SERVANTS OF NATURE

SERVANTS OF NATURE

A History of Scientific Institutions,
Enterprises and Sensibilities

LEWIS PYENSON
and
SUSAN SHEETS-PYENSON

W. W. NORTON & COMPANY
NEW YORK · LONDON

Manufacturing by the Haddon Craftsmen, Inc.

Library of Congress Cataloging-in-Publication Data

Pyenson, Lewis.
    Servants of nature: a history of scientific institutions,
enterprises, and sensibilities / Lewis Pyenson and Susan
Sheets-Pyenson.
        p.   cm.   —   (Norton history of science)
    Includes bibliographical references and index.
    ISBN 0-393-04614-1
    1. Science—Societies, etc.—History. 2. Science—Social aspects—
History. 3. Science—Study and teaching—History.
I. Sheets-Pyenson, Susan, 1949– II. Title. III. Series.
Q124.6.P94 1999
509—dc21   98-53198
                                                                    CIP

W. W. Norton & Company, Inc., 500 Fifth Avenue, New York, NY 10110
http://www.wwnorton.com

W. W. Norton & Company Ltd., 10 Coptic Street, London WC1A 1PU

1 2 3 4 5 6 7 8 9 0

For our children
Nicholas David, Catharine Merrill,
and Benjamin Conrad

CONTENTS

PREFACE

Science, Ernest Gellner has contended, is the mode of cognition of Western industrial society. These words beat a ragged, fading tattoo across the twentieth century. The music asks: why do we seek to know? What is science? Who sees with Western eyes? The pipers form a splendid procession. But will it last? Will we continue to think scientifically? Or does the spectacle mark the end of an epoch? In view of its past, does science have a future?

A reader may ask how science can disappear. With all our modern contrivances and information, how – short of an environment-wrenching catastrophe – could science come to an end? Yet the circumstance has occurred before, when past civilizations like Rome and China expressed little interest in seeking explanations for natural phenomena. They delighted in mechanical contrivances; they celebrated canonical wisdom; they published enduring works of art and literature. But they were not driven to push back the frontiers of knowledge, to use a metaphor associated with European expansion.

The present book has emerged as an enquiry into science as a social activity. It relates directly to the prospect of science in our time. We do not proceed by appealing to the heavy theoretical machinery and the long-distance sentences that are now fashionable in our discipline. We fly no philosophical, political, or methodological colours. We celebrate the observation attributed to writer Marcel Proust, that methodology, when visible in writing, is like a price tag worn on a suit of clothes. We are mindful of poet John Keats's sentence about rejecting poetry that has a design on us.

As we enter a new millennium, the words of Francis Bacon possess a freshness and special pertinence. To invoke his phrase 'servants of nature' is to offer relief from those who would exaggerate or minimise the interpretation of science broadly conceived. This phrase balances the self-satisfaciton, if not the hubris, of some scientists with Bacon's recognition that nature, to be commanded, must be obeyed. Bacon knew that evidence from the natural world comes in many forms; evidence, in some manner going beyond prejudice, can produce general judgments and, on occasion, laws.

Our method will be apparent from the Table of Contents. We identify and elaborate a number of themes that we believe are central to exploring the role of science in society. The themes fall under three headings: institutions, enterprises, and sensibilities. We trace the themes through the recent and more distant past, through Western and non-Western cultures. Our work is grounded in the belief that history may help us see clearly today. We draw inspiration from the great works of French scholarship that celebrate history as a craft based upon a manipulation of concrete particulars, a tradition inaugurated by Marc Bloch and perfected by Theodore Zeldin.

This book is in fact a child of French North America. Conceived in French Canada, where we taught at Université de Montréal and Concordia University, it finally emerged in French Acadiana, where we work at the University of Southwestern Louisiana. While writing it we ran up many debts. We wish to thank our students and colleagues, who heard us elaborate these ideas. We owe special thanks to Eliane Kinsley and Marc Speyer-Ofenberg, and we are grateful to Roy Porter, Crosby Smith, and Bill Swainson for critical comments. Anyone who has lived through a marital collaboration knows what is involved, but children are the most perceptive observers of it. We dedicate this book to our three spirits.

Lafayette, Louisiana
May Day 1996

LIST OF PLATES

1. The 200-inch Hale telescope at Mt Palomar, California, portrayed by Russell W. Porter, 1939.
2. The tube for mounting the 200-inch Hale mirror.
3. Igor Sikorsky's *Ilya Muromets* undergoing a winter test flight near St Petersburg, 1914.
4. Jan van Swinden demonstrates Benjamin Franklin's electrical theory to members of Amsterdam's Felix Meritis Society.
5. University Museum, Oxford, and Keeper's House, about the time of the debate on evolution of 1860 between Thomas Henry Huxley and Bishop Wilberforce.
6. Renaissance surgery, in this case amputation.
7. Toshiko Yuasa in 1933. The first woman physicist in Japan, she worked for the most part in Paris between 1940 and her death in 1980.
8. The Museum of Ferrante Imperato, a natural-history *Wunderkammer*, 1672.
9. The cartographer (literally, 'The Magnetic Polar Stone', seen floating in an enormous basin in the foreground), by Stradanus, sixteenth century.
10. An ideal curiosity cabinet, portrayed by the Amsterdam collector Levinus Vincent.
11. Sixteenth-century print shop, with compositors, proofreader, paper delivery, inker, pressman, office-boy, and owner.
12. Napier power platen press, 1849.

Introduction:
Science and Its Past

Our thirteen-year-old daughter is studying the history of the United States of America. She is memorizing lists of warriors and battles, statesmen and treaties. She sees pictures of people in powdered wigs and frock coats, on horseback and in carriages. Ordinary people wearing rags and buckskin also appear in her books and films. She learns about hopes and fears in times past. Previously, she learned about history from a German perspective. (She can recite the fiercest North Sea storms of the past fifty years.) Our older son learned history first from the point of view of French-Canadian nationalists and then in a traditional English-Canadian vein, before he, too, had to acquaint himself with American facts and foibles. We hope that our children will achieve the level of cultural literacy now being established by prominent intellectuals. If, decades from now, they do not quite recall why George Washington crossed the Delaware or why Chappaquiddick stands for more than an island off the coast of Massachusetts, they will nevertheless retain the notion that what is told about the past is a function of language and politics.[1]

Whatever gaps there may be in our children's schooling, in some sense they will have been educated. Schooling substitutes for travel, for direct experience of distinct cultures. Yet today the distinctness is disappearing. Electronic media and air travel have brought people nearly everywhere in the world into contact with clinically tested drugs, prewashed blue jeans, and the internal-combustion engine. The signs of this convergence have provoked commentary for much of the twentieth century. However the great mixing up is understood, it certainly qualifies as one of the key phenomena of our time.

How did it happen? How did we arrive in our present circumstances? These are the questions posed by historians. They offer many kinds of answer. It has to do with the price of corn over the past 150 years, an economic historian might say. More important are the precedents of Common Law, a legal historian might counter. It is the art of war, thunders a military historian. Everything is family demographics, a social historian counters. Each of these explanations is a splendid room in the mansion of our collective past. But they do not seem to help us understand the form of the objects we use on a daily basis. Does any one of them on its own explain what we see as we go out to purchase the ingredients for dinner or as we watch the evening news on television?

Regardless of the special values that we hold – the religious creed, political persuasion, aesthetic preference, and moral sensibility that together define our character and give meaning to existence – what we experience every day derives from our grasp of the natural and physical world. The following pages investigate how this perception has related to the world of human activity over time. Philosophers and social commentators have argued about how knowledge relates to social dynamics – the regimes of family structure, governmental taxation, religious celebration, and professional obligation that loom large in cultures and civilizations. Perceptions do vary with time and circumstance, but they are not necessarily grounded in incommensurable systems of belief. Microelectronics and molecular biology, for example, which allow all people to share in computer games and biochemical therapy, seem to follow one set of principles everywhere, even though the context of their use varies considerably. The present book explores how knowledge of nature has found a place in society in times past. Sometimes it has transcended language and place. Sometimes it has been anchored firmly in a particular culture.

Our understanding of past knowledge has its own past. Early writings in history of science legitimized a temporal institution, intellectual tendency, or moral lesson. In 1667 Thomas Sprat (1635–1713) promoted the aims of experimental science in his history of the Royal Society of London, the most prestigious association of men of science in the seventeenth century. Joseph Priestley (1733–1804) defended Benjamin Franklin's (1706–1790) notions of electricity against traditional European views in a history of science first published in 1767. And a portion of Georg Friedrich Wilhelm Hegel's (1770–1831) lectures on the history of philosophy, appearing in the 1820s and 1830s, served to instruct readers in his opinions about natural philosophers like Francis Bacon (1561–1626), whom Hegel surprisingly admired. Use of the word history was then as much a synonym for narrative and inventory as a by-word for polemic.[2]

With the nineteenth century, histories of science reverted to the distant and remote past. The impetus came from the crystallization of the historical profession and its installation in European universities. Proponents of the *discipline* of history required a *method* to distinguish themselves from the naive, descriptive narrators of previous generations. The discipline came to centre around the treatment of manuscript documents, which had found their way in large numbers to central repositories like the Bibliothèque Nationale in Paris and the Library of the British Museum in London. The task of the historian became one of transcription, translation, and commentary. Here were the building blocks for synthetic treatments. Historians required their students to master dead languages, old-fashioned handwriting, and ancient chronologies.

The painstaking examination of past science appealed to a number of people throughout the nineteenth century. There were physicians and chemists who, at the end of their career, sought to describe how the method of their own science had evolved. There were mathematicians and astronomers who, having been trained in classics, sought to transcend the ennui of a provincial school or government office by scrutinizing the work of significant predecessors. There were philologists who recognized that the languages of European colonies – notably Sanskrit and Arabic – held a key to a significant literature about ancient and medieval writings on nature.

Nineteenth-century discoveries about knowledge in the remote past are remarkable. The astronomical innovations of medieval Islam, both observational and theoretical, were discussed in the research of Louis-Amélie Sédillot (1808–1875). Euclid's geometry, studied for nearly two millennia as the primary model for clear thinking, received a canonical expression at the hands of Johan Ludvig Heiberg (1854–1928). The notion of medieval Europe as a scientific wasteland came into question through Pierre Duhem's (1861–1916) elaboration of the writings of philosophers at the University of Paris. The deciphering of planetary tables on Babylonian clay tablets by Joseph Epping (1835–1894) established the first reliable chronicle of antiquity.

Learned periodicals arose to circulate findings among the committed band of historians of science. By the last third of the nineteenth century, courses of instruction provided a showcase for the esoteric speciality. As academic philosophy spun out into hundreds of camps and factions, history of science found a practical use in the burgeoning field of epistemology – the philosophical discussion about how we know things. And as the rise of mass education stimulated an interest in teaching methodology, history of science emerged as the most reasonable way to teach physics, chemistry, and natural history. The so-called genetic presentation of scientific disciplines like chemistry and physics, according to which students received an appreciation of old ideas in chronological order, dominated much of science instruction up to the middle of the twentieth century.

The discipline of history of science

Nineteenth-century writings about history of science are grounded in the notion that modern science is a gift of Western Europe. Writers believed that scientific method and practice distinguished the people of the West from the civilizations that the West had conquered. Art, music, and literature were matters of taste; Japanese painting and poetry, for example, could be held only to differ from European painting and poetry. Science, however, was a matter of truth. All peoples, furthermore, could acquire it. A convenient justification for imperialist domination of the world came in the form of instruction in the canons of Western reason. Historians of science were among the firmest apologists for the superiority of European intellect.

The philosopher Auguste Comte (1798–1857) looms as a major figure behind much writing in history of science. In Comte's view, humanity had passed through various stages. Science, using experiment and mathematics to verify theories, would usher in a new age of prosperity and harmony. Comte established a hierarchy for the sciences, with astronomy at the apex and physiology near the bottom. Over time, he believed, all inquiries into nature would become more like mathematical physics. He also outlined how humanity had progressed from a religious worldview to a scientific one. This *positivist* orientation, where qualitative and prejudicial notions fell by the wayside, animated the beginnings of sociology, the quantitative science of human affairs. To bring the new golden age into being, Comte revived the French Revolution faith in Reason and established a church of positivism. Fundamental to the new doctrine was a critical examination of the evolution of science, demonstrating its grand unity and progress. To tell this story, the curators of the Collège de France (the elite institute for research and popular teaching in Paris) appointed Comte's disciple Pierre Laffitte (1823–1903) to a chair of history of science.

Laffitte accomplished little in the course of his long tenure as the world's most visible historian of science; he was entirely overshadowed by Paul Tannery (1843–1904), an administrator in the French state tobacco monopoly who had occasionally taught at the Collège de France. Tannery had published a large corpus on the history of the exact sciences, from classical antiquity through medieval Islam to the Renaissance and on into the nineteenth century. He established a European-wide network of colleagues who shared his passion. Developing a model that would have appealed to Comte, Tannery stated that science originated in Hellenic Greece, passed through Islamic stewardship to medieval Europe, and then blossomed in the seventeenth century. It was quite entirely an affair of the West.

In 1900, on the occasion of scholarly celebrations surrounding the grand Paris Exhibition, Tannery convened the world's first international congress devoted to history of science. He assembled colleagues from Europe and put together an impressive programme. The congress resulted in a permanent commission to plan for future gatherings, establish an international society, and publish a periodical. The organizing epistolary activity (his correspondence was published in many volumes by his widow) contributed to Tannery's dossier as Laffitte's successor at the Collège de France. But politics intervened to deny him the academic position that he merited. Third Republic secularists passed over Tannery, a practising Catholic, in favour of a philosophically inclined disciple of Comte's. When Tannery died of pancreatic cancer in 1904, the newborn discipline lost its most vocal advocate.

Tannery's attempt to form a discipline at the beginning of the twentieth century was one of a number of initiatives for promoting Western civilization. Intellectuals sought organizations and causes that, in spanning the nation states of Europe, could project a common front against barbarism. They proudly pointed to the institution of the Nobel prizes, the creation of the world court in The Hague, and the convening of international congresses in fields of study from mathematics to history. The initiatives depended, however, on funding from national sources. The projection of scholarly and scientific excellence, based not on international assemblies but on national institutions, became a card in the game of diplomacy. Nations tallied up their Nobel laureates, art museums, libraries, and grand research laboratories. During the early decades of the twentieth century, and especially as a result of European wars and political squabbles, the discipline of history of science followed distinct trajectories in various national sectors.

Germany, the land of the research doctorate, contributed dozens of university courses and a number of periodicals. The key figure there was Karl Sudhoff (1853–1938), a medical doctor who in 1905 became director of a privately funded institute for the history of medicine and science at the University of Leipzig. Sudhoff's successor in 1925 was the Paris-born and Swiss-educated Henry Sigerist (1891–1957), who in 1932 became director of the new Institute for the History of Medicine at Johns Hopkins University in Baltimore, Maryland. Great Britain found an energetic patron in the Regius Professor of Medicine at Oxford, the Canadian-born and American-acculturated Sir William Osler (1849–1919). Osler cultivated Charles Singer (1876–1960), who in the 1920s obtained a chair of history of science at the University of London. France continued the philosophically inclined course of Comte with Emile Meyerson (1859–1933) and Gaston Milhaud (1858–1918), which culminated in the Platonism of the Russian-born Alexandre Koyré (1892–1964). And in 1928, Italian-born Aldo Mieli (1879–1950) instituted the International Academy of the History of Science and located it in lodgings in Paris provided by Henri Berr's (1863–1954) Centre International de Synthèse.

All these efforts produced mixed results. Scholarship in Germany was generally compromised by war and political savagery; Sudhoff, in his eighties, willingly embraced Adolf

Hitler's National Socialism. Willy Hartner (1905–1981), one of the brightest lights in Germany, studied at Harvard and became a rare opponent of Hitler in Germany. Singer and his wife Dorothea Waley Cohen (1882–1964) generated scholarly surveys and collections, but they produced few students; Osler's querulous disciplinary successor at Oxford, biologist Robert T. Gunther (1869–1940), found a passion in scientific instruments. Mieli fled Vichy France for Argentina, just as he had fled fascist Italy for Paris; he died there in obscurity, a victim of Perón's wrath. Koyré left Paris for Egypt and then New York; he subsequently received an appointment at the Institute for Advanced Study in Princeton. The change was unmistakable. Talented scholars moved from Europe to the New World.

Notwithstanding his unceasing propaganda in favour of medical reform (he was a consultant for the establishment of socialized medicine in the Province of Saskatchewan in Canada), Sigerist promoted significant scholarship at Johns Hopkins. He found two singular disciplinary fellow travellers in émigrés Otto Neugebauer (1899–1990) and George Sarton (1884–1956).

In the 1920s, Austrian-born Neugebauer was the brilliant student of mathematician Richard Courant at the University of Göttingen. Neugebauer elected to focus his scholarly interest on history of mathematics, rapidly becoming the most accomplished interpreter of mathematical antiquity, from the Babylonian clay tablets to Ptolemy's astronomy. He earned his living, however, as the paid editor of a major journal of mathematical abstracts. Fascism drove him to a position at the University of Copenhagen, and in 1939 to a chair in the history of mathematics at Brown University in Providence, Rhode Island. Neugebauer's unparalleled scientific authority and his access to scholarly resources led to a school of disciples based on mastery of languages, primary and secondary literature, and above all the technical details of the exact sciences.

If Neugebauer was wary of generalizing and popularizing,

Sarton revelled in the broad sweep. Born and educated in Belgium, Sarton used his inheritance to create, in 1913, what has become the leading scholarly periodical in history of science, *Isis*. Following the German invasion of Belgium in 1914, he fled to London with his British wife and infant daughter. There he lived on the brink of starvation until in 1915 he sought his fortune in the United States as a promoter of Tannery's vision of history of science. He found willing listeners in university administrators and capitalists who were interested in appropriating European culture. In 1918 Sarton obtained a salary from the Carnegie Institution of Washington for maintenance at Harvard University's Widener Library as a special researcher. The arrangement continued over the next thirty years, during which time Sarton gradually acquired professorial privileges, edited *Isis*, and produced a grand outline of the history of science up to the Renaissance. He envisioned history of science as the privileged intersection of the natural sciences and the humanities. In his view, history of science was the true record of human achievement.

Sartonian generalizing and Neugebauerist specialization often find expression today in writings about science past. It might be said, indeed, that Sarton's vision lived on for nearly a half century after his death, in 1956, through the enterprise in chronicling the history of science in China conceived and directed by Joseph Needham (1900–1995) at Cambridge. At mid century, however, the emphasis on encyclopaedic chronicle, ponderous biography, and antiquarian curiosity began to recede in favour of methodology. Whereas Sarton and Neugebauer did not often justify the particular focus for their energies (beyond, say, noting something about every science writer in the twelfth century or transcribing and interpreting all known coffin lids with astronomical cyphers), the middle third of the twentieth century demanded relevance. That is, the new generation of scholars found themselves called to consider the end of their vocation. What did Renaissance astronomers or Puritan experimentalists have to do with human suffering and political change? Did the power and prestige of social institutions determine the shape of ideas about nature? How could specialist, scholarly apparatus illuminate the deep structure of human thought?

Physicians in classical antiquity knew that art is long and exhausting, while life is short and demanding. One must gradually build up an intellectual arsenal to attack significant problems. Practical results can certainly be obtained by ingenuous debutants, but even here method and knowledge are everything. Three scholars – Thomas S. Kuhn (1922–1996), Derek J. de Solla Price (1922–1983), and Robert K. Merton (b. 1910) – mixed appropriate quantities of innocence and experience to transform our vision of history of science. They did so by interpreting the prosaic side of scientific endeavour.

Inspiration and method

In the twentieth century, Thomas Kuhn is widely recognized as the most influential commentator on the meaning of science. Derek Price is seen as the guiding spirit behind the quantitative measurement of scientific development. Robert Merton identified the normative criteria for scientific activity and the institutional constraints that guide the life of scientists. Their achievements are the more remarkable because their careers encompassed much else. In an age of specialization, they were polymaths – people able to innovate in diverse ways.

Across the decades spanning the middle of the century, the three interpreters of science share a common *modus operandi*. Most apparently, each is a brilliant stylist. At a time when academic writing discouraged use of first person singular, their intensely personal sentences leap off the page. Kuhn, Price, and Merton each formulated a seminal theoretical overview that was based on a close reading of critical episodes in science. Kuhn examined Copernican astronomy, energy conservation, and quantum physics; Price scrutinized medieval astrolabes, Chinese horology, and measuring and calculating devices from classical antiquity; Merton placed the early years of the Royal Society of London under a microscope and then examined the career of elite American scientists. Each sought general truths by extrapolating from definitive studies of carefully selected examples.

The men who transformed history of science avoided encyclopaedic studies of the kind pursued by Sarton, Neugebauer, and Needham. Common concern with the social dimensions of scientific enterprise led to books about science in society at large – Merton's early treatment of science in seventeenth-century England; Kuhn's analysis of the Copernican revolution; Price's lectures about measuring scientific growth. But their appeal to learned precedent (a distinguishing characteristic of the scholarly life) was based on the advice of personal informants, rather than on systematic bibliographic travail. Lack of scholarly apparatus notwithstanding, each man expressed intense interest in organizing documentation for the next generation of scholars. Kuhn animated the international effort to assemble interviews and private correspondence known as the Sources for the History of Quantum Physics, a project more than any other that has alerted the scientific and scholarly world to the importance of preserving the record of recent science. Price was the most vigorous academic promoter of the quantitative measurement marketed by Philadelphia's Institute for Scientific Information and now used extensively by countless agencies and analysts. Merton collaborated intensively with Paul Lazarsfeld (1901–1976) at the Institute for Applied Sociology in Columbia University, where he assembled sociological documentation on a wide range of phenomena.

They were products of elite universities and all enjoyed the privileges of accumulated honours, but they lent their voices to new institutions and assemblies. Kuhn was entirely

Harvard educated, obtaining a doctorate in theoretical solid-state physics under John Hasbrouck Van Vleck (1899–1980); he sat in uneasy harmony between history and philosophy at the University of California at Berkeley (where an inheritance allowed him to reduce his teaching obligations) until in 1964 he moved to a programme at Princeton tailored to his measure. He helped Charles C. Gillispie (b.1918) steer the signal achievement of American scholarship, the *Dictionary of Scientific Biography*. Price obtained a doctorate in experimental physics from the University of London in 1946. A second doctorate in history of science came from Cambridge in 1954, following which he joined the singular company of scholars in history of science at Yale. He energetically supported new societies and new periodicals. Merton went from Temple University to Harvard, where he obtained a doctorate in sociology. He participated in launching the Center for Advanced Study in Behavioral Sciences at Stanford.

The three found their earliest and strongest voice in the seminal article, rather than the weighty tome, but all mastered general presentations as well as specialized analysis. Following a number of monographs (among which was his brilliant doctoral dissertation), in early middle age Merton produced collections on the sociology of science and social structure. After the profound but general study of Copernicus and the appearance of the *Structure of Scientific Revolutions* (1962), Kuhn issued a searching, technical analysis of the birth of quantum physics. Price published meticulous studies of clocks and calculators as well as collections of essays on measuring science.

Masters in the realm of ideas, from the time of their youth they were no strangers to practical matters. All three men were schooled in the arts of manual dexterity, and all three knew the vagaries of fortune in commerce. Thomas Kuhn's father Samuel was a Harvard-and-MIT educated hydraulics engineer from Cincinnati, a veteran of the US Army Corps of Engineers who established himself in New York City, worked for a bank, and was active in the New Deal recovery. Samuel Kuhn was also a master woodworker, who trained his son in the use of hand tools; a perfectionist, he urged Thomas Kuhn to excel, and the scholar son recalled that he was influenced most by his father and Harvard's James B. Conant (1893–1978). Derek Price's father owned a clothing and haberdashery establishment in London's West End. At the age of sixteen in 1938 Price was appointed physics laboratory assistant at the new South-West Essex Technical College, where in 1942 he received a bachelor's degree. He taught evening classes in adult education while carrying out his own research for an external doctorate at the University of London. Robert Merton, the son of an immigrant carpenter, grew up in working-class South Philadelphia; while a teenage professional magician, he chose the name by which he is now known after considering other variations on 'Merlin'. These tactile sensibilities are common themes among leading twentieth-century intellectuals. In their autobiographies, none less than physicist Albert Einstein (1879–1955) and philosopher Sir Karl Popper (1902–1994) have emphasized the importance of manual training.

Although they were not pacifists, Kuhn, Price, and Merton, like Einstein and Popper, avoided bearing arms during times of conflict. In the Second World War, Kuhn and Price, duly certified with an undergraduate degree in physics, carried out technical research – Price with South-West Essex's Principal Harry Lowery (1896–1967, an expert in musical acoustics) on the optics of hot metals, Kuhn in the American-British Laboratory of the Office of Scientific Research and Development. Merton worked on military-related propaganda and psychological profiles in Columbia University's Office of Radio Research and its successor institute, the Bureau of Applied Social Research.

Each master rose to the summit of the academic world after a short period of ministering *in partes infidelium*. Kuhn arrived at Berkeley in 1956 when the university there was still in the process of overcoming its provincial heritage.

Merton left a lectureship at Harvard to become professor of sociology at Tulane University in New Orleans, from which he fled to Columbia in 1941. A postdoctoral Commonwealth Fund fellowship allowed Price to study theoretical physics at the universities of Pittsburgh and Princeton. From 1947 to 1950 the newly married physicist taught applied mathematics at Raffles College in Singapore, where he was briefly the colleague of the historian Cyril Northcote Parkinson (1909–1993).

Parkinson is famous for studying the flow of people at cocktail parties and for proposing laws regulating bureaucracies, for example, 'work expands to fill the time available for it'. Parkinson's search for social laws and mechanisms finds a counterpart in the life work of Kuhn, Price, and Merton. Kuhn's description of scientific revolutions as an abrupt change from one worldview to another has brought the word *paradigm* (proposed earlier by Merton) to nearly the same level of recognition as Sigmund Freud's (1856–1939) *repression* and Einstein's *relativity*.[3] Price's revival of the term *invisible college* (first used to describe a group of seventeenth-century natural philosophers) and his discussion of *exponential growth* (known to historian Henry Adams [1838–1918] around 1900) have led to a large industry involved with quantifying scientific achievement.[4] Merton's elaboration of *scientific norms* and accumulated advantage (the Matthew Effect), his analysis of anomie in working-class America, and his pioneering development of such notions as the *focus group*, have made him the most widely known sociological thinker of our century.

One question attracted the interest of the three masters in the 1950s. This was the extent to which science depended on the unique contributions of isolated geniuses. They delved deeply into the phenomenon of simultaneous and multiple discoveries. A persuasive argument in favour of a social history of science would stem from the contention that notions or ideas were independent of the personal circumstances of a researcher, that particular interpretations of nature were in some sense the product of their time. The historical evidence the three men had assembled by the early 1960s demonstrated just this contention. The confluence of learned opinion, noted in Price's book *Little Science, Big Science*, marks a turning point in the discipline of history of science.[5]

By the late 1960s, the search for a general understanding of science, stimulated by the work of Kuhn, Price, and Merton, led to renewed interest in quantitative measurement and close scrutiny of scientific institutions and specialties. What were the pathways of authority in diverse disciplines located in special cultural settings? Did all disciplines and technologies function in one way? What were the practical conditions for consolidating a new discipline? How did the constraints on and opportunities for funding generate new research programmes? Do civil unrest and war stimulate or retard the generation of new ideas? Who is qualified as a researcher in science? How have educational institutions set the pattern for scientific research? Careful attention to these questions produced scores of sophisticated analyses and monographs. As the decade of the 1960s closed, history of science, as a learned enterprise, achieved an intellectual vitality envied by many and diverse commentators.[6]

The end of science

In Thomas Kuhn's understanding of science, major changes can occur *within* a particular discipline. That is, new visions of the world do not necessarily destroy the social structures that may have nurtured them. Kuhn also wrote about how disciplines change form, capturing problem areas from neighbouring fields of study and abandoning special pursuits to a new group of researchers. The late twentieth century has witnessed unusual ferment in the evolving taxonomy of scientific disciplines. New disciplines like computing and cognition have taken away large areas of mathematics and psychology. Physicists lament hard times much as classicists complained about declining interest in their discipline a century ago. Most of all, we see the proliferation of an astonishing variety of technical discipline, complete with trade journals and academic programmes. Science is not a victim of its own success, as has been claimed; rather, the times promote technology.

Today science is threatened with absorption by technology, in the way that Hellenistic learning eroded under Roman domination. Roman architects and administrators used existing knowledge to produce monuments of temporal authority -- roads, aqueducts, markets, and public buildings. The durable monuments of Roman civilization, however, were its laws. Romans used what worked to establish what was right. Abstract truth was an affair for Greek tutors.

The early twentieth-century desire to establish truth was an outcome of the Scientific Revolution and the Enlightenment, which led into modern times. But today, echoing Roman sentiment, morality is the watchword. Public servants, for example, are castigated for failing one or another standard of ethical purity. However, since we have no universally accepted notion of goodness, we are surrounded by an appeal to eclectic principles and credos. Eclecticism appears in architecture, with whimsical adornments alluding to diverse precedent. Eclecticism is revealed in the way that orchestras focus on compositions before around 1960, largely bypassing living composers. In the world of letters, all expressions have been called into question. The evidence suggests that we have gone beyond the modern.

The 'postmodern' approach to ideas has extended to science. In postmodernist quarters, it is sufficient to assert that all ideas are expressions of power relations. In the view of postmodernist commentators, scientific writings are merely codes for reinforcing the authority of people in charge. Knowledge, according to Michel Foucault (1926–1984), is 'not made for understanding – instead it is made for cutting'.[7] For Foucault, knowledge is about commanding people incisively. It is about separating things. It concerns morality and values. But it has no privileged claim to truth.

Historian of technology Leo Marx locates postmodernism in the political pessimism of the 1970s. In his view, postmodernists reject the Enlightenment ideal of progress and human perfectability. Sceptical in the extreme, they repudiate all large-scale interpretations of culture and history. The human condition is held hostage by vague, universal forces called power relations, borrowing a metaphor from the Scientific Revolution. But 'unlike the old notion of entrenched power that can be attacked, removed, or replaced, postmodernists envisage forms of power that have no central, single, fixed, discernible, controllable locus. This kind of power is everywhere but concentrated nowhere'.[8] As a result, in Marx's view, postmodernists focus on microscopic manifestations of power. These writers are typically uninterested in long projects that systematically document large populations. Vague, impressionistic surveys share the billing with narrow, idiosyncratic discussions of printed texts.

How did postmodernism find a place in the history of science? Animated by the programme of social history elaborated during the 1960s, the 1970s saw significant works of scholarship and dedication. The innovations of Merton, Price, and Kuhn found concrete application in analyses of eighteenth-century German chemistry, physics in modern Germany and the United States, French scientific institutions, British natural history, and the general issue of the Newtonian synthesis. But this systematic and time-consuming labour took place in a time of growing anxiety about the material survival of the labourers. A long period of economic contraction coupled with demographic changes resulted in a dearth of academic posts for an entire generation of young scholars. Historians of science fared better than linguists or classicists, but the academy groaned under the mass of men and women hired in the flush of the fat 1960s.

The ingenuous assertions of the 1960s – that war is the root of domestic poverty, that racial prejudice and discrimination against women are structural features of capitalism – derived from a perception of social life; to understand the world one had to measure its demography and political economy. By the 1980s, however, mere writings were held to be at once examples and sources of oppression. Postmodernists claim that the ideas and institutions of modern science are irredeemably sexist; that experiment and mathematics, applied to the investigation of nature, are little more than tricks; that science has more in common with styles of clothing than geometric certainty. The assertions appear in the absence of persuasive documentation, for the role of evidence itself is called into question. Indeed, documentation for postmodernists is mere adornment. The content of footnotes or endnotes matters less than the appearance of having appealed to instance and precedent – a matter of legitimizing authority.[9] Sociologist Bruno Latour has published widely reviewed essays about the scientific work of Einstein and Louis Pasteur (1822–1895) without appealing to their scientific publications. Another postmodernist, Latour's sociologist colleague Steven Shapin, alleges in a survey titled *The Scientific Revolution* (1996) that the revolution was a 'non event'[10] – even though his examples persuade a reader of the cause in question.

Latour and Shapin are cavalier about evidence because they hold that all knowledge derives from social interaction. In a sympathetic reading of Latour, philosopher Chris McClellan summarizes the extreme form of this contention: 'everything is actively linked to everything else, while the only form to this seamless cloth comes from the varying durability and strength of the associations that tie it together.'[11] Shapin's unusual approach to evidence and reason lies at the centre of a related essay, *A Social History of Truth* (1994). There he contends that in seventeenth-century England, rhetoric and social standing overwhelmed open discussion of experimental results, and that as a result from its inception modern science has maintained standards and practices at odds with the search for universal truths.

Shapin's sociology of science has generated unprecedented discussion in the pages of the journal founded by George Sarton, *Isis*. Historian Mordechai Feingold observed: 'Shapin's approach is ahistorical. He denies the historian possession of any privileged knowledge of the past. Meanings and intentions in history are forever lost, and all one can do is concentrate on ideals – "publicly voiced attitudes" . . .' Feingold affirmed the importance of assuming that 'there are historical facts that can either sustain or invalidate interpretations', and he insisted 'that a scholar who abolishes boundaries between facts and interpretations must be held accountable'. Feingold again summarized Shapin's methodology: 'Notwithstanding the "elaborate" sources Shapin has gathered, all too often his conclusions are shaped by a confusing and inaccurate discussion of the literature, including citing out of context and the occasional cropping of texts.' Shapin himself replied to the criticism, but without mentioning Feingold by name or providing a reference for Feingold's review. Feingold then patiently reasserted the importance of evidence and Shapin's misleading use of it: 'Thanks to a skillful deployment of rhetoric – copious repetitions intended to drive a message home and the articulation of many key sentences in a subtle and confusing manner – the reader, who has not infinite time to engage in hermeneutics, can easily mistake the conceivable for the actual.'[12] Although we can imagine a flying horse and may deliver orations about it, the image remains firmly in the realm of fiction.

An exchange about the African roots of Western science also reveals the postmodernist style. Sociologist Martin Bernal has contended that much of Hellenic wisdom derived from Egyptian civilization. Bernal believes that 'many cultural similarities that could reasonably be attributed to independent invention in distant communities should not be so explained for those between societies as close in time and place as Egypt and Greece'. But in commenting on Bernal's work, historian of science Robert Palter requires stronger canons of reason. Palter notes that the Egyptians had no mathematical astronomy resembling Greek works, that Egyptian mathematics never attained the sophistication of Babylonian and Greek expressions; and that the traditions of medicine in Egypt and Greece diverged considerably. The point is that Bernal's desire to demonstrate that Aryan civilization derived from black antecedents displaces a concern for evidence.[13] Postmodernist palladins now ride to the rescue of false assertions. In a spirited review of a recent book that criticized propositions advanced by both Bernal and Shapin, the historian of science M. Norton Wise has declined to admit more than that the critics have 'doubtless . . . located some blunders'. Wise prefers to submerge substantive issues in a farrago of unrelated material.[14] By their allegations of wilful misrepresentation, these exchanges are untypical of academic debate in history of science. They point to significant discontent with disciplinary standards.

The word *discipline* carries many meanings, anthropologist Clifford Geertz reminds us, and all of them relate to authority.[15] A leitmotif of the careers of Merton, Price, and Kuhn is a concern with the bounds of authority in science. To explore authority they counterposed the scientific discipline with its complementary social structure, the corporate *institution*. Disciplines function according to general, abstract rules and principles; they attract adherents who earn their living in various ways, profess manifold credos, and pray to diverse gods. Institutions, however, operate by corporate structure and private covenant; they demand allegiance to a chain of command. At the risk of oversimplification, one might say that disciplines exhibit an abstract solidarity while institutions exhibit a more earthy, organic solidarity. Exploring the authority of disciplines and institutions to elaborate the counterpoint of tradition and innovation, in

Kuhn's words, is the project that has animated historians of science since the 1960s. In this book, we begin by considering scientific institutions.

The postmodernist interlude reminds us that generalization is a privilege of experience. The concrete experiences analysed by historians of science – whose number as full-time, dues-paying, certified practitioners is only in the thousands – have transformed our vision of the human condition. They give us new pictures of the ways that people have seen the natural world, and they have added to a long list of misconceived apprehensions. Despite occasional claims to the contrary, the discipline of history of science is indeed regular, cumulative, and progressive.

Recent debates about whether science expresses truths about the world call to mind an observation by a sixteenth-century patron of natural knowledge, Thomas Gresham (1518/19–1579). Councillor of state, founder of the British stock exchange, and endower of a college that served as the nucleus of the Royal Society and persisted into the twentieth century, Gresham proposed a principle of economics that has been epitomized as: 'Bad money drives out good money.' That is, silver currency will inevitably force gold currency out of circulation. The principle applies more generally to governments, trades, and professions. In a parliamentary system of government, the actions of one corrupt delegate can provoke a vote of 'no confidence' that will produce new elections. Gresham's Law suggests why professional corporations are concerned about enforcing standards. If isolated unscrupulous practices shake confidence in, for example, stock brokerage, physical therapy, or dental surgery, people will cease patronizing the enterprise. In the world of scholarship, outrageous or demonstrably false assertions can bring an entire specialty into disrepute. Gresham's Law has found an application in the history of science through the claims of postmodern writers.[16]

An elegy for postmodernism has been written by Frank Lentricchia, professor of English at Duke University and for decades one of the most persistent critics of the notion that ideas have integrity. He confesses that he lived a double life. He read great literature because it transported him with insight and delight. But he taught that 'what is called "literature" is nothing but the most devious of rhetorical discourses (writing with political designs upon us all), either in opposition to or in complicity with the power in place'. There were two of him. 'In private, I was tranquillity personified; in public, an actor in the endless strife and divisiveness of argument, the "Dirty Harry of literary theory," as one reviewer put it.' The contradiction produced a crisis and a response. Lentricchia finally decided that there were writers, clever and dull, whose writings could be read with pleasure and profit. Some writings, he has concluded, transcend the accidental circumstances of the writer.[17] The observation carries over to science. Some of what we see is conditioned by our upbringing, but seminal syntheses of natural knowledge transcend the circumstances of their formulation.

We do not choose our parents, our mother tongue, or the circumstances of our early years. The world is not made for our effortless gratification. Rather, we respond to the imperatives of existence. The latitude of that response – how much we do by choice and inspiration and how much we are instructed to do by way of convention and authority – is one of the most interesting problems for people who study the course of cultures and civilizations. The following pages will have succeeded if they convey a sense of the many ways that we have seen what is all around us.

I

INSTITUTIONS

1

Teaching: Before the Scientific Revolution

Well into middle age the man awoke with a nightmare about honours examinations at his undergraduate college. For years the nightmare took the same form. He was unprepared for the material. Other students streamed towards the classrooms, confident that they had mastered Heine and Heisenberg, Proust and politics, evolution and revolution. He was all at sea, barely familiar with the course syllabi. Before intimations of mortality replaced the fear of inadequacy in the man's sleeping consciousness, the examination dream evolved a more complicated and quite preposterous plot: though the man held a doctorate, he was returning to complete an undergraduate college degree.

Most people have experienced an anxiety dream about school. The reason is clear: schooling is an unnatural and traumatic event. Children are confided to a stranger for instruction in abstractions. They are required to commit great quantities of facts to memory, largely by the intermediary of the written word. It comes as no surprise that some creative minds have questioned the value of traditional schooling, with its emphasis on examinations. Albert Einstein (1879–1955), in one of his earliest popular writings, found little to commend the traditional German secondary school-leaving examination, the *Abitur*. The examination was injurious to mental health precisely because it gave rise to nightmares. Furthermore, a good deal of time in the last year of secondary school was wasted in preparing students for the test.[1] Einstein himself never submitted to the Abitur, although he once failed the entrance examination for the Zurich polytechnical institute, and his lover failed the final examinations there.

Einstein studied in Germany and Switzerland, and he may even have attended school briefly in Italy. He could have affirmed that many nations have a hierarchy of schools where citizens are obliged to receive state-sanctioned training. Knowledge may be imparted anywhere, and skills may be acquired on the job, but an academic institution carries an ethos and acts as a crucible for culture. Most important is teaching manners – the essential, outward features of daily life that distinguish civilization from barbarism. Some academic institutions even instruct about what to say at a cocktail reception, which utensil to pick up first at a dinner party, and how to act *au courant* of the latest intellectual fad. With the eclipse of gentry, priests, and community healers, academic graduates have increasingly been called to officiate in matters large and small.

Whence this prestige attached to the resources controlled by a self-perpetuating guild? The vast majority of academic diplomas no longer lead directly to a post in the workaday world. Today they do not provide evidence, except indirectly, of having mastered the skills required to succeed in business or public affairs. And in an age of sliding-fee structures, social class and family wealth are no longer associated with the crest of a particular institution.

Schools generally are conservative social institutions, and prestige radiates from their traditions, customs, and rituals. They divide the day into class hours and the year into semesters, the calendar of events culminating in colourful ceremonies at which diplomas are conferred. These rituals of formal schooling, which express a way of ordering the world, have entered into the consciousness of a large part of the world's peoples.

School rituals deriving from religious or moral outlooks vary from place to place. Yet all schools subscribe to one common idea. They hold that knowledge may be acquired through diligent study. There are other kinds of knowledge deriving from religious or artistic inspiration. But schools hold that most things can be *learned*. The central notion here concerns a distillation of tradition. Learning about knowledge, largely from books, is what has been called *science* for a thousand years.

In schools, a master imparts knowledge to acolytes, who may eventually create something new beyond their lessons. Whether scholastic lessons are abstract or practical, esoteric or mundane, schools prepare students for a place in society. That place is generally keyed to facility with the written word, which has been the most secure means of transmitting knowledge from one generation to another. In fact, it is not unreasonable to imagine that schools *invented* writing, and hence that schools are the prime mover of history – the science of knowing the past by its documentation.

In this chapter and the next one we examine how schools of higher learning have been involved with scientific tradition and change. We shall see that academia has both promoted scientific innovation and also stifled it. One of the challenges facing universities in the new millennium will be to implement new ways of breathing relevance into the accomplishments and promises of the past.

The Mediterranean world

What we know about science education in antiquity derives from a variety of documents: a few hundred clay tablets from several sites in Mesopotamia; a few treatises written on papyrus; and diverse histories and texts recopied and reprinted in Chinese, Greek, Arabic and Latin. To this must be added inscriptions on stone, masonry, coins, and pottery, precious castings and carvings, and the accumulated wisdom of archaeology. Because our knowledge of the distant past derives from fragmentary sources, it has sometimes been said that the study of antiquity appeals to people who like mastering a small, fixed syllabus. The sources, however, are much more abundant than commonly imagined.

Clay tablets allow us to conclude that schools existed in Mesopotamia, and that they coincided with the earliest representations of the Sumerian language about 3100 BC. Among the documents of Old Sumerian, which existed until about 2500 BC, are school exercises – lists of signs and words. At the time of the Semitic invasion of Mesopotamia, about 1700 BC, we find a compendium of celestial omens called the *Enuma Anu Enlil*. These omens concern the moon's eclipses, halos, and conjunctions with fixed stars; solar eclipses; weather and earthquakes; and planetary stations. They held special importance for those who believed in astrology, a system of correspondences constructed between celestial phenomena and terrestrial events. The celestial phenomena must have been catalogued over centuries and at diverse places by trained observers. These circumstances suggest an early social pairing of priestly and scholastic functions.

Many of the Sumerian calculations we possess treat practical measuring problems, often involving land area. (In modern terms they reduce to complicated algebraic equations, often cubic or even quadratic expressions.) The problems are sometimes formulated with what we may call malice of forethought (correct answers are integral numbers), and sometimes they have absurd proportions (lengths stretching more than a thousand kilometres or food for an impossibly large army). We have problem sets both with and without solutions, and some solutions feature elementary mistakes. We must conclude that the corpus relates to instruction in schools. The techniques were no doubt useful for keeping track of state assets, but it seems more reasonable to imagine that this specialized knowledge served better to discipline young minds.

The presence of codifying abstruse calculations (whatever their ostensible, practical referent) implies the existence of schools, even if we cannot say much about scholastic organization. Egyptian mathematics, for example, is based on unit fractions – fractions where the numerator is always one. It is possible to speculate about the origin of such a convention (in terms of family structure, inheritance practices, land tenure and taxes), but there can be no disagreement about the ultimate impracticality of the convention for advanced mathematics. Among the few surviving compendia of Egyptian mathematics, we find calculations dividing the contents of a jug of beer into minuscule parts, obviously a school exercise by its lack of utility.

A new kind of teaching emerged in the fifth century BC, and it left its mark on learning in all cultures with access to the Mediterranean world. The innovation related to a group of Greek teachers known as Sophists. They were private professional pedagogues (like later-day itinerant lecturers) who operated in a free-market economy. They would teach by contract whatever people wanted to learn. Their syllabuses suited individual tastes, and their pitch seems to have been a mixture of affable cultivation and practical skills designed to propel a citizen forward in his city.

Their innovations notwithstanding, Socrates (ca.470–399 BC) and Plato (ca.427–347 BC) were teachers in the Sophist tradition, even though they distinguished themselves by their strong claim to methodological precision and systematization of knowledge. Plato's Academy occupied a large athletic facility long used by teachers like him. Aristotle (384–322 BC), who might have succeeded Plato, created his own school at another athletic facility, the Lyceum. Aristotle's chosen successor Theophrastus (372–287 BC) produced written anthologies of his pre-Socratic predecessors in addition to general manuals and new works. He purchased land near the Lyceum and donated it in perpetuity to his colleagues for a school, although the Lyceum's library left Athens for Anatolia as a result of an ideological schism. Later the library returned to Athens and eventually found its way to Rome (as spoils of conquest), where it received wide notice. Permanency of place and syllabus, coupled with the international and public nature of instruction, produced a search for *certainty* rather than, as with the Sophists, mere expediency.

The Academy and the Lyceum were institutions of higher learning. They departed from the smorgasbord of Sophist offerings whose heritage we find, today, in undergraduate liberal-arts curricula. Young people associated with these schools absorbed particular truths as well as the spirit of the place, and then contributed to the discourse; it pleased some men (we have no clear record of women scholars) to stay on for part or all of a lifetime. The excitement of scholarly discussion and the presence of libraries, where knowledge was collected and stored, made such a choice attractive. We possess no diplomas from antiquity because the world of Greek learning was so small as not to require them. A quick conversation would be enough to establish a person's credentials.

State funding ensured the contemplative life of these colleges, which continued in some form for many hundreds of years. At least at the beginning of the Hellenistic period, academic contemplation related directly to political involvement. Because the end of all learning was to train better citizens, scholars often applied themselves to statecraft. The goal was to produce someone like Henry Kissinger or, more optimistically, Woodrow Wilson, each of whom was a distinguished academic before entering politics.

The Big Three – Socrates, Plato, and Aristotle – closed the Greek golden age. In the far-ranging conquests and the subsequent *Hellenizing* process initiated by Alexander the Macedonian, these and other thinkers of contemporary renown received tremendous exposure. What distinguishes the sequence of the Big Three is not speculative moral or political philosophy, but rather a tradition of collective enquiry into nature. They also sought explanations rooted in experience and capable of standing up under sustained, reasoned debate. Whatever the philosophical colour of knowledge-seekers in Hellenistic times (the philosophies came in dozens of hues), their accomplishments depended on libraries and secular centres of higher learning.

Institutions with a teaching function began to take shape, emphasizing the search for knowledge of nature, with the result that the contentious ethical-political side receded into the background. A pupil of Aristotle's successor Theophrastus, Demetrius Phalerius (ca.345–293 BC), deposed as dictator of Athens, went to the Egypt of one of Alexander's generals turned potentates, Ptolemy I Soter; there Ptolemy, acting on Demetrius's advice, founded the institute for advanced study known as the Museum of Alexandria. The name suggests a secular temple for receiving inspiration by the muses, the nine avatars of arts and letters (including astronomy) in classical antiquity. Though under the direction of a priest (until Rome imposed a supervisor) and with their material needs overseen by curatorial staff, the Museum's fellows were free to study what they liked. They lived sumptuously at the king's expense. They had outdoor galleries and lecture theatres for learned discussions, and they ate in a large dining hall. Attached to the Museum were a botanical garden and what became the largest library of Mediterranean antiquity, the Serapeum. The prestige of the Museum made it a magnet for scientists throughout Hellenistic and Roman times – Euclid (fl. ca.295 BC), Apollonius of Perga (fl. ca.200 BC), Aristarchus (ca.310–230 BC), Eratosthenes (ca.276–ca.195 BC), Archimedes (ca.287–212 BC), and Hero (fl. AD 62) all resided in Alexandria for longer or shorter periods. Museum fellows could and did take on pupils – the grammarians Dionysius Thrax of Alexandria (fl. AD 40) and Apion (fl. AD 30) are traditionally held to have studied there under Didymus (b. 63 BC). Scholars generally found it a safe haven from political storms. The Museum was the nerve centre of a cultural community that we would find today in places like the Cambridges.

The Museum inspired copies at the administrative centres of Antioch, Ephesus, Smyrna, Seleucia, and Rhodes. The Attalids of Pergamum in Anatolia (in modern Turkey) imitated the Alexandrian example by creating a medical school and magnificent library, an environment of learning that centuries later nurtured the Pergamum native, the famous physician Galen (ca.129–ca.200). A second-century contemporary of Galen's, the great thinker Claudius Ptolemy (ca.100–ca.170, not related to the royal family) held a professorship at the Alexandrian Museum, part of the small number of chairs in philosophy that Egypt's nonresident monarchs, the Romans, had financed. After AD 200, however, the Museum began to lose some of its intellectual centrality, despite the extraordinary achievements of Ptolemy. Galen's writings suggest as much, because he visited the Museum and wrote disparagingly about its physicians. Alexandria's Museum – with its hundreds of thousands of rolls of books and its heritage in speculative philosophy, with its tradition of high-table meals and sparkling dinner conversation – is a distant mirror of twentieth-century universities. It is difficult to say how much was left of the library and its intellectual circle when Caliph Umar, following a tradition of book burnings stretching from the pre-Socratics through the early Christian zealots, ordered a perhaps largely symbolic purification by fire in AD 646.

Although the ancient museums appear much like the best of our universities today, their line to the present is broken. The medieval arts and philosophy faculties in Europe were not exactly corporations for generating new knowledge; indeed, they owe more to secondary-school instruction in antiquity than they do to the academies and museums. In their final form the seven liberal arts (the quadrivium: arithmetic, geometry, astronomy, and harmonics or music; and the trivium: grammar, rhetoric, and dialectic or logic), which formed the base of medieval university instruction, may be traced to schools of the first century BC. By the imperial Roman period, however, in the schools that retailed these liberal arts, literary studies overwhelmed natural sciences. Like their European successors, Hellenistic and Roman engineers, surveyors, and sailors learned their craft apprentice-style.

The schools of higher learning at Athens, Rome, and elsewhere (or rather, the collection of professors of grammar, rhetoric, law, and medicine at these locations) continued into the sixth century AD, when they were extinguished by Christian fanaticism or barbarian neglect. But the classical tradition nevertheless survived for a thousand years, in Constantinople. Between 425 and 1453, diverse classically inspired schools provided the administrative elite of Byzantium.

The warriors of the Fourth Crusade turned their attention to the conquest of Byzantium. They sacked Constantinople in 1204 and then set about to conquer the outlying provinces. The first Latin emperor of Byzantium, Baldwin I, asked Pope Innocent III to send professors from the University of Paris to found a Latin institute in Constantinople. Innocent agreed to the plan. Also in the thirteenth century, Paris founded a Collegium Constantinopolitanum, designed to lodge and train a score of Byzantine clerics. When Michael Paleologus recaptured Constantinople in 1261, he revived higher learning by appointing George Acropolita (1217–1282, a politician, general, and historian, whom he had freed from prison) to the chair of Aristotelian philosophy. Acropolita also served as ambassador to Rome, effecting a reconciliation of sorts between the eastern and western churches. Twelfth-century Europeans knew about classical learning thanks to hundreds of years of translation from Arabic, but Aristotle entered the fledgling European universities on the tide of Greek learning that issued from Byzantium. It is possible that the notion of European faculties of higher learning – variously guaranteed by church and state – derives from Byzantine precedent.

Eastern cultures

Learned colleges appeared in other ancient civilizations, such as South Asia. The end of the Vedic period in India, about 500 BC, saw the emergence of a wandering brotherhood of secular teachers, the *vadins*. They codified their teachings when imaginative literature began to appear in writing, which until then had been used for administration, commerce, and music. The vadins were in some measure South Asian Sophists, and their activity led to the great schools of Jainism and Buddhism. Jainism, founded by Vardhamana Mahavira, and Buddhism, the teachings of the fifth-century BC Gautama Buddha, both questioned the polytheistic divinities and hierarchical social structure of Vedic traditionalists. For both religious teachers, enlightenment resulted from individual study. Jainist asceticism spread by mass education, while Buddhist thought was concentrated in monastic orders.

With the progressive expansion of Buddhism came the revival of Sanskrit – the language of the Vedic commentaries – as a learned *lingua franca*. The fusion of Buddhism and Vedic traditions around 1200 led to classical Hinduism, with three kinds of educational institution. First (and especially in northern India) were the Gurukula schools, small groups of pupils gathered around a private teacher; astronomy was part of the curriculum. Second were the Hindu temple schools of southern India, inspired by the Buddhist monastic seminaries and supported by land grants; natural sciences seem not to have figured in the syllabus, but because the temple schools had hospitals we may imagine that they incorporated medical instruction. Third were the agrahara centres designed to spread Brahmanic learning. These Hindu schools were pale reflections of the Buddhist colleges that had functioned within grand monasteries since the fifth century. Nalanda (located south of Patna in Bihar, eastern India), one of the most famous of these monasteries, had 10,000 inhabitants at the end of the seventh century; of these as many as 1500 were teachers and about one third were students. It was at Nalanda in the seventh century that the Chinese scholar I-hsing (672–717) copied 400 Sanskrit texts.

Natural sciences in South Asia found their firmest supporters not in schools, but in family-controlled guilds. Astronomical knowledge, for example, was a guild secret. The restricted nature of certain kinds of natural knowledge also coloured science instruction at Chinese colleges. Insofar as we have certain knowledge of them, Chinese institutions of higher learning may be traced to the philosophical schools formed at the time of the Warring States, from 475 BC to 221 BC, when kingdoms large and small contested for supremacy. Teachers were required to train and discipline a civilian bureaucracy, and states naturally competed to recruit teachers who could transform administrative norms into ethical principles. The resulting philosophical free-for-all is known as the time of the Hundred Schools. In terms of the multiplicity of sectarian doctrine, the Hundred Schools seem not unlike the late Hellenic period. A handful of the Hundred Schools survived a period of internecine warfare and continued to have an impact long into a time of imperial rule, indeed, up to the present: the Confucianists, the Legalists, the Mohists, the Taoists, the Logicians and the Naturalists.

The Confucianists, followers of Master Kung, held that virtue could be acquired by learning, although his disciples, from Mencius to Xun Zi, differed about how much education might do for people. Legalists, under Han Feizi, believed in the literal interpretation of legal canons and the inflexible application of jurisprudence, a procedure offered to make law both equitable and independent of executive privileges. Mohists, followers of Mo Zi, proclaimed a religious vision of love and encouraged technological improvements that would defend the weak against the strong. Taoists, tracing their origin to the teachings of Lao Zi, advocated the dissolution of reason in ascetic spirituality; their disengagement from the mechanism of statecraft translated into an antipathy for mechanical contrivance, but their quest for a state of grace led Taoists to experiment with therapeutic regimes for extending and improving life. Logicians, followers of Hui Shih and Kungsun Lung, emphasized a search for generalized concepts transcending the ephemeral particular. The Naturalists elaborated the theories of the two forces (Yin and Yang) and the Five Elements (water, fire, wood, metal, earth), attributed to their master Tsou Yen; in contrast to the other schools, they actively sought to advise heads of state.

As in Hellenistic times with the schools of Athens, the Hundred Schools came together in a secular institution of higher learning, the Academy of the Gate of Chi, located in the capital of the State of Chhi. Founded by King Hsüan about 318 BC, and perhaps inspired by one of its fellows, the naturalist Tsou Yen, the academy assembled scholars of many persuasions and from diverse states. These included Taoists, Mohists, and the great Confucian scholar Mencius. Fellows wore special, flat caps and apparently had no obligations beyond advice-giving; they could aspire to the title of grand prefect.

The Academy of the Gate of Chi – the Chinese counterpart to the Museum of Alexandria – did not survive the imperial unification that ended the Warring States period. The grand victor, Chhin Shih Huang Ti, organized an imperial bureaucracy, brought the defeated aristocracy to heel at his court, expanded public works, maintained a large army, and engineered the great northern wall. As part of his codification of laws and rites, he ordered the destruction by burning of all books except his own archives and treatises on medicine, divination, and agriculture. Along with purging wrong words, the new potentate executed wrong-thinking scholars. Chhin Shih Huang Ti died about 210 BC, barely fifteen years after unifying China; his successor, a usurper son, lasted four years more before the Chhin empire (and its academy) dissolved in social disintegration and revolt.

Liu Pang, an escapee from death row, emerged from the ruins of the Chhin to found the Han empire in 202 BC. His dynasty invented 'classical' China. Genuine concern for preserving what the Chhin had condemned (and not entirely eradicated) is found in the establishment of an imperial school (Ta Hsüeh) in 124 BC, with various chairs (occupied by professors, *po shih*); its aim was to produce functionaries. The Han school produced scholars for the imperial regime, and they were selected by examination. Students received honorary titles commensurate with their test results; the best of them landed positions in the central bureaucracy. (The whole process was sped by the invention of paper, traditionally attributed to Tshai Lun late in the first century AD.) Various accounts describe an impressive campus, with entry restricted to the sons of noble or administrative families. Although students paid no fees, they were required upon arrival to offer gifts to their professors.

Buddhism made its appearance in China by the third century AD. Its ascetic and non-aggressive doctrine found popularity at the time of material dislocation surrounding the collapse of the Han empire into competing kingdoms. In disunited China there were significant attempts at promoting institutions of higher learning, but the instaurations all seem to resemble the various ephemeral and unsuccessful universities of medieval Europe. Around the beginning of the fifth century, for example, the Northern Wei established an imperial school in their capital; the name soon changed to the Central Book School, reflecting its concern with the Confucian classics, for which an anthology, or codex, had recently been prepared.

Chinese civilization emerged from divisions and rivalries to create a golden age under the autocratic Sui and then the Thang. About 583 the first Sui emperor, Wen Ti, revived the nobles' school (Kuo Hsüeh), a school for meritorious commoners (T'ai Hsüeh), and a preparatory school (the Four Gates School), each of which had five professors; he also created for the first time a separate mathematics school with two professors. The purpose of higher education under the Thang was still to prepare students for a government position, and this could be attained by success in a national examination. An inflexible form of this system emerged much later, in the Yüan, when the mandarinate drew exclusively from students who had mastered the Confucian classics. The system did not entirely ignore natural knowledge (from the Thang onward there were separate mathematics examinations), but science undoubtedly constituted the lowest path to success.

A later Thang emperor, Hsüan Tsung, assembled an independent group of high officials to advise him in scholarly matters – the Hanlin (literally, 'Forest of Pencils'). The Hanlin Academy, as it came to be called, emerged as the premier learned authority in China. Awarded the title of Learned Scholar in 738, Hanlin associates – men who were practical as well as erudite – became, by the middle of the century, China's court society of government advisers. Hanlin academicians were charged with emending and authenticating the Confucian corpus that served as the basis of the civil-service examinations. By the Ming period, membership was an exclusive prerogative of senior and accomplished scholars. The Academy extended its authority straight through the Chhing (Manchu), and it expired only in 1911.

The Hanlin Academy regulated orthodox scholarship. Furthermore, the genre of scholarship to be regulated – the Confucian classics – offered scant place for treatises in natural knowledge. The Hanlin did, however, directly supervise an advanced imperial school, revived in the middle of the eighth century, and over the next five hundred years there are persistent intrusions of extra Confucian discourses into diverse state schools. In part this reflects the syncretic evolution of devotional thought, where Buddhist and Taoist notions were incorporated into Confucianism; in part it was a desire to train adepts in medicine, agriculture, and possibly also geography. The time of the Yüan, under the Mongols, again saw the introduction of foreign ideas, the expected result of an empire that stretched from Budapest to the Pacific Ocean. Interest in things Islamic continued with the establishment of the Ming dynasty in 1368. As Chinese traditions merged with those of the Mongols, it becomes appropriate to turn to the institutions of higher learning in medieval Islam.

Islam

A little more than a hundred years after the death of Muhammad in 632, Muslim rule in the form of the caliphate (the successors of the prophet) extended from Samarqand to Barcelona, stopped only by the Byzantines and the Franks. After a century or so of imperial rule, the caliphate devolved into a number of autonomous kingdoms and regimes, the periphery seceding first, organized under a diverse spectrum of caliphs, sultans, maliks, emirs, wazirs, and so on. The notion of a pan-Islamic world survived internecine wars and foreign invasions. Islamic rights were not restricted by political regime, and they entailed no national citizenship. All Muslims were equal before Quranic law in any Muslim jurisdiction, and this equality received continual reinforcement from trade and from the experience of the *hadj*, the pilgrimage to Mecca made by pious Muslims.

Because there was no Islamic pope to decide doctrinal matters (and disputes about dogma precipitated a number of schisms beginning with the earliest caliphs), the teaching of Islamic law became a practical necessity. By the ninth century, nonresidential law schools, or *masjids*, retailing Islamic knowledge in the context of everyday problems, emerged in association with mosques in most large Islamic centres; students lived in *khans*, nonprofit Islamic hostels for pilgrims and transients. From the masjid and the khan came the *madrasa*, the signal educational institution of Islam. It dominated learned life from the end of the tenth century until the nineteenth century.

Masjids and madrasas owed their existence to the charitable donation, or *waqf*. The usual inspirations for charity – piety and pride – lie behind the endowment of madrasas, but Islamic law provided special encouragement for it. A waqf donation, made in person or in a will, circumvented the divisions of an estate among a man's sons, which resulted in the dissolution of private fortunes. By analogy with today's philanthropies, an Islamic waqf could prevent fortunes from being taxed. Furthermore, the donor exercised complete liberty about the conditions of his waqf, provided that he did not contravene Islamic law. He could, for example, purchase or construct an institution, endow it, install himself as director, and specify that direction pass to his descendants. The waqf was inviolate, and it could be broken only if its object was heretical or uncharitable. It comes as no surprise that breaking a waqf – like breaking a modern will – was a regular occurrence.

The madrasas were waqf-endowed colleges for Islamic wisdom, complete with buildings, libraries, curators, service staff, dormitories, and (one imagines) dining commons. Professors and fellows, appointed by terms of the waqf, taught students in numbers from a dozen to more than a hundred. The madrasas had no corporate identity beyond the terms of the waqf, however; Islamic law gave rights only to individuals. There were, then, no corporate diplomas. A disciple received a written commendation from an individual master, his madrasa professor. By implication, madrasas had no sinecures. A professor was paid not to write books, but rather to train students in the art of debating Islamic truth. If he had no students, he could not receive a waqf-endowed salary, and the exercise of dazzling rhetoric was the way to attract students away from hundreds of competing madrasas.

The individualistic approach to higher learning (the lack of which in modern universities educators so often decry) extended to the matter of documentation. A madrasa student aspired to a certificate of mastery signed by a professor. The competent authority – always a man – authorized the acolyte to teach law or issue legal opinions. This licence to teach was a unique development. The Islamic certification, it may be argued, is the origin of the *facultas ubique docendi* – the authorization to teach a particular subject anywhere – issued corporately by professors or by the church at the early Christian universities in Europe.

The madrasa curriculum generally excluded the so-called ancient sciences, the inheritance from the schools and museums of the Hellenistic-Roman world, which in the ninth century, under the patronage of caliphs Harun al-Rashid and especially al-Mamûn, had been translated into Arabic. The exclusion has been seen as a conservative rejection of heretical, or at least contentious, doctrines. Yet the madrasas do seem to have instituted just the method of disputation that dominated the Hellenistic schools, survived into the late medieval period at Constantinople, and formed the basis of scholarly interchange at Christian universities in medieval Europe. Despite contempt for and amusement directed at the ancients, classical works in science did not suffer the opprobrium of a universal ban. Students informally read treatises in natural science and medicine with madrasa professors. Twelfth-century Iberian-based Ibn Rushd (1126–1198), known in the Latin world as Averroës, is illustrative. A professor of Islamic jurisprudence, he wrote major treatises on astronomy and medicine. Although his philosophical works were anathematized and burned at Córdoba in Spain, his writings on the ancients suffered no indignities.

Law – whether natural, conventional, or supernatural – requires a record of opinions, and for this reason large libraries were also a familiar feature of the ninth-century Islamic world. The most famous of these was the Bayt al-Hikmah, or Hall of Wisdom, founded by the caliph al-Mamûm at Baghdad, but it was by no means the only one – in Baghdad or elsewhere. In antiquity and the medieval world, libraries were places for all activities related to books, whether reading, copying, or convening seminars and debates. A library in tenth-century Basra even had a professor-in-residence who gave courses. We see something of this tradition today in the broad sponsorship of cultural activities by the world's great libraries. Influential scholars

and historians of modern times – Lucien Herr and Philippe
Ariès in Paris, George Sarton in Cambridge, Massachusetts,
and Daniel Boorstin in Washington, DC – published their
work as associates of a library.

The Middle Ages

The thirteenth century, the century of great cathedral construction, was a time of unusual organizational ferment.
One of its achievements was the emergence of modern universities. Over the preceding centuries, Christian Europe
had been studying Roman law and wrestling with foreign
wisdom, variously Byzantine and Arabic. Law became
important as the Catholic church contested with civil authorities for control of temporal realms. Teachers of law and
related matters – the grammar, rhetoric, and logic of the
classical trivium – were in great demand at urban centres.
There, overlapping and competing jurisdictions exerted by
the church, nation, county, town, or guild, each with its
rights and privileges, divided up the citizenry. At the same
time cities stimulated the formation of groups of like-
minded people into corporations with clearly defined responsibilities and fields of action. Students and masters of
higher learning fell into line. They took the term *universitas*,
a legal entity with powers extending beyond the individuals
who composed it, much as other trades could have taken
the term.

Universitas was at first always qualified to indicate
whether the academic guild was one of students who named
a rector (as at Bologna, Salamanca, Leipzig, the first Cracow
university, or the law faculty at Montpellier and Prague),
one of masters (as at Paris), or a power-sharing arrangement
of students and masters together, as at Louvain. A large
class of students were also masters, especially those in the
higher, professional faculties who had completed a teaching
licence in the lower arts faculty. (The system continues
today in universities like Yale, where undergraduates are

taught by graduate-student faculty, and in Oxford's Christ Church college, where the masters are called 'students'.)

Student-masters or the senior professors – the *doctores* – could hold courses under a wide range of institutional shelter. These shelters derived from the 'nations' – the protective associations for foreign students that were loosely affiliated with regional origins. By the late medieval period, the shelters were variously called *fraternitates, societates, congregationes, corpora, paedagogia, contubernia, regentia, aula, collegia,* or *bursae*; these fraternities, halls, and colleges had as principal or rector a master who was accredited by the university and was responsible for organizing instruction and overseeing living accommodation. In many cases the halls emerged as an act of charity with stipulations (regarding who might join, for example) reminiscent of stipulations involved in the Islamic charitable trusts that endowed madrasas. The system encouraged a division of the student body into hierarchies of wealth and privilege.

Universities, indeed, were confederations of constituencies – the faculties, the colleges, and all those from maids to apothecaries, copyists, stationers, and later book printers who came under academic protection. Students in many cases ran the show. In southern France, Iberia, and eastern Europe, students not infrequently controlled university offices, notably those of councillor and rector. At Bologna, the student nations had proctors, bursars, and beadles, and they managed considerable amounts of cash. Indeed, masters at Bologna and Padua (both institutions were known as students' universities) organized into doctoral colleges just to defend their interests; in the Paris arts faculty, the masters controlled the nations – and their treasuries. The constituencies took diverse forms across Europe, but they were everywhere at the organizational centre of things. The familiar name of the University of Paris, the Sorbonne, derives from a college founded by Robert of Sorbon (1201–1274), royal chaplain and canon of Notre Dâme Cathedral.

The collegiate structure continued for a long time at a

number of universities. The most famous examples are the colleges still at the base of the universities of Oxford and Cambridge, but colleges figure in the life of other universities as well, notably Pavia. As late as the middle of the eighteenth century, Paris had ten residential colleges. The ambiguity attached to the word 'college' – certainly a learned connotation but otherwise unspecified with regard to *level* – persists up to our time. North-American colleges function variously as faculties or dormitories at universities, while in the Latin world a college is the usual designation for a secondary school. One of Montreal's most distinguished private schools, Lower Canada College, now instructs a class of six-year-olds; Montreal's most distinguished university, McGill, includes Victoria College, a women's dormitory. In Paris and Mexico City there are national colleges with professors who devote most of their time to research. At Rome and Washington, special colleges convene from time to time to choose a new pope or president. In all these instances we find the notion of a common pursuit.

The waning of the Middle Ages led to vesting ultimate academic authority in the faculties on the one hand, and to a levelling of the student body on the other hand. The masters appropriated the collegiate model for their own ends. But the medieval legacy is not hard to spot today. In addition to faculty senates and directors of residential life, we have university fraternities and privately endowed student societies, residential colleges and dining clubs, concessions (bookstores, presses, tailors), and a bewildering hierarchy of professional bargaining units – trade unions of professors, teaching assistants, janitors, and cafeteria workers.

The greatest medieval legacy is that of academic freedom – and not merely for the masters. Medieval students enjoyed considerable privileges. These sometimes included the right to strike as well as protection against cruel and unusual treatment by civil authorities. The privileges were eroded when, over the fifteenth and sixteenth centuries, the masters claimed control of corporative activity, but certain student rights have persisted into the modern world. At some universities today, students exercise a decisive role in the choice of new professors, in matters of professional promotion, and in curriculum reform. Universities continue to discipline members of their own corporation. Even institutions that bear little resemblance to medieval guilds – certain state or provincial universities in North America – tread carefully around the matter of allowing municipal police or soldiers on their campus.

Following the model of early thirteenth-century Paris, law, theology, and medicine were the recognized 'higher faculties'; arts, the fourth faculty, was a grab-bag of skills deemed preparatory for professional careers. The object of university study was to acquire knowledge and be able to teach it, and the course of study was open to any qualified person – that is, any man of the right faith and class. Although the medieval university was a fast track to the three professional guilds, it did not directly prepare students for earning a living productively. There were, it is significant to note, no faculties of engineering, architecture, navigation, or commerce.

By the late medieval period, European Christian universities issued various certificates: the baccalaureate, for competence in teaching certain subjects under supervision; and a master's or doctorate, awarded after a public examination, for admission to the corps of masters. These diplomas persist to our own time, albeit with modifications. German university faculties came to offer, by the eighteenth century, only a doctorate, by which time Oxford and Cambridge awarded, as earned degrees, little beyond a bachelor's. (The nineteenth-century honours degree at Cambridge was held to be equivalent to a German DPhil, according to the polymath John Theodore Merz [1860–1922], who had intimate experience with both systems.[2]) In all cases the diploma signified that the holder came from a background of wealth and ease, and it augured (but did not promise) a career in law, medicine, one or another church, or government.

The universities functioned until the eighteenth century in the absence of a coherent system of secondary education, although something in this line came to be provided by Jesuit colleges and English public schools, among other institutions. For this reason the lower faculties – arts (frequently divided later into letters and sciences) or in Northern Europe, philosophy – continued to provide basic, or remedial, services. Professors of many sciences, then, were from the beginning under continual pressure to lecture far below the level of the research front. The pattern persists to the present day. Medical students learn about the latest diseases, drugs, and instruments; prospective lawyers study last year's legal opinions; future theologians receive the party line from clerical conclaves. But a great many science students never get beyond rational mechanics of the Baroque and thermodynamics of the nineteenth-century Industrial Revolution.

This is not to say that research into natural phenomena and laws did not occur at universities. Medieval university philosophers at Paris, Oxford, Valladolid, Cracow, and elsewhere laboured to elaborate Aristotelian notions of motion, both terrestrial and celestial, as well as Galenic medicine – for these pagan texts had been translated from Arabic and Greek into Latin by the thirteenth century. Investigators committed to understanding the laws of the world including Nicole Oresme (ca.1320–1382), John Buridan (ca.1295–ca.1358), Albertus Magnus (ca.1200–1280), and Roger Bacon (ca.1219–ca.1292) all taught at universities for longer or shorter periods of time. Then as now, however, a professor's freedom to navigate by his conscience depended on the secular and ecclesiastical winds, even after medieval universities acquired self-policing statutes.

A central paradox of institutions of higher learning has always been their vulnerability to ideological or political repression. The burning of academic libraries in classical antiquity and medieval Islam is exactly matched by conflagrations over the past five generations – at Strasbourg, Louvain, Madrid, Königsberg, Tokyo, Beirut, and Kuwait. The condemnation in 1927 of anarchists Nicola Sacco and Bartolomeo Vanzetti by officers of Harvard University and the Massachusetts Institute of Technology, a *cause célèbre* of the 1920s, echoes the condemnation of the subversive Joan of Arc by the University of Paris.

Universities do not make society. They teach what people want to learn, and they give voice to what people prefer to hear. But because they are keepers of tradition and accumulated wisdom, their response time is slow. This allows universities to become authorities for what we know. Relative isolation from prosaic concerns provides a unique environment for encouraging new knowledge about the world. The tension between tradition and innovation is a fundamental characteristic of the European university, and it is central to the enterprise of modern science.

Teaching: From the Time of the Scientific Revolution

Henry Adams, who as a Harvard University professor brought the history seminar to North America from Germany, pondered a thousand years of European culture and proposed, early in the twentieth century, laws for what he saw. In his view, the civilization of western Europe had reached a crisis, as the foundations of medieval faith sank into the shifting sands of technological change. Changes occurred at an ever increasing pace. Knowledge grew and events accelerated. Even with the finest tutors, a person could not keep up with all that was new. Cast adrift in the modern age, Adams dropped his anchor at the cathedral of Chartres, France. From this mooring, he reckoned the meaning of the world, and he calculated its demise in the year 1921. Adams (1838–1918) lived almost from the advent of electromagnetism through the observational verifications of general relativity; he himself measured his life by the technological inventions that he had experienced. He called himself a child of the eighteenth century who struggled to come to terms with the twentieth.

The literate speculations of Henry Adams – who contemplated regularities in the development of Western culture – spawned scientometrics, the science of measuring science. Derek de Solla Price, a firm advocate of the new science who found inspiration in Henry Adams, proposed that the rate of scientific change, however one measured the rate, obeyed a law first formulated by Alfred Lotka (1880–1949). The number of discoveries, periodicals, pages of print, individual researchers, and so on, all grow exponentially for a time until the growth levels off at a plateau. This S-shaped curve, in Price's view, reflected a basic fact of civilization.

The take-off point for Price's exponential curves occurred around 1650. At this time, the institutions of science – whether educational facilities, scientific societies, or scientific journals – blossomed. A host of new ideas, from the heliocentric universe to the circulation of the blood, shook the foundations of Western thinking about the natural world. This constellation of institutional and intellectual factors has been called the Scientific Revolution, a term that describes a period of rapid and radical change.

The Scientific Revolution

The Scientific Revolution of the sixteenth and seventeenth centuries developed to a considerable extent outside the universities, which were bastions of scholasticism and Aristotelian thought. When the Catholic canon Nicholas Copernicus's (1473–1543) book on the revolutions of the heavens appeared in 1543, universities could trace their traditions and prerogatives back more than three centuries. Yet a large percentage of contributors to the new natural philosophy (however it may be defined) were employed by universities, and by far the majority were university alumni. Over the latter half of the sixteenth century, university lecturers at Wittenberg (Georg Joachim Rheticus [1514–1574] and his colleagues Erasmus Reinhold [1511–1553] and Kaspar Peucer [1525–1602]), Tübingen (Michael Maestlin [1550–1631]), Oxford (Henry Savile [1546–1604]), and possibly Cambridge (Henry Briggs [1561–1630]) constructively criticized and otherwise promoted Copernicanism. Salamanca permitted, by statute, Copernicus's thought to be taught. Although by 1600 only a dozen men had lined up solidly behind heliocentrism, the new doctrine was widely disseminated at various universities.

Without labouring the point, it is well to mention some

among the architects of the Scientific Revolution with significant university connections. Copernicus attended universities at Crakow, Bologna, Padua, and Ferrara; in Italy he studied medicine and canon law. Andreas Vesalius (1514–1564) learned medicine at Louvain and Paris and then taught surgery and anatomy at Padua. Galileo Galilei (1564–1642) went to Pisa for medicine and then at the end of the sixteenth century taught mathematics at Pisa and Padua. William Harvey (1578–1657) studied medicine at Cambridge and Padua. René Descartes (1596–1650) received instruction in (among other things) Galileo's telescopic discoveries from the Jesuits at La Flèche and read law at Poitiers. Christiaan Huygens (1629–1695) attended the University of Leiden. Gottfried Wilhelm Leibniz (1646–1716) went to Leipzig, Jena, and Altdorf (where he took a doctorate). Isaac Newton (1642–1727) took a BA at Cambridge and then became Lucasian professor there. Their innate conservatism notwithstanding, universities have indeed served as crucibles for new ideas in natural knowledge.

As the example of Newton indicates, the universities did respond to the 'new science'. Experimental and mathematical natural philosophy at once transcended and underlay the professional interests of the three traditional, higher faculties. The faculties of arts and sciences (or as they were known in northern Europe, faculties of philosophy) were the natural home for this learning, for they had long harboured professors of astronomy, mechanics, and mathematics. Furthermore, by the sixteenth century, schools to prepare students for the university assumed increasing importance, building on a tradition found in several of the medieval English Public Schools (Winchester and Eton) and the Dutch teaching order known as the Brothers of the Common Life. In St Paul's, Shrewsbury, Westminster, the Merchant Taylors', Rugby, and Harrow (all sixteenth-century English creations), and in the profusion of Jesuit colleges in Western Europe generally, adolescents could acquire the basic skills – languages and mathematics – that had previously been retained by university professors of the liberal arts. This preparation freed at least some arts-and-sciences professors from elementary instruction and allowed them to spend more time on the latest word. Clever professors in Italy, the Netherlands, and Germanic Europe were increasingly able to transmit the news and add to their income by attracting interested students. Since the seventeenth century, the prestige of a university has related to the situation of its professors on the research front.

The liberation of natural philosophers in the universities is not unrelated to a general climate of tolerance for diverse religions and credos. This openness governed the golden age of the Dutch Republic (1581–1795), offering a haven to giants like René Descartes and Benedict de Spinoza (1632–1677). Dutch universities were much frequented by foreigners, notably the British, in the seventeenth and eighteenth centuries. By the eighteenth century, Leiden featured an unusually strong corps of science professors, including Herman Boerhaave (1668–1738), Willem Jacob 's Gravesande (1688–1742), and Petrus van Musschenbroek (1692–1761, originator in 1746 of the electrical capacitor known as the Leiden jar). The brilliant Dutch expositors of experimental science had at their command the unparalleled Dutch instrument trade. They were stimulated by the daily arrival of colonial exotica on the one hand and the deadly struggle against the North Sea on the other hand. In their hands, the dissertation for the doctorate became what it is today – a passport in the world of science. Indeed, from the eighteenth through the twentieth centuries, the Dutch doctoral dissertation – much longer than its German or French counterparts – has set the standard for the unwieldy tomes that now issue from the hands of aspiring scholars around the world.

An atmosphere of tolerance also characterized late seventeenth-century and eighteenth-century Scotland. There, as in the Netherlands (and unlike in England, France, or

Spain), a student's religious views were his own business; and like the Netherlands, Scotland enjoyed close contacts with both Lutheran Germany and Catholic France. Aberdeen, Glasgow, and especially Edinburgh cultivated mighty traditions in medicine and natural philosophy. James Gregory (1638–1675) and Colin Maclaurin (1698–1746), prominent Newtonians, taught at Edinburgh, as did the three anatomists called Alexander Monro (father, son, and grandson). Monro *primus*'s physician father had studied at Leiden where he formed a friendship with fellow student Boerhaave. Monro *primus* (1697–1767) cultivated the friendship and brought Edinburgh to rival Boerhaave's Leiden as a medical school. Chemists William Cullen (1710–1790) and Joseph Black (1728–1799) both taught at Glasgow and ended up at Edinburgh. By the middle of the eighteenth century Edinburgh and Glasgow variously featured David Hume (1711–1776) and Adam Smith (1723–1790). The glow of a Scottish scientific education lasted through the nineteenth century – the tenures at Glasgow and Edinburgh of physicist William Thomson, Lord Kelvin (1824–1907) and physician Joseph, Lord Lister (1827–1912) reflect the brilliance of eighteenth-century predecessors. English speakers from the time of Charles Darwin (1809–1882) have gone down to Oxbridge to make social connections and up to Scotland to learn the sciences that were ancillary to medicine.

Autocratic theocracies of the seventeenth and eighteenth centuries, France and Spain, did not encourage freedom of thought in independent academic institutions that might ultimately threaten their own stability. The seventy-odd educational institutions of higher learning in French and Spanish lands (adding in engineering and mining schools as well as large colleges to the list of 'universities', properly speaking) did not blaze with scientific learning. In France, the universities receded before large establishments for scientific research created by royal patronage: the Paris Academy of Sciences and the Paris Observatory in the seventeenth century and their eighteenth-century off-spring, the Paris botanical gardens and the natural history museum. Spain did not see comparable royal research institutions until the mid eighteenth century, and by then there were not many of them (notably the Madrid and Cadiz observatories and the Barcelona Academy of Sciences), but it maintained a string of institutions of higher learning in its colonial possessions.

The rise of the German university

Scottish universities possessed drawing power and brilliant professors; Dutch universities had these attributes as well as a tradition of publishing science doctorates. German universities were something of a question mark. 'The total annual matriculations in the German universities averaged 4200 from 1700 to 1750,' writes historian of science John Heilbron, 'and then declined almost linearly to about 2900 in 1800.'[1] Why, then, do we associate the modern research university with Germany?

Part of the answer relates to a medieval and Renaissance heritage that left Germany with a large number of institutions of higher learning. In the eighteenth century there were four times as many German-language universities as Dutch (five) and Scottish (four) together. The smallest of the German universities, Herborn and Duisburg, shrank to virtual extinction (sixty and eighty students respectively), but nearly all of them awarded a philosophical doctorate. German professors had their hand in scientific research from the very advent of printing, around 1450, and in old-style Jesuit universities and colleges (where philosophy still preceded professional studies), there was adequate employment for science researchers.

Part of the answer relates to historical accident. Only three Dutch universities (plus Louvain and Ghent, which the Netherlands lost to Belgium in 1832) survived the Napoleonic interregnum; it was not until after the middle of the nineteenth century that Leiden and Utrecht began to benefit again from Dutch colonial prosperity. Then, too, the Scottish medical faculties overwhelmed arts and sciences, which never succeeded in organizing a doctoral programme. Some of these conditions also applied to Germany, of course, which lost a good number of universities to Napoleonic reorganization. But eighteenth-century Germany neverthe-less pioneered a new kind of university, where priority went to the philosophy faculty, and this is the image we see everywhere today, when we are accustomed to 'doctor of philosophy' degrees in such unphilosophical subjects as nursing, engineering, and agriculture.

The German universities benefited from competition among the various German states in attracting students and generally building up academic prestige. The dominant late seventeenth-century universities were at Leipzig (belonging to Saxony) and Jena (belonging to Weimar). Prussia then founded Halle in 1694 to siphon off talent from nearby institutions. Hannover founded Göttingen in 1737 to remove the shine from Halle. Maria Theresa revived the moribund universities under Austrian care beginning in the 1750s, banishing Aristotelian scholasticism in favour of experimental physics; her reorganization affected Freiburg im Breisgau, Graz, Innsbruck, Prague, and Vienna, and it had a notable impact on collegiate-structured Pavia, in Aus-trian Italy. To make their mark, these new universities were charged by their state to teach and inspire by propagating and contributing to the stock of knowledge. This notion appeared early in the eighteenth century under the Leibniz-ian natural philosopher Christian Freiherr von Wolff (1697–1754), who lectured and wrote from Halle and Mar-burg, consulted widely across Europe, and turned down a dozen or so university calls.

Emphasizing research in a teaching climate followed the rationalist precepts that had taken Europe by storm in the seventeenth century – notably those of Descartes, Newton, and Leibniz. Uniting research with teaching fitted well with the emphasis on facts and experience that radiated from the writings of the foremost proponent of the new science, Francis Bacon. For the most part the innovation occurred earliest in universities without a medieval pedigree. The receptivity of institutions to change is related inversely to their entrenched traditions.

The professorial research function was opposed by privileged members of scientific societies, who received state emoluments for innovating without having to lecture. But the new style universities were adamant about encouraging research, and they made producing new knowledge a condition of professorial appointment, as, for example, with Johann Tobias Mayer's (1723–1762) chair of physics at Göttingen in 1750. The condition extended to all fields of learning, and universities that ignored it – for example Basle, which at the time chose professors by lot from a slate of three men who usually belonged to the local patriciate – did so at their peril. It gave rise especially to the earliest institutional union of research and teaching known as the philological seminar.

The eighteenth-century university seminar was a key development, and it emerged from the discipline of comparative philology. Two hundred years of European expansion had stockpiled an astonishing variety of tongues. The literature in some of these was sophisticated and not completely foreign to European minds. Sanskrit – the Latin of India – found great appeal among scholars at the new universities, who set out to relate it to everything else they knew, dead or alive. The puzzle had endless parts, each one of which was ideally suited for a doctoral dissertation. The programme demanded specialized libraries, which would be increased from one generation to the next; it required a home and a budget, which university authorities then (no different from now) grudgingly provided. The doctoral seminar was thus born in a room surrounded by dictionaries and reference works. It has remained there ever since.

The doctoral seminar did not extend easily to France.

Napoleonic Europe, focusing on grand state institutions, was no friend of independent corporations with a royalist heritage. In the wake of the French Revolution, Napoleon created a score of pyramidal educational authorities, each one consisting of faculties, *lycées*, and elementary schools, all ultimately responsible to functionaries in Paris. This University of France continued through the nineteenth century, recruiting teenagers to become schoolteachers and, later in the century, becoming a motor of regional economic growth. Higher scientific learning was transmitted in special *grandes écoles* outside the university. The most important of these early in the century was the Ecole Polytechnique, governed then (as now) by the Ministry of War and designed to produce military engineers. There was also the Ecole Normal Supérieure, the national school that set norms for schoolteachers, which at mid century, under the inspired direction of Louis Pasteur, became a privileged conduit to a scientific career. By the twentieth century there were a score of these *grandes écoles*, which recruited by competition and which promised graduates a civil-service posting in diverse technical fields. The French universities have never received their place in science, but a comeback of sorts was made at the end of the nineteenth century in direct response to developments *outre-Rhin*.

Beyond the borders of France, Napoleon engineered the end of a number of universities in the Netherlands and German-speaking Europe. German rulers used the occasion of their new independence to open new universities in propitious administrative seats like Berlin, Breslau, Bonn, and eventually Munich. The notion of pure learning, or *Wissenschaft* (a neologism from the German Enlightenment intended to denote scholarship and science), lay at the centre of the reorganized and the new universities, especially in Prussia. The research spirit permeated the University of Berlin, created in 1810 with the guidance of historian Wilhelm von Humboldt (1767–1835), brother of the naturalist Alexander von Humboldt (1769–1859). Over the next generation research became a way of life for German university professors, as councillors of the various kings, princes, dukes, following a long tradition, competed for prominent men of science.

The German research university in context

The research ethos, already displayed at the larger eighteenth-century German universities, became rooted in nineteenth-century academic life. Germans believed, along with the poet Heinrich Heine (1797–1856), that the promotion of culture through education was the path to national regeneration. Eighteenth-century courtly life, lacking the means to indulge in the profligate dissipation that characterized Paris and London, was nothing if not intellectual. The courtly ideal of *Bildung* – an appreciation of the world combined with self-realization – was achieved by serious study.

Education became a German passion early in the nineteenth century, and the university reforms were connected with a new system of primary and secondary schools. The guiding light for educational curricula was a romantic invocation of classical antiquity which was known as neohumanism. The path to *Bildung*, then, required a large detour through Greek and Latin, the knowledge of which (attested by a diploma, the *Abitur*, issued by a classical secondary school, the *Gymnasium*) was held to be a prerequisite for study in any university faculty. This much was also true of French and English education, although to a lesser degree. The signal characteristic of the German educational synthesis lay in grafting research on *Bildung*.

Accomplishment in research, certified especially by having received a doctorate from a philosophy faculty, signalled that a man was the right sort for instructing German youth. By extension, the battery of examinations instituted to certify young men as customs agents, mine inspectors, and *Gymnasium* teachers went far beyond the practical knowledge necessary for the job. Gymnasium mathematics teachers, for example, had to acquit themselves in many subjects not taught in the secondary schools, such as celestial mechanics. The requirement is less bizarre when it is recalled that the examiners for civil-service ratings were university professors, whose material interests included persuading future civil servants to attend their rather arcane – not to say useless – lectures. In this way, professors circumvented the inconvenient tradition by which philosophy faculties awarded no diploma except the doctorate; the civil-service rating examinations defined a kind of undergraduate degree (comparable to the *licence* in France or the ordinary BA at a Scottish or English university).

Bildung, like Wissenschaft, was practically irrelevant. A cultivated person was unprepared for greasing the wheels of commerce and industry, just as a master of Wissenschaft had no sense of how to turn his knowledge to lucrative gain. University learning, at least in the philosophy faculties, was intentionally abstract. As the German states required engineers and manufacturers, notably for their armies, they looked to France and adapted her solution. They set up civilian copies of the Ecole Polytechnique. By mid century these schools evolved into institutes of technology, independent of universities, called *technische Hochschulen*. Although these institutes did not create the German industrial revolution of steel, chemicals, and electricity (sometimes called the Second Industrial Revolution), they provided thousands of unusually well-prepared engineers (among them Albert Einstein's uncle, Jakob Einstein [1850–1912]) who stewarded the revolution into the twentieth-century age of gigantic industrial firms.

Institute professors, like those who taught Albert Einstein in Zurich, were infected with the research ethos of their university counterparts. Publishing brilliant work was one way to move up from institute to university, a common career move for many scientists (Einstein bucked the trend, resigning a professorship at the German University of

Prague for one at the Zurich Institute of Technology). An institute diploma, however, had nothing of the cachet of a university doctorate. The Zurich Institute of Technology in fact had a special arrangement with the local university, whereby institute alumni could become university doctoral candidates (Einstein twice failed to obtain a doctorate in this way). The engineers clamoured for parity with university graduates to achieve respect at dinner parties and to obtain state privileges bestowed on cultured university alumni. This they progressively (and slowly) achieved. The right to award a doctorate of engineering came in 1900, and more than fifty years later came the honour of calling themselves *technische Universitäten*.

In the nineteenth century, German education was generally a battleground between practical studies, or *Realen*, and impractical Wissenschaft. Chemistry provided the first demonstration that pure learning, left to its own devices, could turn a profit. The demonstrator was Justus von Liebig (1803–1873), who mass-produced chemists from his laboratory at Giessen. He revived the felicitous notion that certain kinds of science, especially chemistry, were teachable less by magisterial lectures than by hands-on experience. He expanded on the lecture-demonstrations of the Enlightenment by extending to natural sciences what had been common for Sanskritists or Provençalists – the seminar. Liebig's students, at first instructed in his home, were largely apothecaries. He taught a technique, synthesizing chemical compounds, that could apply to all nature. He retailed the technique with a goal that no government could dispute, increasing agricultural production. State authorities showered riches on him: a title of nobility and a well-appointed laboratory for teaching and research. The disciples of organic chemistry established themselves across Germany when the great boom of synthetic dyes began, guaranteeing the discipline an independent home in the philosophy faculties. We shall see later that a similar evolution characterized nineteenth-century physics.

The idealism of the Gymnasium movement did not completely extinguish an eighteenth-century emphasis on secondary instruction in Realen – notably modern languages and mathematics. Municipalities and occasionally states also sponsored a variety of trade and commercial schools designed for people who could not afford the luxury of higher learning. Following a natural tendency among academic institutions to seek greater privileges for their graduates, some of the trade schools developed a curriculum and a diploma to rival the Abitur. These advanced trade schools took the name *Oberrealschule*, and their graduates (having learned French and English in place of Greek and Latin) could go on to study at the technischen Hochschulen. To satisfy practical students who wanted a bit of classical gloss, as well as impractical students disabused of the significance of Greek for modern times, a third school emerged at mid century, the *Realgymnasium*, which offered Latin and modern languages and whose diploma (after much soul-searching on the part of university professors) allowed entry to certain professional studies.

The absurdity of preventing future organic chemists from learning modern languages and advanced mathematics before the age of nineteen continued to the very beginning of the twentieth century, when the privileges of the Abitur (including university entry and preferential treatment generally by the state bureaucracies, including the army) extended to graduates of all three kinds of secondary school. But the prestige and unifying force of Gymnasium education – experienced by everyone from Karl Marx and Friedrich Nietzsche to Otto von Bismarck and Max Planck (1858–1947) – has continued to the present day.

The reputation of the Gymnasium, the great scientific engine of the doctorate of philosophy, the vaunted emphasis upon professorial research, the fabled encouragement of independent thought – all these things produced imitators and adaptations around the world. Not everyone, however, accepted the German model uncritically. Boston's Henry

Adams graduated from Harvard University in 1858 and headed for Berlin. He found the university law lectures there depressing and useless. To work up facility in German, he then spent a number of months as a special student with thirteen-year-olds at a Berlin Gymnasium. Of this experience he recalled half a century later:

> The arbitrary training given to the memory was stupefying; the strain that the memory endured was a form of torture; and the feats that the boys performed without complaint, were pitiable. No other faculty than the memory seemed to be recognized. Least of all was any use made of reason, either analytic, synthetic, or dogmatic. The German government did not encourage reasoning.
>
> All State education is a sort of dynamo machine for polarizing the popular mind; for turning and holding its lines of force in the direction supposed to be most effective for State purposes. The German machine was terribly efficient.[2]

That inhuman efficiency, flying the colours of neohumanism, found many theatres of operation over the next century.

The Gymnasium also had unabashed admirers. One of the most illuminating accounts of science at nineteenth-century German universities comes from John Theodore Merz, the British-born and German-educated entrepreneurial and intellectual wonder. He recalls of his Gymnasium days in Darmstadt during the 1850s:

> All my teachers, with perhaps one exception, were, in my opinion, very superior and earnest-minded men, who performed their duties very conscientiously and certainly did not shirk work. They expected the same from the boys, and I believe succeeded largely in securing this. I remember only few instances of serious punishments either for laziness, insubordination, or untruthfulness. Only twice during the six years of my attendance was a boy caned before the class for telling an untruth.

Merz, a polymath in an age of specialization, flourished in Darmstadt's neohumanism. Latin and Greek poetry,

> which we were expected to commit to memory, made the greatest impression on me, and I learnt many passages and whole poems without any special effort simply by hearing them read and repeating them to myself. Many of them I have carried with me through my whole life, and they have been sources of great enjoyment to me in lonely hours.[3]

The nineteenth-century German university is known for its principles of *Lehrfreiheit* and *Lernfreiheit* – the freedom of staff to teach whatever they liked and the freedom of students to attend any course they desired. Tenured instructors called *Privatdozenten* (because their income derived only from student fees, not state salaries) did indeed lecture on subjects of arcane interest (Merz attracted an audience of three when he was a Privatdozent at Bonn). Though salaried professors could do the same, they generally gave large introductory lectures to supplement their income. Students moved freely from one university to another and attended lectures ranging from philosophy to physics. A doctor of philosophy could apply to become a Privatdozent. This process of *Habilitation* meant submitting a dissertation for the right to teach, the *venia legendi* of medieval origin that allowed a professor to teach at any institution of higher learning. University faculties, controlled by tenured professors, were naturally extremely careful not to dilute their ranks (and their earning power) with a large number of Privatdozenten.

Rules were meant to be broken. The *venia legendi* could be revoked by the corporation (the university faculty) that issued it. From time to time professors and Privatdozenten were unceremoniously dumped as political or social liabilities. The most famous of these were the Göttingen Seven, removed from their posts in 1837 for associating themselves with political reform, although other *causes célèbres* included the exclusion of physics Privatdozent Leo Arons (b. 1860) from Berlin in 1898 for his membership in the Social-Democratic Party and the firing of Berlin physiology professor Georg Friedrich Nicolai (1874–1964) for his pacifist activity during the First World War.

Despite the refrain of academic freedom that resounded everywhere in Germany before 1933, university lecturers, much less professors, had to be the right sort of people. Jews, as Max Weber (1864–1920) noted, might take their cue from the motto written over the gates to Dante's hell ('Abandon all hope, ye who enter here'), but they became part of academia anyway. Women, although by the twentieth century not formally excluded, were almost completely absent as German university professors. This situation was not unusual in western, continental Europe, where women professors were phenomena. Physicist Marie Curie-Skłodowska (1867–1934), one of only two women professors at the Sorbonne before 1940 (the other was the organic chemist Pauline Ramart-Lucas [1880–1953]), obtained an appointment as a foreign-born, Nobel laureate, professor's widow; Emmy Noether (1882–1935, daughter of a university mathematics professor) taught mathematics at Göttingen during the First World War only because most young men were serving as soldiers. Discrimination extended to neighbouring lands. In the Netherlands, where women had been earning medical diplomas since the middle of the nineteenth century, the first woman university professor did not begin lecturing until 1917. She was Johanna Westerdijk (1883–1961), who occupied the chair of plant pathology at the University of Utrecht with great distinction.

More significant exceptions to this situation are found beyond the western part of continental Europe. By the last quarter of the nineteenth-century, the Russian-speaking and the English-speaking worlds had created separate, university-grade colleges for women – complete with women professors. St Petersburg featured a women's university, a women's medical faculty, and a women's normal school, all with women science professors. Barnard (at New York's Columbia University) and Radcliffe (at Harvard University) followed the model of women's colleges at Cambridge; colleges like Bryn Mawr, Mount Holyoke, Wellesley, and Vassar, founded independently of male institutions and staffed largely by women, offered advanced degrees; and around 1900 coeducational sectarian colleges such as Oberlin, as well as a host of universities from Sydney to Manchester, signed on women as lecturers of various sorts.

Universities elsewhere

During the Third Republic, from 1871 to 1940, French administrators tried to borrow features of the German universities. Of all nations they were slowest to make the desired improvements, the research doctorate firmly establishing itself in France only in the late 1920s. But what of laissez-faire England?

For nearly two generations, the Scottish pressure valve accommodated the enormous demands for scientific education which had been generated by the First Industrial Revolution of steam, coal, iron, and textiles. The valve became insufficient by the 1820s, when Oxford and Cambridge still discouraged entry from religious nonconformists (the last of the religious 'tests', required for obtaining a diploma, were swept away only in 1871) and offered nothing approximating advanced scientific or even medical instruction. The reform of British education occurred over the middle quarters of the century. Generations after dissenters had established schools for languages and science outside the pale of the Church of England, English Public Schools were renovated with French, German, and mathematics under remarkable headmasters like Shrewsbury's Benjamin Hall Kennedy (1804–1889), Charterhouse's William Haig Brown (1823–1907), and Rugby's Thomas Arnold (1795–1842), the father of poet Matthew Arnold.

Early in the nineteenth century London was a bit like Prussia's Berlin had been: a seat of government without a university. Unlike Berlin, London generated scientific colleges piecemeal. The Benthamite-inspired University College founded in 1828 and the establishment King's College in 1830 eventually became the larger installations of a huge organization, the University of London, which countersigned diplomas (by examination) at many domestic and colonial locations. London colleges offering scientific and technical instruction multiplied: the Royal College of Chemistry (founded in 1845 by students of Justus von Liebig), the Royal School of Mines (founded in 1851 along the model of the Ecole des Mines in Paris), and the City and Guilds of London Institute (founded in 1878 and located in the storied, sixteenth-century Gresham College).

The earliest of the so-called 'red-brick' universities in England's industrial north also grew by collegial accretion. Durham revived its Cromwellian university in 1832, added a college for physical science at Newcastle-Upon-Tyne (of which John Theodore Merz was for many years the guiding spirit) and a medical school, and then picked up affiliated colleges in places such as Barbados (Codrington) and Sierra Leone (Fourah Bay). Manchester, growing from Owens College to a university in 1877, had within its orbit colleges at Liverpool, Leeds, Birmingham, and Sheffield, although these branches declared institutional independence within a generation. University affiliations were marks of prestige and avenues to power at a time when the old Scottish and English universities sent members to Parliament and enjoyed the privilege of conducting courts of common law with the prerogative of imprisoning women for morals offences.

The prosecution of research in English universities was something of an inconsistent accident: Cambridge's Cavendish Laboratory rising to world prominence under its first four directors (James Clerk Maxwell [1831–1879], John

William Strutt, Lord Rayleigh [1842–1919], Sir Joseph John Thomson [1856–1940], Ernest, Lord Rutherford [1871–1937]) and Oxford's magnificently appointed Clarendon Laboratory (founded with money originally willed for a hippodrome) sinking into desuetude under its *fainéant* directors Ralph Bellamy Clifton (1836–1921) and Frederick Alexander Lindemann (1886–1957). To a certain extent science in England lived vicariously from imperial recruits, Rutherford's trajectory (from Christchurch, New Zealand, to Cambridge to Montreal to Manchester and back triumphantly to Cambridge) being a paradigmatic illustration. Before 1918 the preparation of scientists did not generally include a doctorate, the British having marked this diploma (as the Russians also reserved it) as a *laudeo* for illustrious professors. (Rutherford's DSc came courtesy of McGill University after he had been appointed Second Macdonald Professor of Physics there.)

Unlike England and France, the United States responded with enthusiasm to notions Germanic. Early in the nineteenth century the new nation had religiously-affiliated colleges of the English kind, public universities financed by individual states, and a diverse collection of privately endowed institutions of higher learning – there being generally no governmental restrictions on recognizing institutions that variously styled themselves colleges, academies, and high schools. There was, indeed, no clear distinction between secondary education and higher learning, high-school and college diplomas being roughly equal in number across the nineteenth century. Americans adopted French engineering schools as soon as the Germans did, and with more felicitous results. Until 1850 science was best acquired at the West Point military academy and at nearby Rensselaer Polytechnical Institute, both modelled on the Ecole Polytechnique. Diverse polytechnics since then, such as the Massachusetts and California institutes of technology, Case, Carnegie, Armour, Rice, Stevens, and Drexel, established themselves as temples of science and technology, the Americans never having separated (as the Germans, the French, and the English separated them) the two distinct traditions.

About 1870 a number of high-minded American educators introduced the German philosophy doctorate under its Latin cognomen (*Philosophiae Doctor*, or PhD), even though not one of them oversaw an institution with a philosophy faculty. The innovation spread through refurbished religious institutions, like Yale and Harvard, older private institutions like the University of Pennsylvania, state universities like the ones at Berkeley, Ann Arbor, and Madison, and newly endowed institutions of learning like Johns Hopkins, Vanderbilt, the University of Chicago, and Stanford. When universities like Princeton and Duke upgraded themselves, they expanded in the direction of 'graduate' studies. To distinguish the research function from the usual propaedeutic mandate, American universities invented the 'Graduate School' as one of their constituent divisions.

In Europe the university was a corporate entity with state prerogatives – a guild structure – rather than a self-contained and contiguous physical plant. Sixteenth-century and seventeenth-century transplants in places like Santo Domingo, Quito, Puebla, and Manila followed the European model, lodging professors and students wherever room could be found in the neighbourhood of ecclesiastical monuments. Although certain private corporations continued the European pattern (notably in dense, urban settings like Philadelphia and New York), by the eighteenth century, the colleges and universities erected beyond Europe had made use of their greatest asset – land – as a privileged domain. The university *campus* proliferated. The College of William and Mary is emblematic. It is situated at one point of an isosceles triangle, the other apexes being the colonial Virginia legislature (House of Burgesses) and the governor's residence. The granting of land has subsequently figured in the foundation of new universities, and even the Europeans came to embrace the principle. The finest extended example of nineteenth-century German academic architecture, in fact, is the splendid campus of the Université Louis Pasteur, erected as an imperial German university at Strasbourg shortly after the Germans conquered Alsace in 1871.

The nineteenth-century campus was designed as a bucolic retreat, incorporating sylvan glens, conspicuously vacant fields, and arboreta, all of which might be seen as compensation for the lack of state privileges. As science helped to propel the Second Industrial Revolution, scientists were able to withdraw into specially constructed temples of limestone or marble in pristine settings.

Whether on or off a campus, specially designed research laboratories graced universities for the first time during the latter part of the nineteenth century. At the beginning, form was everything – allusions on the building's portals reminding immature minds about the long tradition of science and the great power it represented. By the last quarter of the century, form was only skin deep. Science laboratories circled around large lecture halls for elementary courses (especially the ones frequented by medical students, who by 1880 were often required to go outside their faculty to learn about physics). Next to the lecture halls was an array of teaching laboratories, special halls for diverse kinds of professorial research (including massive stone plinths set independently of the building's foundations to provide vibration-free working surfaces), rooms for housing steam and electrical generators, and seminar rooms and libraries. Larger physics institutes like the one at Berlin, inaugurated in 1877 on piers driven into a canal, had apartments for the director (Hermann von Helmholtz [1821–1894]) and his family, the assistants, and the maintenance staff of mechanics and maids. Like the one at Berlin, enormous structures arose at Leipzig and Zurich – veritable kingdoms under the command of a director. European institutes – massive, multistoried structures with several wings – dwarfed older academic buildings. The specialized laboratories eclipsed even the jewel of humanists, the library.

Harvard's Widener Library, the world's largest university-owned book repository at its dedication in 1915 and an imposing memorial to a young man who went down with the *Titanic,* was not a great deal bigger than the Berlin physics institute.

By the end of the nineteenth century, nation states were bankrolling prestigious empires of science at independently administered universities (or, in France, at *grandes écoles*). Bureaucratic response followed swiftly: the state constructed its own scientific laboratories and funded them even more handsomely. Some scientists abandoned universities for the new settings (Helmholtz's presidency of the Physikalisch-Technische Reichsanstalt represented the leading edge of the new wave), but most sat on the fence between the traditional prestige (and independence) of a university position and the vast resources (with strings attached) of the new federal research centres. The usual arrangement was to divide time between (and accumulate emoluments from) university and state laboratory. A clever scientist could play off each patron against its competitor. This is what Albert Einstein did when he went to Berlin in 1914. Appointed to a salaried chair at the Academy of Sciences (positions in the same vein had been funded by the academy for many years), he received a courtesy appointment at the local university (allowing him to supervise doctoral students) and a titular directorship of an institute for theoretical physics in the federal laboratory structure known as the Kaiser-Wilhelm Gesellschaft. Einstein used the academy position for publishing rapidly and circulating reprints free of charge; the university post for staying abreast of bright young talent and new scientific ideas; and the Kaiser-Wilhelm post for privileged access to its industrialist, financier, and politician patrons. Certification in research nevertheless remained a university prerogative. German universities continued to award doctorates throughout the twentieth century, and the doctorate became a *sine qua non* of scientific life elsewhere – even in England, where today a certain prestige still attaches to a scholar who, like Lawrence Stone (b. 1919), Quentin Skinner (b. 1940), or Simon Schama (b. 1945), may not sport an earned doctoral degree.

Universities in the United States grafted the doctorate onto an existing structure, the undergraduate college, whose standards approximated those of a French lycée or German Gymnasium. Seeking the grail of appropriating European wisdom, American professors (complemented by a large number of European imports) taught specialized courses to students registered for an advanced degree. This new structure – departing from the freedom to choose courses which was enjoyed by European students – slowly but inexorably increased the time required for obtaining a doctorate and inflated the length of doctoral dissertations. As higher learning experienced an uneven course in Europe under the excesses of fascism and Stalinism, the modified American model provided a new standard for research training.

From the end of the nineteenth century, foreigners were astounded by the material resources of American universities. The English mathematician James Joseph Sylvester (1814–1897), Swiss naturalist Louis Agassiz (1807–1873), and German biologist Jacques Loeb (1859–1924) held significant university posts in America; by the end of the century, an American lecture tour was obligatory for leading scientific lights, like Englishman Thomas Henry Huxley (1825–1895), German Felix Klein (1849–1925), and Austrian Ludwig Boltzmann (1844–1904, who ironically referred to his tour as a voyage to *El Dorado*). Immigrant talent educated in the United States – physicists Albert Abraham Michelson (1852–1931) and Michael Idvorsky Pupin (1858–1935) – rose to the heights of their discipline. But all comers did not stay. Max Abraham (1875–1922) took the measure of a physics chair at Urbana in 1909 and then returned to Europe, where he had no comparable position. Einstein's first scientific collaborator Jakob Laub (1882–

1962) declined to fill Abraham's Urbana chair, opting instead for one at La Plata in Argentina. Shortly after the turn of the century, Ernest Rutherford would not forsake McGill University in Montreal for Yale (although he did leave when Manchester beckoned). The United States of the 1890s held no permanent attraction for young Bertrand Russell (1872–1970), fresh out of Cambridge and married to an American Quaker. For scientists at the peak of their career in Europe, the preferred arrangement was a visiting lectureship, like those liberally endowed before the First World War. Under this arrangement, physicists Hendrik Antoon Lorentz (1853–1928) and Max Planck taught at Columbia University. After 1918, Einstein was lured to the California Institute of Technology for months at a time. As these examples suggest, by the first decade of the twentieth century, it was normal for German or French professors to take leave from their universities in order to occupy positions abroad, notably in the New World. There were even world-ranging, extramural professorships. In 1914, for example, geophysicist Gustav Angenheister (1878–1945) became a special professor who split his time between Göttingen and the capital of Western Samoa.

Technology has made commuting professors an established feature of academic life. In the 1920s, theoretical physicist Wolfgang Pauli (1900–1958) commuted by train from Göttingen to his lectureship at Hamburg. The possibilities of commuting coincided with the end of the university science institute as a personal empire, presided over by the professor and his wife. The institute or laboratory became a university monument, rather than (as it was during a brief moment, between approximately 1870 and 1910) a living part of a professor's aura. Only the president's mansion, often conspicuously located on the campus of a new university, allowed state or private overseers to place an administrator on public display. But because the presidential office served as an obvious focus for student discontent, the mansion sometimes became a white elephant. Today, the president of the University of Southwestern Louisiana lives happily on campus, but the gothic presidential mansion of the University of Tokyo stands vacant – the victim of student protests a generation ago.

Along with the end of the university institute came the rise of the university *department*. By 1900 professors and lecturers sometimes organized sequences of courses, assigning responsiblity for all the parts of a domain, but the spectacular fragmentation of knowledge led to a hierarchical structure for managing it only in the United States. There, the arrangement extended to a military command structure, with a department chair, professors, associate professors, assistant professors, and a host of supporting staff. The departmental innovation coincided with the rise of the department store and the departmentally structured industrial firm. The inspiration is found in the administrative units of the federal government. With the model of academic departments in science, American universities distanced themselves from the European tradition where a professor taught what he liked. Science instruction became highly organized and goal-oriented. In the nineteenth century, European academics were traditionally able to take advantage of fast-breaking developments in neighbouring disciplines; in the twentieth century, innovative American academics spent much time and energy breaking out of disciplinary confinement.

Both geographical decentralization and interdisciplinary innovation have become watchwords in academic science. Electronic information-processing to some extent obviates the necessity for a scientist or scholar to reside at an ancient college of learning. Universities everywhere have adapted to new socioeconomic conditions by expanding curricula. They have always responded in this way, although never as quickly as their critics would like.

Measured and deliberate innovation is one of academia's heavy burdens. It is also a great strength. Emerging fields of knowledge become new scientific disciplines only after they have found a secure place in universities. We look to universities for an authoritative word about the latest innovations. New scientific ideas emerge in a variety of settings, but they become the common heritage of humanity only when processed by an institution for advanced instruction like the modern university.

Sharing:
Early Scientific Societies

Above the deafening cacophony of a dozen screaming four-year-olds, a daycare teacher admonishes, 'Now share!' The concept of sharing a toy – of sacrificing individual possession for a communal experience, of deferring pleasure until others have taken a turn at gratification – is altogether foreign to the toddler, whose universe heretofore has been entirely self-centred and unabashedly selfish. It is seen as an important measure of maturity when the child is able to transcend the universe of 'me and mine', and to begin to entertain the idea of a greater social imperative.

The development of science seems to recapitulate the odyssey of every individual as he matures from infantile egotism to participation in the universe of social give-and-take. In the ancient and medieval worlds, learning about the natural world proceeded by fits and starts. People recorded intriguing theories and thoughts, constructed ingenious mechanisms and monuments, and even established schools. There existed, however, no special notion of a common mission to uncover new truths about nature, no clear idea that a division of labour could prove especially conducive to the rapid accumulation of knowledge. Earlier thinkers tended to guard and keep secret what they knew, fearing that good ideas might be stolen by a rival.

With the Scientific Revolution of the mid seventeenth century, the cultivation of natural knowledge ceased to be solitary and introspective; it became shared and communal. By working *together*, according to this new outlook, philosophers could accomplish more than they could by working separately; the cumulation of individual efforts by sharing would result in more gains to science than the summing of its isolated parts. Furthermore, what contemporaries labelled the 'new science' – signified by a corporate or composite effort – also aimed to replace words with deeds, the library with the laboratory, and systems with facts. This emphasis on activism, experiment, and experience stimulated the establishment of scientific societies, special associations where individuals could congregate and cooperate in advancing the new science.[1] In this chapter we examine the anatomy of the new societies.

These institutions for sharing became the dominating and distinguishing feature of science during the second half of the seventeenth century. Scientific societies were an essential component, not a mere by-product, of the Scientific Revolution. They became a vital instrument for formulating and transmitting scientific norms and values. They transcended the pedagogical tradition associated with universities and established a new routine, inspired by everyday circumstances. Scientific societies held meetings at regular intervals; they elected officers and set up committees. Such daily activity led to the establishment of 'a seasonal calendar of ritual: the first formal meeting of the year, periods of election, ordinary meetings, breaks for religious and state holidays, public meetings, vacation' and so forth.[2] Scientific societies may have exalted the tedious and the dull, but they enshrined a secular calendar for these mundane affairs – an essential figure of modernity. In other words, time was organized without the traditional appeal to sacred celebrations or agricultural cycles.

What led natural philosophers to embrace a new ideology associated with *sharing*? Certainly they did not think that invention would cease to be the fruit of one mind and would become a collective procedure. They were, after all, proud of their own discoveries. Rather, they saw advantages to associating with a group of like-minded people. The form of their association departed from medieval guilds. Associations for promoting the new science ignored matters of faith and livelihood. Nor did scientific associations seek to train apprentices. They were an avocational service club – the seventeenth-century equivalent of the Odd Fellows or Rotarians.

The learned society or academy of the seventeenth century incited and rewarded independent work. It also provided an avenue for communicating the results of scientific investigations, at first by means of the private correspondence of a secretary, and later through formal minutes and journals. Scientific societies housed books in their libraries, displayed specimens in their museums, and collected instruments in their cabinets, all these services assisting the investigations of individual members. Groups were naturally better able to purchase the costly tools required by the new science, whether telescopes, microscopes, or burning-mirrors. In this way, scientific societies made the matériel for conducting science accessible in a convenient and relatively inexpensive form. By the end of the seventeenth century, any man of scientific reputation and accomplishment belonged to a learned society or academy.

Nascent scientific organizations fulfilled less obvious functions, as well. Just to be associated with these enterprises conferred prestige on a member. This has been true virtually from the beginning, and 'FRS' (Fellow of the Royal Society of London) or 'membre de l'Institut' (member of one of the national academies of France) is today a coveted designation. In addition to this honorific function, periodic meetings of societies provided a forum for individuals to meet and discuss their work. Universities had no real place for the exchange of ideas among equals (there were neither faculty clubs nor professorial offices), but in the halls of the academy, controversies could be aired, alliances forged, and criticisms vetted.

Whence the notion for these associations? Some of them found inspiration in an invocation of Platonic free assembly and corporate activity, beyond political control. Others looked back to the Renaissance, when learned men came together under the influence of a particular patron or court. Yet, as we shall see, the Royal Society of London represented a novel departure: For the first time, individuals united together in a public body dedicated to the corporate prosecution of scientific research.

Engines of the Scientific Revolution

The Royal Society of London, founded in 1660, promoted 'a cluster of disciplines concerned with natural and mechanical phenomena to the exclusion of others, linked by common methods'. It aimed to advance the realms of natural philosophy and natural history (roughly equivalent to our physical and biological sciences), and distanced itself from discussions of theology or scholastic philosophy, which it perceived as sterile. The Society's devotion to the production of knowledge, rather than to its dissemination, sets it apart from other contemporary institutions. Its importance and prestige was confirmed by royal incorporation at the hand of Charles II.[3]

Sir Francis Bacon, a lawyer and chancellor to James I, became the patron saint of the Royal Society and of many other scientific societies as well. Bacon's scientific contributions were unremarkable, but he enjoyed tremendous posthumous influence as the principal polemicist for the new science. In the *New Atlantis* (1627), he called for the creation of research institutions to accommodate the new learning. There he described 'Salomon's House' – a collaborative effort dedicated to 'the knowledge of causes, and the secret motions of things; and the enlarging of the bounds of Human Empire, to the effecting of all things possible'. Bacon maintained that only by combining the efforts of individuals could humankind hope to tackle the enormous range of questions that should be raised about the natural world. This programme formed one of the components of his projected *Great Instauration*, a work incomplete at the time of his death, and it complemented the inductive approach sketched in his *New Organon* (1620).[4]

Baconian ideology infused the creation and early years of the Royal Society. As the Society's apologist Thomas Sprat put it, Bacon's writings contained 'the best Arguments, that can be produced for the Defence of experimental Philosophy, and the best Directions, that are needful to promote it'. Bacon's views not only permeated Sprat's official *History of the Royal Society* (first published in 1667), but they also found expression in the Society's charters, diffusion in the *Philosophical Transactions*, and reiteration in the writings of fellows like Robert Boyle (1627–1691) and John Evelyn (1620–1706). Baconianism so well reflected the motivations of diverse associations of scientifically inclined amateurs in England that historians still try to identify the group that led directly into the creation of the Royal Society. Depending on which historian's arguments one believes, the Royal Society may be traced to a gathering of gentlemen associated with Gresham College in London, to a less pragmatic network of London philosophers and social reformers, or to a collection of natural philosophers who eventually came to reside in Oxford.

The first of these, Gresham College, had been founded in 1597 by a legacy of the London merchant Sir Thomas Gresham to provide a series of educational lectures on a variety of topics for the local townspeople. Gresham also established resident professorships in astronomy, geometry, and medicine. His former townhouse provided a natural meeting place for scientifically inclined men, including sometime lecturers Robert Hooke (1635–1702), Christopher Wren (1632–1723), and Isaac Barrow (1630–1677).

A second London group of Puritans and Parliamentarians, who flourished during the 1640s and 1650s, was attracted by the millenarian zeal exuded by Continental collaborators Jan Comenius (1592–1670), Samuel Hartlib (d. 1662), and Theodore Haak (1605–1690). John Dury (1596–1680),

William Petty (1623–1687), and John Evelyn numbered among the reformers who viewed the association of scientists in a scheme by Hartlib for an 'Office of Address' as a mechanism for practical improvement and social advancement. The 'office', motivated by Protestant fervour, collected information about utilitarian discoveries and inventions.

Still another group – including Seth Ward (1617–1689), Thomas Willis (1621–1675), and William Petty – went up to Oxford from London because their mentor John Wilkins (1614–1672) had assumed the wardenship of Wadham College. Wilkins, brother-in-law to Oliver Cromwell, made the remarkable transition from Puritan divine to Anglican bishop. His followers were part of the Royalist exodus from London (and Gresham College) that had occurred during the upheaval of the Commonwealth period, when the Puritans assumed the reins of government. Robert Boyle's move to Oxford attracted others to the quiet college town, including architect Christopher Wren and experimenter Robert Hooke. This small group of natural philosophers organized weekly meetings to perform and conduct experiments. Some scholars contend that this was the incipient Royal Society – an association that had existed as an 'invisible college' under the Puritans and even previously during the reign of Charles I.

Whatever its historical antecedents, the creation of the Royal Society of London for Improving Natural Knowledge was assured when twelve men of diverse backgrounds – from Royalist to Cromwellian – gathered at Gresham College during the early days of the restoration of the monarchy, in 1660. They resolved to meet weekly to discuss and advance natural philosophy. Two years later, Charles II granted the group a royal charter. A second charter of 1663 established the operating rules and procedures of the Society. These actions bestowed upon the group of 115 scientific virtuosi a corporate status comparable to the one enjoyed by lawyers in the Inns of Court and by medical doctors in the College of Physicians. The incorporation of the Society itself meant that it could own property, employ officers, possess a seal and coat of arms, and license its own books.[5] These were significant legal privileges at the time.

In his book *The Great Instauration* (1975), Charles Webster suggests that questions about the Royal Society's origins and true character can be resolved by determining the Society's *active* members. Webster identified twelve fellows – among them Boyle, Evelyn, Petty, and Wren – whose activity dominated and sustained the fledgling Society during its first two and a half years. Webster concludes that preliminary meetings were held in London during the closing years of Cromwell's republic and that 'diversity of outlook and experience' brought a remarkable advantage to the group. He contends that it is 'superfluous' to ask whether the nucleus was Puritan or Anglican, Parliamentarian or Royalist. The early Society evolved continually in terms of its composition and interests, just as religious beliefs and political convictions fluctuated beyond its confines.

The diverse religious and political composition of the Royal Society set a premium on limiting activity to natural philosophy. The exploration of experimental and mathematical problems concentrated the energies of early fellows and minimized more fundamental differences of opinion. In this way, the Society's work remained unaffected by the collapse of Cromwell's republic and the restoration of the monarchy. In Webster's words, 'scientific work was insulated from ideological friction'. Science, according to this view, is an anodyne for social dislocation.

The Royal Society dedicated itself to 'the advancement of the knowledge of natural things and useful arts by experiments, to the glory of God the creator and for application to the good of mankind'. It was governed by a president and a council of twenty-one fellows, from whose ranks were elected a treasurer and two secretaries. The Society employed at least two Curators of Experiments, obtained the cadavers of criminals for anatomical demonstrations, and built quarters for its assemblies in London. Fellows had to be elected by the general membership and upon election had to pay an admission fee, in addition to an annual subscription.

Although the Royal Society may be considered an organization that rewards the achievements of a scientific elite, its membership down from the early days has been relatively large, especially when compared with the size of other national scientific organizations. From its inception, the Society included a large proportion of virtuosi from the leisured classes, men whose interests have encompassed historical, literary, artistic, and archaeological studies. To the more avid scientific practitioners in the Society, the concerns of this element (who were needed for their wealth and social status) appeared aimless, unfocused, and obscure. The virtuosi also gave the Society a tendency to devolve into a social club for gentlemen. (When this current took hold in the Society during the early nineteenth century, it was ironically a member of the aristocracy, the duke of Sussex, son of George III, who reformed the Society and restored its learned purpose.)

Historian of science Marie Boas Hall has recently shifted attention from the organization's origins and sociological composition to what actually occurred at its meetings. She has been particularly interested in the extent to which experiments were performed by the Society's paid employees, both curators and operators, during its early years. Empirical discussions and demonstrations of experimental results seemed to offer a respite from potentially divisive political or religious issues. The airing of hypotheses, says Boas Hall, in contrast, led to 'disputes and wranglings' inappropriate to a 'quiet atmosphere of learned debate'.

Boas Hall concludes that although the early Society paid lip-service to the promotion of 'Physico-Mathematicall Experimentall Learning', early enthusiasm soon gave way to the mere reading of papers and discussion *about* experiment. Although a small core of virtuosi maintained interest in the demonstration of experimental phenomena by *operators* (the title is significant) like Robert Hooke, most fellows sensed that the descriptions of experiments in the Society's *Philosophical Transactions* possessed more enduring value than demonstrations. In the words of A. Rupert Hall, the Royal Society became 'a place of report rather than a research institute'. Rhetoric and the prestige that flowed from association with eminent names like Isaac Newton and Robert Boyle nevertheless ensured that contemporaries and historians alike have linked the Royal Society with the new experimental philosophy.[6]

The early Royal Society's fulfilment of the Baconian imperative depended entirely on individual initiative, whether Operator Robert Hooke's enthusiasm for performing experiments or Secretary Henry Oldenburg's (ca.1618–1677) prosecution of the plan for creating a universal natural history. Oldenburg's zeal for the task led to the publication of some 'histories' (more properly, narratives) of trades in the *Philosophical Transactions*. These experiential accounts derived from Oldenburg's queries addressed on a regular basis to correspondents all over the world; by 1668, the annual volume of incoming and outgoing letters supervised by Oldenburg generally exceeded 300. James McClellan characterizes the Royal Society as encouraging 'a vaguely defined Baconian empiricism that meshed well with the format of its meetings and the looser interests of its members'. He also sees the outward turn away from a dedicated Baconian core as the mechanism that propelled the Society to become the most important learned society of the second half of the seventeenth century.[7]

Part of the Royal Society's Baconianism may have been rhetorical. The society encompassed a heterogeneous membership and tended to create myths about its cohesiveness when it was under attack. And attack its critics did. In *Gulliver's Travels* (1726), for example, Jonathan Swift ridicules the futile projects pursued in the 'Academy of Lagado', inspired by the research undertaken by members of the Royal Society. Historian Martha Ornstein is so persuaded of the rhetorical use of Baconianism that she sees the imagery of 'Salomon's House' as fulfilling for learned societies what the Communist Manifesto did for socialism.

Other scientific societies did not trace their inspiration so directly to Bacon. Galileo wielded enormous influence over scientific developments in Italy, and he was a member of Rome's Accademia dei Lincei, founded in 1603. Like Rome, many Italian cities housed learned societies, more properly Renaissance academies that promoted a range of subjects: Bologna claimed an Accademia degli Affidati (1548) and Naples an Accademia Secretorum Naturae (ca. 1560) and later an Accademia degli Investiganti (ca. 1650). Unlike other Renaissance academies, however, those in Bologna and Naples concerned themselves with the cultivation of natural knowledge, rather than literature or the arts.

The foremost among the Italian academies was the Florentine Accademia del Cimento (Academy of Experiments), founded in 1657. The small society of nine members – including the important naturalists Giovanni Alfonso Borelli (1608–1674) and Francesco Redi (1626–ca.1698) – depended on the patronage of Prince Leopold de'Medici and answered to his whims. It assembled a fine collection of scientific instruments to effect its sole purpose: conducting experiments. Members tested the theoretical work of Galileo and his disciples and recorded the results anonymously in the Academy's *Saggi di naturali experienze*. Despite the group's pronounced commitment to empiricism and their rejection of all speculative theorizing, Academy members fell victim to the conservative backlash of the Inquisition and Counter-Reformation. It also suffered through the centrifugal force of members' personal quarrels, resulting in disbandment for ten years until they settled their differences.

Even seventeenth-century Germany, in its state of political fragmentation and economic torpor, could claim scientific societies. In Altdorf, a Collegium Curiosum sive Experimentale was created in 1672 with twenty members, after the model of the Accademia del Cimento. Some twenty years earlier, an Academia Naturae Curiosorum had been founded, whose principal function was to publish an annual volume of contributions by its physician members, the *Miscellanea Curiosorum*. But it was only with the creation of the Berlin Academy in 1700, at the urging of Gottfried Wilhelm Leibniz, that Germany could claim a society along the lines of the Royal Society or France's Académie des Sciences. The society was to be funded by the proceeds from the monopoly on printing calendars owned by the elector (the future Prussian king, Frederick I). Part of the Berlin Academy's programme involved the advancement of German technology and nationalism, giving particular attention to improving the German language. Leibniz's activism also led to the creation of the St Petersburg Academy of Sciences in 1724.[8]

In France, academies could be found in provincial towns like Caen, Rouen, and Montpellier. These included not only learned societies as such, but also other kinds of educational institution, including schools of manly exercise, classical languages, and oratory. The capital city (as in England) dominated scientific life at this time. One of the earliest informal circles in Paris – dating back to the 1630s – was organized by the Minim monk Marin Mersenne (1588–1648), himself devoted to the physical sciences. Mersenne, who had studied mathematics with Descartes, translated some of the writings of Galileo into French and popularized the work of Blaise Pascal (1623–1662). After Mersenne's death in 1648, a successor to his academy was organized by nobleman Habert de Montmor (ca.1600–1679), which adopted a formal constitution in 1657. Weekly meetings took place in Montmor's house; mathematician and cleric Pierre Gassendi (1592–1655) presided over them. But the Montmor Academy became as much a social club for the highest levels of Parisian society as a forum for disseminating the new science.

It was through the Montmor Academy that the Royal Society began to influence the future shape of science in France. Members of the two organizations were linked by correspondence and personal visits; some individuals, like the Dutch scientist Christiaan Huygens, belonged to both. The French admired the new spirit of critical enquiry exemplified by the English cultivation of empiricism and experiment. It remained unclear, however, how the English model of cooperation among men of different social backgrounds, political persuasions, and religious convictions might be applied in the French milieu. Personal rivalries – fuelled by competing philosophical doctrines like Cartesianism and experimentalism – helped to spell the collapse of Montmor Academy by 1665. The instability brought about by its indifferent financial support strengthened pleas by Melchisédech Thévenot (ca.1620–1692), Adrien Auzout (1622–1691) and Pierre Petit (ca.1594–1677) for the creation of a *subsidized* society for experimentation.

Jean-Baptiste Colbert, minister to Louis XIV, responded sympathetically to the advances of the former Montmorans. He adapted the plans put forward by Thévenot and his friends, in the end calling for fifteen salaried academicians, hand-picked from among the most distinguished scientific names of Europe. The positions were divided between two categories or classes: 'mathematicians' (also including astronomers) and 'natural philosophers', made up of chemists, physicists, and anatomists. (The decision to emphasize the physical sciences resulted from Colbert's concern to minimize conflict with other established bodies, such as the Faculty of Medicine in Paris.) In contrast to the Royal Society, members were expected to specialize in a particular area of study. Their first meeting was convened in the Royal Library in 1666. Subsequently, meetings were held twice a week: mathematicians met on Wednesdays; natural philosophers on Saturdays.

There were strings attached to this act of royal munificence, especially on the part of the mercantilist Colbert. The

Académie des Sciences joined the Académie Française in the Sun King's intellectual firmament; at the very least, it was intended to proclaim, affirm, and reflect his glory. Academicians, in addition, were expected to deliver on the experimentalists' utilitarian promises, which linked scientific investigations with advancement in industry, trade, and military prowess.

As a result of being given a clear mandate from the government, the early Académie des Sciences appeared to embrace the Baconian programme of cooperative research in at least two concrete ways that the Royal Society did not. The establishment of the Observatoire de Paris in 1699 allowed Academicians to carry out a continuous programme of observing the heavens and mounting scientific expeditions, with these undertakings ultimately leading to the solution of navigational and astronomical problems. The Académie also required its members to cooperate on a regular basis in order to adjudicate the merit of technical processes and to bestow patents on worthy inventions. The practice of the early Académie des Sciences suggests that cooperative efforts were more effectively applied to evaluating new ideas than to creating them.

The workings of the early Académie des Sciences remain somewhat obscure, at least until a total overhaul occurred in 1699. Before this date, the Académie had possessed neither rules nor constitution. Colbert himself had selected the first academicians, foreign as well as French, the most distinguished being the Dutch natural philosopher Huygens. Later appointees to the working membership of fifteen pensionaries – rigidly divided according to scientific speciality (geometry, astronomy, mechanics, anatomy, or chemistry) – included the astronomer Gian Domenico Cassini (1625–1712) and the polymath Leibniz. The Académie possessed, in addition, ten honorary positions. Somewhat surprisingly, Cartesians were excluded in this, the home of Descartes; activists like Auzout and Thévenot were marginalized. At this early stage in its history, the Académie des Sciences functioned under Baconian inspiration, with a small membership undertaking joint experimental investigations on a range of topics. It was an elitist association, limited in size with an exclusive admissions policy.[9]

To some extent the early Royal Society and the Académie des Sciences may be seen as typifying the English and French scientific traditions. The Royal Society grew out of individual initiative and received royal recognition only after the fact. From its inception, it drew heavily upon the landed gentry for its membership and treasury; as a result, the breadth of its interests wandered away from the narrowly scientific. The Académie des Sciences, by contrast, functioned more as a branch of the French civil service, with a high degree of regimentation and control exercised from above. It remains difficult to assess the relative merits of the two scientific systems: the French, with its strong stamp of centralization and control, versus the English tradition, which cultivated individual self-reliance, perhaps as a direct result of the lack of state support. Whatever the advantages of either system, we see here the first crystallization of national differences in scientific traditions. The rise of nation states in the nineteenth century enhanced these distinctions.

Science flourished in Britain during the last half of the seventeenth century, despite the collapse of earlier humanitarian projects and the cynicism displayed by the king. Any decline in membership in the Royal Society was more than counterbalanced by the rise of new provincial centres of scientific activity, for example, in the creation of philosophical societies at Dublin and Oxford, both founded in 1683. As Michael Hunter has explained, seventeenth century English society showed a penchant for establishing *public* bodies, as opposed to impermanent, highly mutable structures dependent on personal whim.

France, on the other hand, failed to emerge as a centre of scientific excellence, despite the elaborate designs of enlightened despotism which had brought the full support of the state to a host of scientific projects. By the late seventeenth century, these programmes fell afoul of political and economic contingencies. The increasingly extravagant ambitions of Louis XIV, ushering in an era of prolonged warfare with England, meant a decline in financial support for science. A period of domestic intolerance, inaugurated with the Revocation of the Edict of Nantes, further contracted opportunities for the free exchange of scientific ideas, and Protestant intellectuals like Henri Justel (1620–1693) were marginalized.

The rise of the scientific correspondent

The creation and persistence of the new institutions attests the strength of the scientific movement. An additional 'barometer of intellectual health', in the words of Harcourt Brown, was the 'exchange of news, books, and journals' among these organizations, particularly through official or unofficial representatives. Operating from the Place Royale in Paris, for example, Mersenne circulated information to an informal network of French natural philosophers, including Descartes, Gassendi, Pierre de Fermat (1601–1665), Gilles de Roberval (1602–1675) and Blaise Pascal. Mersenne constructed an unprecedented system of scientific communication, with nearly eighty participants. An even more elaborate correspondence network was established by the Royal Society's Henry Oldenburg, who as secretary from 1662 until his death in 1677, exchanged information with Mersenne and Henri Justel, secretary to Louis XIV. Modern science began as an international undertaking.

Justel disseminated English scientific news and books across continental Europe. For nearly thirty years, until his death in 1693, he was Henry Oldenburg's most important link with Europe; he lent incalculable assistance to advancing the Royal Society's reputation. Justel channelled information through a circle of intimate acquaintances who attended his 'conferences' in Paris, as well as through a more widely ranging network of contacts with the leading intellectuals of Europe. French members of his circle included Pierre Daniel Huet (1630–1721), founder of the Caen Académie des Sciences and the Abbé Charles, one of the editors of the *Journal des sçavans*. Despite Justel's illustrious collaborators, his correspondence has been seen as valuable not for its coherent exposition of a particular point of view, but for 'the mass of dissociated facts and opinions . . . conveyed'.[10] Even a cursory examination of the letters exchanged between Oldenburg and Justel reveals how much useful scientific information could be gleaned from what appears to be, on the surface, just delightfully candid gossip.

Intelligencers like Justel and Oldenburg depended upon travellers and diplomats to transmit their parcels and letters. A network of courtiers, statesmen, and civil servants scattered across the Continent, the Near East, and the New World provided Oldenburg with the machinery for collecting information and gaining new foreign agents. Oldenburg's contacts, who introduced him to local virtuosi, sent summaries of new books, reports of experiments, and simple accounts of everyday scientific activity. Formal relations between the Royal Society and foreign academies were merely polite and sterile; virtually all news of Continental science went to Oldenburg from Justel or from Englishmen abroad.

The importance of these connections suggests that the rise of scientific societies has depended on the emergence of the apparatus of the modern state. Departing from traditional Marxist arguments by which science is driven by economic need, the demands of capitalism failed to dictate a set of problems to seventeenth-century researchers. Rather, the expansion of trade and commerce associated with the rise of capitalism provided a means of collecting and amassing valuable information. Groups in one geographical location could be brought into communication with like-minded individuals elsewhere. Essentially, seventeenth-century mercantile developments nurtured and sustained the evolution of learned societies.

Eighteenth-century expansion

A century after the creation of the Royal Society and the Académie des Sciences there were around two hundred societies devoted to science or technology. In France alone, twenty-five provincial academies appeared by the eve of the French Revolution. Generally speaking, these societies stimulated research and provided for the diffusion of that research through their publications. The appellation of 'literary society' – characteristic of eighteenth-century societies – refers less to their cultivation of *belles lettres* than to their concern with scientific literature.

Over the course of the eighteenth century, learned societies emerged, as James McClellan puts it, as '*the* characteristic form for the organization of culture' throughout the Western world and its spheres of influence. A host of subsets of these societies might be discussed, but for our purpose those exclusively or even partially devoted to science (along with literary studies or technology) are the most interesting. Their exponentially increasing number outdistanced other institutional forms of scientific activity, whether botanical gardens, observatories, or universities. No leading scientist was without an affiliation to one of them. Not only did they sponsor publications, but they endowed prizes and funded expeditions. McClellan understands the flourishing of scientific academies during the last half of the eighteenth century as 'an unprecedented development in the organizational and institutional history of science'. As he demonstrates, by the end of the eighteenth century, scientific societies extended 'from Philadelphia and Kentucky in the west to Saint Petersburg (or arguably Batavia, the East Indies) in the east, and from Trondheim (Norway) in the north to Sicily and Haiti in the south'.

The establishment of learned societies during the

eighteenth century became an international movement, reaching its peak in the 1780s. These institutions were concentrated in European urban centres, particularly in France. Few nations failed to support scientific societies; only the European capital cities of Spain and Austria were without them. They were – alongside churches, courts of law, and universities – manifestations of high culture, with all its implications of exclusiveness. Only during the next century would this fundamental characteristic of scientific societies be altered; no longer would they be the exclusive prerogative of a learned and powerful elite.

Eighteenth-century developments may be categorized according to the two dominant models for scientific organization established during the seventeenth century. One was that of the Paris Académie des Sciences, the generic 'academy', frequently found on the Continent. The other, the 'society' model exemplified by the Royal Society, emerged in the less stratified societies of Britain, the United States, and Holland. Both types are united by their possession of chartered corporate status and written rules. They convened regular meetings, appointed officers, and elected a restricted number of fellows. In addition to official quarters, they often claimed libraries, collections, botanical gardens, and observatories.

Important distinctions, however, may also be drawn between the academy and society models. Academies, more so than societies, tended to be state-supported institutions; the state accordingly extracted its due by controlling their duties and responsibilities. Societies enjoyed much more autonomy and independence, but because they lacked a clearly defined mission, they tended to be less productive. The internal structure of the two forms of scientific institution differed significantly, which may be illuminated by comparing the Académie des Sciences with the Royal Society.

The Académie des Sciences possessed a restricted, yet heterogeneous membership, stratified in a strict hierarchy.

Its officers were drawn from its two constituent classes, the regular and the honorary members. At the top of the scientific core of regular members (*pensionnaires*, who were paid pensions for their services) were eighteen individuals, three of whom represented each of the Académie's six sections: mathematics, astronomy, mechanics, anatomy, chemistry, and botany, in addition to the permanent secretary and treasurer. Below them in the hierarchy were twelve associate and twelve adjunct members. Nonresident members who did not have to attend meetings but who were excluded from decision-making came next: twelve members from the provinces, eight distinguished foreign scientists, and seventy corresponding members. On average, the Académie could claim just over one hundred and fifty members at any point during the eighteenth century, with fewer than fifty among the resident scientific core. In total, only 716 men belonged to the Académie over the course of the century.

The Royal Society – 'larger, less professional and exclusive, and more homogeneous', in the words of James McClellan – was no match for the success of the Académie des Sciences, where scientific accomplishment, finally, was the currency of admission. The Royal Society averaged 325 fellows, seven times the size of the core group in Paris, with nonscientists outnumbering scientists two to one. Election was decided by the membership itself. Without any internal differentiation of its membership into categories or classes, the society became too unwieldy to conduct administrative matters, let alone prosecute any kind of joint scientific endeavours, at its weekly meetings. During the eighteenth century, a twenty-one member elected council, led by an increasingly powerful president, assumed all administrative responsibilities and became the 'guiding force' of the Society.

Both the Paris Académie and the Royal Society spawned imitators elsewhere. Academies in Montpellier, Turin, and Mannheim, for example, imitated Paris's example. Boston's

American Academy of Arts and Sciences and Philadelphia's American Philosophical Society copied the Royal Society. A new hybrid form introduced during and characteristic of the eighteenth century was the 'universal' society devoted to both science and the arts. The Royal Society of Edinburgh contained literary and scientific sections; the Royal Irish Academy was divided into science, *belles lettres*, and antiquities. The typical French provincial society dedicated itself to science, *belles lettres*, and the mechanical arts. German academies often focused on *wissenshaftlichen* disciplines. Those in Göttingen and Prague had sections for physical, mathematical, and historical sciences. As McClellan summarizes this diversity, when resources were scarce, the 'multi-area' institution was adopted; the 'single field type' emerged where resources were plentiful.

The most important scientific societies of the eighteenth century were official institutions, legally recognized by their respective governments. This legal status conferred important privileges on the societies, including technological consulting, control of the scientific press, and self-government. McClellan argues that according to this arrangement, institutions and governments 'struck a deal', whereby institutions received 'recognition, funding, and privileges in exchange for technical service and advice'. In essence, societies and academies sold their expertise and knowledge for the power to control the practice of science within their own cultural milieu. The emerging nation states of Europe supported scientific societies as a gesture of alliance with the forces of rational enlightenment, progress, and modernization.

Scientific associations, coming in many shapes and sizes, also may be arranged according to a pyramid of importance. An elite group of national academies in capital cities belong to the top of the hierarchy: the Royal Society, the Paris Académie, the Berlin Academy, the St Petersburg Academy, and the Royal Swedish Academy. Almost all were devoted to scientific pursuits exclusively; they received generous support and powerful privileges, often dating back a century or more. At the next level fall a host of institutions founded in large urban centres and provincial capitals; these include societies and academies in Edinburgh, Montpellier, Göttingen, Bologna, and Philadelphia. They received only modest financial revenues, they tended to be founded later in the eighteenth century, and they cultivated nonscientific subjects alongside science. The scientific accomplishments of this more heterogeneous group were less uniform and less sustained. The base of the pyramid rested on institutions that never built a reputation; these include societies at Marseilles, Barcelona, and Rotterdam, for example. Many were founded in smaller towns and cities late in the century and did not obtain official recognition for years. They cultivated a range of disciplines and possessed undistinguished memberships.

What makes the eighteenth century unique for the institutionalization of science is that individual organizations – big or little, national or local – interacted to forge a larger institutional network. As the *Memoirs* of the Medical Society of London, founded in 1773, stated: 'The principal part of our knowledge must be ever derived from comparing our own observations with those of others. In this view the utility of societies, which afford an opportunity for the mutual communication of our thoughts, must be sufficiently apparent.' Sending memoirs and soliciting exchanges became a routine activity. This meant that scientific research and information could henceforth be circulated through regular channels. At issue here is something other than publication, which had already been inaugurated through the system of official journals; rather, academic publications found an assured venue of distribution. In a word, the academies began to 'market' science, having done their utmost to create an audience.

Nineteenth-century consolidation

Once firmly established in the collective consciousness, scientific societies and academies became arbiters of science. With the French revolutionary zeal to abolish privilege in all of its manifestations, it is hardly surprising that the Académie des Sciences became a prime target. It was an institution, even an instrument, of the king, and it was a bastion of elitism. Myths were perpetuated about how self-taught artisans presented their inventions to a jaded academy, only to be rebuffed and humiliated. Not only did the Académie represent an intellectual aristocracy, but it contained a special class of honorary members selected from the social aristocracy. It met for the last time on 21 December 1792; it re-emerged in 1795 as the First Class (or division) of the Institut de France. (With the restoration of the monarchy, the former title of Académie was likewise restored.)

Even before this reincarnation, the Académie des Sciences had begun to turn its back on its Baconian heritage, particularly the collectivist imperative. Rather than acting to generate scientific knowledge, the Académie emerged as an adjudicator, passing judgement on the merits of its members' contributions, in pure and in applied fields. Its imprimatur became a coveted sign of national, or even international, prestige. Election to an academy seat became the crowning achievement of a life's work; appointment to a professorial chair seemed trivial by comparison. Unlike the case during the eighteenth century, all academicians (at least in theory) were equal, since junior ranks had been abolished. By making election a process of 'filling dead men's shoes' – whereby leading contenders most closely approximated the deceased (or soon to depart) academician – the Académie defined the shape of science in France. As Maurice Crosland explains, it was almost as if the subjects included in the First Class of the Institute chose the academicians.

The Académie des Sciences inaugurated new functions

during the nineteenth century, such as semiannual public meetings. The *Comptes rendus*, created in 1835, brought the proceedings of the Academy and the eloquent *éloges* of its deceased members to the attention of an international community. Crosland argues that the centrality and comprehensiveness of the *Comptes rendus* tended to relegate all extramural efforts to oblivion. Responsibility for publications belonged to the permanent secretary, elected for life and given a comfortable annual salary of 6000 francs (about 300 pounds sterling).

By the mid nineteenth century, the Royal Society had also forsaken parts of its earlier scientific mission. Its statutes finally recognized that its role in experiment was more passive than active, more imagined than real. Regulations stated simply that the Society's purpose was 'to read and hear letters, reports, and other papers, concerning Philosophical matters'. In Boas Hall's words, the atmosphere of meetings changed from 'an atmosphere of lively discussion and debate and the frequent display of experiment' to one that was 'determinedly formal and lifeless'. This change did not signify that experiment was held in low regard, or that fellows had ceased to be good experimentalists, nor that nineteenth-century experiments could not be demonstrated. Rather, it indicated that the Society's conventions had changed, placing new emphasis on results rather than processes. Papers might derive from experiment, but they were no longer accompanied by experiential demonstration.

No one disputed that experiment formed the centrepiece of the Society's activities, only that this approach offered an imperfect and impartial view of the natural world. This was precisely the criticism of seventeenth-century natural philosophers, who believed that experimentalism offered an insufficient replacement for general principles, frameworks, and even theories. The complaint resurfaced among the spiritualists, vitalists, and theologians of later centuries, who found their concerns excluded by a materialist Royal Society. In 1878, the British geologist John Jeremiah Bigsby

(1792–1881), for example, lamented the fact that in the Royal Society 'Belief in no God and no Bible is openly paraded'. His protégé, Canadian paleobotanist John William Dawson (1820–1899), concurred that the religious scepticism of its leaders was 'eating the heart' out of science.

The emergence of specialized societies

Since its inception in the seventeenth century, the scientific society has sought to represent a range of philosophical interests. Sometimes art and antiquities were included to accommodate the interests of aristocratic virtuosi; certainly members' investigations into any part of the natural sciences (and their applications) were welcomed. With the growth in size of the scientific community over the course of the eighteenth century and with the expansion of its interests, organizations devoted to the sciences in general no longer commanded attention. Scientists began to occupy themselves with a more restricted range of human experience, seeking, as well, to associate themselves with others who held similar concerns. As a result, the specialist society – one based on what we would recognize as the contents of a particular scientific discipline – began to emerge. Organizations like the Geological Society of London, founded in 1807, became known for the camaraderie and conviviality exhibited by its members, in contrast to the stiff formality displayed in the proceedings of the Royal Society.

James McClellan sees the creation of specialist societies around the turn of the nineteenth century as an accentuation of a tradition in existence decades earlier. He admits, however, that with the foundation of the Linnean Society of London in 1788, the single-discipline society became 'less the institutional oddity, and more the norm'. In England, the Geological Society of London (1807), Zoological Society of London (1826), Royal Astronomical Society (1831), and Chemical Society of London (1841) followed in relatively quick succession. Henceforth the tendency in scientific organization was a coalescence around disciplinary interests.

As it turned out, the partial solution to specialist interests provided in the sections of the Académie des Sciences simply meant that societies restricted to certain scientific disciplines were created, on average, about a generation later in France than in England.[11] French academicians did not perceive the establishment of these societies as a threat to their hegemony, since, in their view, the Académie contained the most distinguished practitioners in any particular speciality. Academicians often accepted (with some degree of condescension) senior positions in these societies, a procedure intended to elevate the new organization's status. Unlike the Académie, specialist societies in France acted to diffuse the study of one particular science to a wider audience.

Jealousy towards rival scientific organizations was not an unreasonable reaction on the part of established societies, particularly when *new* fields of knowledge were represented. The danger was that specialized societies might become associated with the vanguard, and general societies with the rearguard, of the scientific enterprise. Indeed, in the case of Paul Broca (1824–1880) and the Anthropological Society in Paris, the new society offered the means of establishing the legitimacy of the nascent social science of anthropology. The formal organization attracted attention to and supplied a power base for the discipline's founders and promoters.

The complexity of the relationship between established national societies and new specialist ones is revealed in the interactions between the Royal Society and the Geological Society. As Joseph Banks (1743–1820), the powerful president of the Royal Society, expressed his fear about the incipient importance of the geological and other London societies: 'these new fangled Associations will finally dismantle the Royal Society and not leave the old lady a rag to cover her'. Geologists, for their part, felt that their interests commanded little respect in the eyes of the older society.

One aspirant to membership was cautioned that 'unless a geological paper be of high merit it does not meet in the Royal Society such acceptance as one in terrestrial magnetism, electricity, [or] chemistry'.

It is hardly surprising that the Royal Society should have felt some jealousy towards its younger, more lively sibling. Since its foundation, the Geological Society grew more fashionable and scientifically significant. It was composed, wrote the distinguished Cambridge geologist Adam Sedgwick (1785–1873), of 'robust, joyous, and independent spirits, who toiled well in the field, and who did battle and cuffed opinions with much spirit and great good will'. Charles Babbage (1792–1871) lauded the Geological Society in his generally gloomy treatise on the decline of science in England, and no important geologist refused to join the organization. Furthermore, governments and universities referred geological matters not to the Royal Society but to influential members of the Geological Society. The rolls of the Society listed distinguished fellows by the 1830s – peers, members of parliament, landowners, and bankers; both Charles Darwin and the comparative anatomist Richard Owen (1804–1892) joined during that decade. Leading scientists filled positions on its Council: Roderick Murchison (1792–1871), Charles Lyell (1797–1875), and William Whewell (1794–1866) served as president; secretaries included Henry De la Beche (1796–1855) and Darwin.

At the same time that they inspired others to copy them and as they accommodated their hegemony to specialist interests, scientific bodies also fuelled petty feuds and disputes, particularly from those who had been excluded. Sometimes the jealousy remained merely isolated, negative, and remote; on other occasions, it assumed a more positive role, by uniting the dissatisfied and bringing them together to form rival institutions. Even the Canadian Sir William Dawson, whose interests had been badly served by establishment science, so esteemed the Royal Society that he modelled Canada's national scientific society after it. A range of alternative scientific organizations – some broadly conceived, some specialist in focus – were spawned from the late eighteenth century onwards. They generally sought to democratize the scientific enterprise and to extend the benefits of membership to a larger circle.

Watching: Observatories in the Middle East, China, Europe and America

On a clear summer night walk as far as you can beyond the electric colours of urban life. Leave the shimmering rivers of hot air, as they snake above pavement and monument, causing the stars to twinkle. Go to where you can smell no exhaust, hear no human noise. Go at dusk and look up as the stars come out.

What may be seen? First there is the spectrum of the sky, yellow to red to faint-green and on to indigo. There may be birds, insects, and bats. There are condensation trails from high-flying jet aircraft, rapid transits of orbiting satellites, and shooting stars. Depending on one's eyesight and location on the globe, the night sky reveals between one thousand and two thousand points of light. Located in a narrow band among these fixed stars there are seven objects that trace cyclical patterns. Until quite recently, the accidental trajectories of these seven objects – the sun, the moon, Mercury, Venus, Mars, Jupiter, and Saturn – found a central place in many civilizations. The stars have never reliably predicted the outcome of commercial, military, or personal initiatives, but their regular movements have nevertheless had an impact on our lives.

One among the seven moving stars is of critical importance. Biological cycles of growth and renewal reflect the apparent periodical motion of the sun – the solar year. We reckon age by solar cycles, not lunar ones, even in societies where the calendar is closely tied to the moon. This is so because the moon's periodicity will not in itself predict spring inundations or winter rains, the return of migratory birds or fishes, or the best time to plant or harvest. Periodical changes in the moon's aspect, linked with the slower, uneven velocity of the sun's changing position in the sky, can be made to establish a yearly calendar of twelve months (each beginning with new moons) and a rather large fraction of leftover days. Astronomical science has traditionally focused on how to take care of the fraction. Once a calendar (months with a fixed number of days each) was in place, astronomical observations could be kept reliably. Records made possible the identification of cycles for the five remaining planets, the precession of the equinoxes, and predictions of such things as solar eclipses (or the possibility of them).

The existence of a calendar must not imply that we have direct access to events noted by it. Establishing a reliable chronology of antiquity – a goal sought by Europeans since medieval times – was possibly the greatest achievement of the broader historical discipline in the nineteenth century, and this occurred following a meticulous analysis of planetary records on Babylonian clay tablets. All calendars require intercalation of some sort (ours today supplies the odd day or second to round out the apparent solar year). The corrections may follow a formula or, more empirically, a celestial observation. The advantages of a determination of days and years by first principles is apparent to any head of state. Indeed, the state has generally supported astronomical observation – perhaps even (as some interpreters of Stonehenge contend) from paleolithic times.

Until Galileo Galilei pointed his telescope skyward, the seven stars that change their relative positions in a cyclical pattern were the givens of scientific endeavour. Predicting the movement of these jewels and orbs provided an arena for mathematical virtuosity, a justification for maintaining libraries, a reason for establishing schools of advanced learning, and an excuse for international collaboration. Because the patrons of this apparatus demanded practical results in the way of reliable calendars, astronomers devoted effort toward studying persistent empirical trends, such as the precession of the vernal equinox, the change in the stars behind the sun on the first day of spring.

Patrons demanded a great deal of their star gazers. Astronomers were called upon to pronounce on occasional spectacular events, such as eclipses. Through the twentieth century, astronomers have addressed meteorology – the corruptible, sublunar domain of Aristotelian physics named after the blazing objects in the sky, *meteors*, that were apparently as ephemeral as the rain. Astronomers were charged with telegraphical signals and radio broadcasts. They measured fundamental physical quantities in gravimetry (the gravitational constant identified by Isaac Newton) and optics (the speed of light, first calculated by Ole Christensen Römer [1644–1710]). Occasionally they chronicled the flight of migratory birds and assembled demographical statistics. They addressed whatever depended on a sharp eye and a head for figures. Until the twentieth century, astronomers were the practical masters of the realm of numbers.

The Islamic observatory

Astronomers differed from casual stargazers in that they required a special place for making observations. Observing in a grand observatory required a team of people. They had to be ready for the right moment and hope that a cloud did not intervene. In practice, this implied a support staff of servants and some form of lodging for the observers. Understanding the data required a library and calculating devices – whether pen and paper, abacus, clay tablet, or sand table. Apprentice observers had to be trained. Instruments had to be maintained. Regular reports about celestial omens and calendars had to be produced. Four thousand years of astronomical practice are continued at today's enormous, mountain-top research installations.

We have seen that the endowed, residential college, or madrasa, was an innovation of medieval Islam. It is also to Islamic civilization that we owe the invention of the astronomical observatory. This occurred under al-Mamûn, early in the ninth century. A great patron of learning, al-Mamûn financed major astronomical complexes at Damascus and Baghdad. These possessed modifications of the instruments mentioned by Ptolemy, including an armillary sphere of concentric circles for tracking the stars, a marble mural quadrant (a graded quarter-circle mounted on a wall) for observing the height of stars above the horizon, and a five-metre gnomon or stile. The observatories assembled a group of perhaps as many as a dozen talented astronomers, one of whom was Ptolemy's commentator al-Farghânî (Alfraganus, fl. 850), who constructed tables, or *zijes*, based on observations. Astrological interest, especially as it related to solar eclipses (for which Ptolemaïc data had to be corrected), was undoubtedly the motor of al-Mamûn's astronomical patronage.

Knowledge may naturally tend to disaggregate, pooling here and there, channelling along one or another stream, evaporating into the air. The disaggregation is present in Islamic astronomy. During the Abbasid golden age, al-Mamûn's observatories were distinct from the learned academy at Baghdad, the Dar al-Hikmah or House of Wisdom, which had been founded by Caliph Harun al-Rashid. The academy functioned as a collector and filter of learning from all sources, east and west. Greek and Indian texts, and possibly also Hebrew ones, were recovered and translated into Arabic. Among the most notable academicians was Abu Jafar Muhammad ibn Musa al-Khwarizmi (fl. 830), author of the first Arabic text on algebra (based on both Greek and Indian sources) as well as a work on Indian numerals. Al-Khwarizmi also composed a treatise on Hindu astronomy, recalculated much of Ptolemy's data for the seven planets, and provided tables for calculating eclipses as well as trigonometrical functions. He certainly knew about the work conducted at al-Mamûn's observatories, especially on establishing the obliquity of the ecliptic, but he chose not to incorporate the new results.

Al-Mamûn's observatories did not survive his reign (he died in 833), but they established a precedent for observing nature. Over the next centuries, Islamic observatories extended their programmes to all the planets. The institutions became characterized by grand instruments (sometimes surpassing in size those at European locations up to the eighteenth century) and the staff (more numerous than European staff) to manoeuvre them. Observatories acquired legal status and operated under the eye of a director. The astronomical work and instrumental innovations of the polymath Ibn Sînâ (Avicenna, 980–1037), based on observations taken early in the eleventh century at an observatory financed by the amir of Isfahan at Hamadân, followed the earlier pattern. But the institutional evolution occurred unevenly. Distinguished observers, such as al-Battânî (Albategnius, fl. 880) and Ibn Yûnus (late tenth century), seem not to have availed themselves of a *permanent* observing facility, even though they were much concerned with astronomical innovation. Ibn Yûnus, for example, invented something akin to the method of transversals.

European commentators have traditionally celebrated Islamic savants as transmitters of Hellenistic learning; less time has been spent detailing Islamic scientific innovation. But there is no doubt that in astronomy, Islamic observations expanded and became more sophisticated. The crucial tasks of an Islamic observatory related only to the sun and the moon. One needed to establish dates of religious observances (for the Muslim lunar calendar) and times of daily prayers, keyed to sunrise and sunset. With the accessibility of Hellenistic texts, precise measurements of the sun led to interest in anomalous motions, such as precession of the equinoxes, and eventually to concern with the five remaining planets. Indeed, programmes to observe the five smaller bodies provided a *justification* for the permanent endowment of an observatory. It takes about thirty years of watching the sky to document all planetary regularities, and this is the working lifetime of an astronomer. Among observatories with a long-term programme was the one founded by the late eleventh-century Seljuq sultan Jalal al-Dîn Malikshâh at Isfahan; its staff of as many as eight men included al-Khayyami, the mathematician and astronomer known for his poetry as Omar Khayyam (ca.1048–ca.1131). The astronomers at the Malikshâh Observatory were the first to emphasize to their patron that it would take thirty years to record changes in the sky; from their time forward the generational argument became an astronomical watchword.

The slow pace of institutional development reflected uncertainties about using large measuring devices. One principle has dominated astronomy since antiquity: the larger the measuring device, the more accurate the observations. During the Islamic period large azimuthal rings installed on the ground to measure points on the compass were cast in copper (notably one five metres in diameter at the early twelfth-century al-Afdal al-Bataihî Observatory in Cairo), and large mural quadrants were cut into the ground and faced in marble. The moving parts of these instruments were usually made from wood – indeed, wood was preferred to brass for mural quadrants and even sextants up to the eighteenth century. But the wood warped with time and weather, especially as the large instruments were normally open to the elements. Heavy moving parts – the arm on one of Ptolemy's rulers, for example – had to be suspended in such a way as to minimize creep. One reason for the slow growth of early observatories is that many astronomers, among them Ibn Yûnus, actually favoured small devices – even portable ones – that could be manipulated by one observer.

The peak of Islamic observatory-building took place during the thirteenth century, and its exemplar was the one founded at the city of Marâgha, south of Tabriz in present-

day Iran, by Mangû, brother of the Muslim conqueror Hul-
âgû. Mangû, by all accounts a convinced patron of learning,
seems to have first thought about inviting the most distinguished astronomer among his new Islamic subjects, Nasîr
al-Dîn al-Tusî (1201–1274), to found an observatory at Beijing or possibly the Mongol capital of Qaraqurum. Indeed,
during the Mongol period there was renewed intellectual
interchange between East Asia and Central Asia. Several
accounts refer to an Islamic astronomer, with his instruments, visiting China at just this time, and Chinese astronomers certainly travelled west. Nasîr al-Tusî may have gone
east, but he certainly supervised the construction of the
Marâgha Observatory, beginning in 1259. The inspiration
for the Marâgha observatory, it is reasonable to assume,
was Mangû's familiarity with the Chinese tradition of constructing a new calendar for a new sovereign. To keep his
hand in traditional, Chinese star-reckoning, Mangû brought
Fao Mun-Ji to Marâgha at the onset of the enterprise.

To insure the life of the observatory beyond his own reign,
Mangû provided it with a *waqf* endowment – the first
known application of applying to astronomy the mechanism
for endowing madrasa and hospital. The resulting revenues
financed the observatory during the reigns of subsequent
rulers until the dissolution of the Mongol state about 1316.
Nasir al-Tusî's sons succeeded him in directing the observatory, and it may be that he and they were the waqf administrators. The charitable endowment allowed the observatory
to become an institution for instruction in the secular, or
ancient sciences – the natural sciences excluded from the
madrasas. In this, too, the observatory followed the pattern
of Islamic teaching hospitals.

The Marâgha Observatory had a main building sur-
mounted by a dome, through a hole in which the sun could
be observed. It included an enormous library (by one
account more than 400,000 volumes) and housed terrestrial
and celestial globes. Many of the observatory's rooms were
excavated caves. (Astronomical observations are often made

when a star passes overhead at the zenith, and for these altitude measurements, an excavated trench with a mural quadrant is fine.) Among its instruments were a fixed armillary sphere with five rings, a mural quadrant, a solar armilla, an equinoctial ring, and a parallactic ruler. The instruments went to produce a set of zijes, the so-called Ilkhâni Tables which provided data for all seven moving stars.

Marâgha formed a precedent for Mongol astronomical patronage. In the fifteenth century, Ulugh Beg (grandson of Timûr, feared in Europe as Tamerlane) erected the most magnificent of Islamic observatories at Samarqand. He became an expert astronomer, apparently constructing his observatory around an existing madrasa. He initiated astronomical instruction at the madrasa and drew talented astronomers, notably Ghiyâth al-Dîn al-Kashî (d. 1429), to the observatory, which he endowed with a waqf. The observatory apparently survived Ulugh Beg's reign (he was murdered by his son), settling into a slow decline over the succeeding century. The Islamic tradition of grand astronomical institutions continued into the sixteenth century, with the construction of an observatory at Istanbul under the direction of Takiyüddin al-Rasid (1526–1585). It functioned for several years before being dismantled in 1580, at the request of the sultan who founded it. The last of the great Islamic observatories came in South Asia early in the eighteenth century, courtesy of Maharaja Swai Jai Singh II.

The Islamic tradition may be placed in perspective by introducing the late sixteenth-century astronomical fiefdom of Tycho Brahe (1546–1601), granted by King Frederick II of Denmark and financed by Tycho's inherited fortune supplemented by Frederick's largesse. Tycho had workshops for his instrument-makers, a mill for producing paper, and a printing press. The main house of Tycho's Uraniborg functioned as a chateau, complete with running water, kitchen, chemical laboratories, workrooms, library (housing a five-foot-in-diameter celestial globe) and bedrooms. Tycho had large instruments mounted on the top of the house. A separ-

ate observatory building had more instruments set in subterranean rooms equipped with plinths. With its generous
royal endowment and its massive, innovative instruments,
Uraniborg was nothing other than the European counterpart of the Istanbul Observatory. Tycho's observatory was
an inspiration for Francis Bacon's invocation of the notion
of a House of Salomon, which in turn became a model for
the Royal Society of London. In a sense, we may trace
modern scientific institutions to medieval Islam.

It is doubtful that Tycho had first-hand information about
the observatories of his Islamic predecessors. His instruments followed Ptolemy's instructions, which he adapted
and added to on the basis of European tradition – just as medieval Islamic astronomers began with Ptolemy and constructed innovative measuring arms, armillary spheres, and
scales. Medieval Islamic observatories, located as they were
between Western Europe and Eastern Asia, nevertheless suggest questions about interchanges between West and East.

When we look west, we find little direct Islamic inspiration for the organization of astronomical activity. The area
of closest contact between Islam and Christianity, Anda-
lusia, was insulated against the urge to construct grand state
observatories. The insularity derived from the effective independence of Western Islam and especially the diverse
Spanish emirates and kingdoms. This is not to say that intellectuals in Spain were less interested in the stars than were
people in Central Asia. Certainly the eleventh-century
group of astronomers around al-Zarqâli (d. 1100) who compiled zijes that became known in Europe as the Toledan
Tables undertook significant observations, but the work was
apparently accomplished without a permanent observing
facility.

Chinese astronomy

What about the East? Can it be that the inspiration for
Islamic observatories came from China? There is no doubt

that various Chinese governments maintained astronomical offices, and with them the means of making sophisticated astronomical observations, for at least seven centuries before a similar spirit infected Islamic authorities.

Knowledge of the sky was an imperial prerogative from the time of the Han. The heavens were held to have conferred a mandate on the imperial house, and reading the stars was a way of learning whether terrestrial policies found divine favour. *Portent* astrology (where one sought divine instruction from the sky by reading celestial signs) rather than individual *fate* astrology (the notion of a preordained future that suffused western Eurasia) dominated the court institutions of Chinese astronomy. New rulers and new regimes, in fact, promulgated new calendars as a practical sign of their celestial mandate. In the Han, astronomy went under the Office of the Grand Historian, for it combined the functions of archivist and omen reader. About 90 BC, the head of this office, Ssu-ma Chhien, compiled the dynastic history known as the *Shih Chi*, which had chapters devoted to calendar construction and astrology. This tradition continued (with the same kind of ups and downs that characterize institutions of higher learning in the Mediterranean basin) for two millennia.

The Chinese dynasty at the time of the rise of Islam, the Thang, received ambassadors and merchants from Byzantium, Persia, and elsewhere. Among the foreigners living in China under the Thang were Indian astronomers. In the seventh century there are indications of Brahmin astronomy being translated into Chinese. Beginning around 650, three families of Indian astronomers held positions in the imperial astronomical bureau. Of these, astronomers of the Gautama family found their calendrical work officially adopted. Chhüthan Shi-Ta or Gautama Siddhartha (*fl.* 718), the greatest of the clan, became director of Thang astronomy and wrote a major mathematical work in 729 which featured the zero symbol, division of the circle into 360 degrees (the Chinese circle traditionally contained 364.25 degrees), and sexagesimal minutes and seconds. No doubt the resident Indian astronomer families made use of trigonometry, then unknown in China. Despite internecine disputes about astronomical secrets (the Chinese Buddhist monk and brilliant mathematician I-hsing became involved in some of these disputes), the Indian families produced an officially accepted calendar, calculated solar eclipses, and wrote an astrological treatise.

The observatory where the Indian astronomers lived and worked was large, even by modern standards. Two grand astrologers supervised the Astronomical Bureau in Thang China, an institution that combined features of observatory and college. They operated one of the largest astronomical schools of any time. In the bureau's astrological department, 2 professors supervised 5 observers and 150 students; one professor of calendar-making oversaw 2 technicians and 41 students; 6 professors of time-keeping had 37 technicians, 440 clerks to handle various bells and drums that signalled the hours, and 360 students. Separate from the Astronomical Bureau was the Divination Bureau. Divination concerned foreseeing the future on the basis of traditions ranging from the *I-Ching* (Book of Changes) to geomancy (the favourable attributes and aspects of land that still inspire architectural design in Asia), and it followed the art of Yin and Yang (the qualitative masculine and feminine spirit that resided in all things). The director of divination had 2 vice-directors, 2 professors, 2 assistant professors, 37 technicians, and 45 students. On the twelve-rung scale of the civil service, the astrological directors held posts fifth from the top; experts in calendar-making ranked ninth, and experts in timekeeping apparently had no rating at all.

The apprentice system in Thang astronomy led into middle-management positions. The enterprise departed from a strict technical meritocracy, because directors were parachuted in from outside the bureau. And as foreigners came to carry out many of the calculations, there was little interest at the top or at the bottom in accuracy, fidelity, or innovation. With the exception of the foreign calculators, this institutional structure, modified and diminished in size, also took root in eighth-century Japan, where astronomical knowledge became the domain of a few families and where the dominant Chinese focus on calendar-making ceded to portent astrology.

The structure of astronomy at Chinese observatories separated calendrical mathematics from practical problems of terrestrial mensuration. Chinese maps followed a grid, for example, but unlike Ptolemy's geography the grid was not keyed to astronomical measurements. There was a small mathematics school founded during the Thang period. Its professors did not rank high in the civil service, and the students were not destined for administrative posts. Sons of minor officials and commoners, the students did not have access to other schools; with their 'Master of Mathematics' diploma, they anticipated a career as land surveyor. To an extent even greater than in medieval Europe, Chinese society separated mathematical scholars and mathematical craftsmen.

In both Chinese and Islamic civilizations, the motivation for observing the skies related to legitimizing state authority, which promoted (or at least guaranteed) a faith. Both Chinese and Islamic rulers had heavenly mandates, and it was only natural to read heaven's signs in the stars. The reign of a Chinese potentate often began with a new, star-informed calendar. The prayers of an Islamic caliph were regulated by the sun and moon, and his life was foretold by the remaining planets. Notwithstanding a divergent interpretation of celestial signs, observatories provided essential information both East and West.

We may identify a progressive evolution of techniques at both Islamic and Chinese observatories. There was in fact persistent interchange of techniques between the two civilizations: al-Khayyami reinterpreted what he thought were Chinese mathematical techniques, and Nasîr al-Tusî received an invitation to Beijing or Qaraqurum. Nevertheless, foreign knowledge (such as the Persian, Manichaean, and Nestorian texts that were translated into Chinese during the eighth century) eventually disappeared. One finds, for example, no trace of Ptolemaïc notions in Chinese texts. Chinese astronomers may have been instrumental in setting up one or another Islamic observatory, but we see nothing of Chinese norms in Central Asian astronomy.

Why? Because astronomy was a state secret and a clan monopoly, foreign astronomers found an ephemeral place in China. The astronomical sciences – astrology, navigation, cartography – could be prosecuted for the most part only under imperial authority; data, methods, and calculations were not available in the public sphere.

In the record of those times when astronomical innovations came to Asia and the Islamic world, however, we see another part of the answer. Innovation in the sciences of observation occurs in the context of aggressive expansion. When a civilization is actively assimilating foreign peoples and exotic cultures, traditional notions of all kinds are subject to modification. The scientific fruit of this expansive vision appears in Hellenistic Alexandria, tenth-century Salerno, thirteenth-century France, Renaissance Italy, Restoration England, eighteenth-century Scotland, and twentieth-century America.

Innovation in instruments

What were the innovations in astronomical instruments between the ancient observers of Stonehenge and the comparably majestic observatory on Hven where Tycho Brahe brought classical, Ptolemaïc astronomy to its apogee? Among the innovations of the Istanbul Observatory was a mechanical clock based on a European design. (Tycho, it may be noted, did not consider mechanical clocks reliable for astronomical work.) Clocks of all kinds flooded the Ottoman world during the sixteenth century, even though they were ill-suited for indicating prayer times, just as they streamed into East Asia as goodwill offerings of European ambassadors and missionaries. Only with Christiaan Huygens's pendulum clock and the precise chronometers of the eighteenth century did regular, mechanical timepieces enter the observatory.

The astrolabe, perfected in medieval Islam, became a useful navigational device, and its precise scales – stamped and engraved on brass – could be employed for determining planetary positions, as could various wooden cross-staffs of European origin. Brass was also worked into armillary spheres, which allowed for a simultaneous measurement of celestial latitude and longitude. The sphere itself, more cumbersome than useful, could be reduced to a two-dimensional circle or part of a circle. When of large proportions, like the six-foot-radius quarter circle used by Tycho, the instrument could be fixed to a wall and adapted for taking altitude and meridian transits simultaneously. The gradual evolution of instruments, pioneered by professional astronomers, led directly to heliocentric, celestial mechanics: Johannes Kepler (1571–1630) began his reformation of astronomy because he focused on an 8-minute-of-arc discrepancy between Tycho's observations and traditional calculations.

Observational practice, in particular the use of meridian transit instruments, guaranteed that Tycho's notion of an observatory would continue to the end of the nineteenth century. The Paris Observatory, for example, took shape in the late seventeenth century as a residential mansion where quadrants, octants, and the new telescopes perambulated to an outdoor terrace. Telescopes went on the roof from the beginning, and wings were added for additional telescopes. In the nineteenth century, advances in metalworking made possible lightweight movable domes, which could enclose permanently mounted telescopic leviathans.

Small instruments evolved slowly and continually at least since the time of Ptolemy, but the large ones changed hardly at all. Innovations derived from star-watchers who needed to determine time and place. The astrolabe, invented in the

Mediterranean around the fourth century, responded to the requirements of sailors and astrologers and especially to the men of affluence who underwrote the voyages and horoscopes. With its plates for various latitudes, the astrolabe provided a picture of the fixed stars in stereoscopic projection, and its obverse served to sight the altitude of celestial objects. A serious student of the stars, realizing the limitation by latitude of such a calculating device, would seek to generalize it; the universal astrolabe appeared by the eleventh century in Toledo. It would be obvious to a frequent user, furthermore, that one needed only a quarter-circle for determining time and place; as we have seen, quadrants of this kind enjoyed popularity by the fourteenth century.

The utility of devices like the astrolabe depended on the precision of their lines and scales and the regularity of their moving parts. Precision related to the rise of a craft tradition that eventually led to the emergence of professional instrument-makers. With the Renaissance, precision replaced figurative allusion as a rhapsody for people in the workaday world, and the prime measure of precise movement, the stars, attracted increasing attention. The growth of commerce and banking brought number – and its various transformations from one to another currency or system of weight and measure – to wider circles. Marine commerce with Asia and the New World generated demand for maps and navigational instruments. Calendrical reform assumed crisis proportions as feasts and anniversaries no longer corresponded with the seasons. Astrology became important in daily affairs as religious heterodoxy conveyed doubt and uncertainty about humanity's place in the cosmos.

The European Renaissance experience featured a strong desire for improvement in observational precision and an interest in carrying out correlative arithmetical calculations. But this desire is one of the West's greatest debts to the East. European familiarity with a handful of Islamic data-tables pales before a multitude of Islamic zijes, which provided detailed information about astronomical time-keeping. The invention and employment of astronomical instruments in Islam and China point to a persistent desire for precision. The East has been no less quantitatively oriented than the West.

We have seen that the principal research institution, West and East, has been the observatory – an organization mandated by the state to collect data and issue forecasts. The observatory has traditionally been self-replicating, the mastery of its arcane techniques being achieved through a long, practical apprenticeship. Instruments to record the motion of the stars have been a central feature of these institutions. Data has been assembled over generations. There has been the implicit requirement of a patron to finance all parts of the operation.

Ptolemy's description of his instruments provided a model for Eurasian observatory-builders until the time of Tycho Brahe. As precision related directly to size, large instruments could be built only under patrons with grand resources, and in this relationship is written the history of big science. The instruments themselves – whose grandness could be an object of popular veneration – came to represent the authority and magnificence of the patron. But the daily, mundane operation of these instruments provided relatively little increase in patronal prestige. One finds consistently that observatories declined after several generations of existence, if changes of regime did not spell their immediate extinction.

The observers who flourished beyond the lifetime of their patron are those who diversified into cognate fields of learning. During the eighteenth century, the Paris Observatory, for example, established expertise and secured continuing royal patronage through a focus on geodesical questions – from determining the shape of the globe (a crucial question for deciding between Cartesian and Newtonian physics) to mapping the Kingdom of France. On the eve of the French Revolution, France was the only nation to have been mapped completely by astronomical methods, although the focus on terrestrial measurement (which led naturally to the invention of the metric system) deprived French astronomy of leadership in telescopic astronomy. The Paris Observatory, indeed, had been designed without provision for making measurements with permanently anchored instruments. Even with a turn to the skies during the nineteenth century, the observatory embraced ambitious research programmes in meteorology and experimental optics.

Time and prediction

By far the largest secondary industry for observatory staff concerned establishing the time of day. Time moves inexorably toward a natural end, we think today, its deliberate momentum sweeping aside everything in its path. But until medieval times the clock struck off continually varying units. Day and night were divided into an equal number of hours (usually twelve in the Mediterranean world), and these were marked on a sundial. Not only did time vary with the seasons, then, but also with latitude and longitude. The regulation of the hour was the natural province of state astronomers, who established calendars on the basis of the sun's regularity. Since solar days did not vary nearly as much as shadow hours, however, the utility of equal-time hours would have been apparent to state preceptors. We cannot say when the earliest of the equal-time clocks – the clepsydra based on water dripping at a constant rate – appeared, although their use is noted throughout Eurasia in the Hellenistic period. Derek Price has shown that mechanical representations of the heavens, probably powered by water and activated by gears, were constructed at the Hellenistic-Roman interface, the Tower of Winds in Athens being a convenient reference point.

Local time may be determined most readily by observing the shadow of a marker at midday. Terrestrial latitude can be established by measuring the elevation of the pole star (whichever star that may be), or the elevation of the sun at a particular day, say, the vernal equinox, and then comparing elevations at diverse places. With a clock for measuring local time at a constant rate, it is possible to determine longitude differences if two distant observers record the time of a celestial event, for example, when the moon eclipses a star. With longitude established, eclipses can be predicted.

Mariners and their financial backers have always been sensitive to the catastrophe of overshooting an island destination. It is not surprising, then, that mariners supported the development of accurate chronometers late in the eighteenth century. Their practical requirements led observatories to standardize the local time of day. By the nineteenth century, the daily exercise of setting clocks became essential in ports. Time synchronization usually took place at noon to allow for a direct observation of the solar transit. The time was signalled by firing a canon or lowering a large sphere attached to a mast. The daily ritual provided a justification for observatory directors to reach into the public purse for an accurate pendulum clock (a seventeenth-century invention) to mark time between astronomical verifications. With the adoption of error analysis early in the nineteenth century, the possession of accurate clocks led to requests for purchasing powerful transit telescopes. This ratcheted approvisioning accounted for the dramatic growth of observatories from Greenwich to Sydney. Observatories denied this opportunity for growth during the nineteenth century, for example those at Montreal, Batavia (Jakarta), and Valparaiso, developed local research programmes in astronomy with great difficulty – if they managed to do so at all.

Astronomy and related disciplines

Mastery of time led astronomers to reassert their traditional authority in all areas that required the reporting of global natural phenomena. Measuring variations in the intensity

of the earth's magnetic field, a worldwide programme conceived by Göttingen astronomer Carl Friedrich Gauss
(1777–1855) and supported by the British Admiralty, led
to setting up magnetical recorders next to accurate clocks.
The astronomers' claim to supervise terrestrial magnetism
found confirmation with the discovery that magnetical variations depended on solar flares. Magnetical variations also
depended on local geology, as did gravity measurements,
and astronomers became involved with geophysical surveys. With the development of reliable and sensitive instruments for recording earthquakes, astronomers offered
shelter to seismologists. Synoptical meteorology, requiring
daily reports from hundreds of stations across a continent,
also depended on timing. It is not surprising that in the
middle of the nineteenth century, astronomer Urbain Jean
Joseph Le Verrier (1811–1877), who is credited with discovering the planet Neptune, organized a national meteorological service at the Paris Observatory.

The strategy of institutional aggrandizement based on
mastery of time did not succeed everywhere. Gravimetrical,
magnetical, and seismological recording involved delicate
instruments requiring full-time maintenance; whoever was
in charge of them could not also be expected to scan the
night-time skies. Furthermore, early in the twentieth century, each of the geophysical specialties achieved the status
of a distinct discipline governed by international com-
missions with central offices and periodicals. The geophysical disciplines left their heavenly quarters following the
course set by the most ambitious of the astronomical deriva-
tives, meteorology, which had declared its independence in
the middle of the nineteenth century. National weather
offices emerged in most North Atlantic countries by about
1860. Even an institution with the prestige and acquired
rights of the Paris Observatory could retain control of the
weather only until 1878. In very few instances did geophysical sections provide more than a grand headache for men
and women who had given their hearts to the stars.

The effective independence of geophysical observatories related on the one hand to their grand mapping schemes – for example, traversing North America, India, Brazil, or Java with rod and chain, and measuring magnetical declinations on a grid across the Pacific Ocean. Because of the purported economic benefits of these surveys, federal and imperial authorities financed them and kept their human assets on a short leash – often under military discipline. Geophysical independence also related to the low cost of outfitting an observatory, at least in comparison with the enormous outlay required for large telescopes. By the last third of the nineteenth century, major geophysical observatories adorned cities like Strasbourg and Tokyo that had no significant astronomical installations. At other locations – Padua, Göttingen, New York, and Pasadena – geophysics was a distinguished complement of astronomy.

The diverse imperial, national, and colonial weather services managed enormous budgets for producing useful climatological surveys (where to plant vineyards and rubber plantations) and forecasts (notoriously imprecise before the advent of Vilhelm Bjerknes's [1862–1951] front theory, about 1919). The other parts of geophysics have also lent their expression to costly orchestration of widely separated observers. This multinational cooperation found a useful precedent in Gauss's international magnetical association, which set standards for observatories from London to Hobart Town. The cooperative precedent produced significant issue in other disciplines. During the latter half of the nineteenth century, chemists met to establish standards for nomenclature, precision metrologists exchanged standard metre lengths and kilogramme masses (and the machines for comparing them with various copies), and astronomers finally convened (in Paris, naturally) to apportion out the stars for systematic, telescopic inventory. Geophysical observatories were also prominent in another institutional innovation: the central bureaucracy with permanent foreign dependencies.

From the middle of the nineteenth century, formulating geophysical theories and predicting the weather required a network of widely separated stations reporting to a central authority. As the soil and the weather do not respect the accident of state frontiers, clever researchers naturally sought to accumulate data from foreign locations. French authorities in Paris, especially, lent expensive equipment to and formulated protocols for collaborators around the world. In the absence of retaining foreigners on salary (something all nations were loath to do outside their imperial domains) the incoming information was always received with a certain scepticism.

A strong initiative to establish client geophysical observatories came from the world's largest institution devoted to research in physics, the Department of Terrestrial Magnetism of the Carnegie Institution of Washington. The department began life as one division of Andrew Carnegie's Baconian research institute, endowed in 1901. Under its director Louis Agricola Bauer (1865–1932), the department circled the world, mapping magnetical intensities on sea and land. Not content with measuring ephemeral values at distant locations, Bauer financed observatories at Watheroo (Western Australia) and Apia (Western Samoa) to record magnetical variations over many years.

Bauer, a native American with a doctorate from Berlin, may well have been inspired by the original Apia Observatory, founded in 1902 by geophysicists from the University of Göttingen. Bauer knew about client observatories built by United States astronomers in South America – Harvard's observatory at Arequipa (Peru), the University of California's observatory at Santiago de Chile, the San Luis (Argentina) station set up by the Dudley Observatory, William Joseph Hussey's (1862–1926) simultaneous directorship of observatories at Ann Arbor and La Plata (Argentina), and the long-term Yankee direction of the Córdoba Observatory, also in Argentina. In his extensive expeditions aboard the wooden yacht *Carnegie*, Bauer saw

United States administration of observatories in Cuba and the Philippines; he contemplated the French administration of observatories at Santiago de Chile, La Plata, and Shanghai, as well as French scientific authority in the technically independent countries of Morocco, Tunisia, and Lebanon; he knew about the international research institutions at Naples (in marine biology, founded by Anton Dohrn [1840–1909]) and on Mont Blanc (in astrophysics, founded by Pierre Jules César Janssen [1824–1907]).

In the 1920s large branch observatories of major universities (Harvard, Yale, Columbia, the University of Michigan, the University of Leiden) came to South Africa. In the 1930s, National Socialist Germany contemplated an observatory at Windhoek (the former capital of German Southwest Africa, today Namibia, had been suggested as the site of an observatory as early as 1909), and the Soviet Union angled for an observatory in New Zealand. The imperializing precedent – erecting expensive observatories on foreign soil – accelerated during the second half of the twentieth century, as enormous telescopes came to Cerros La Silla (Chile), Mauna Kea (Hawai), Las Palmas (Canary Islands), Calar Alto (Spain), and at last Namibia.

By the twentieth century, then, observatories had for the first time in more than a thousand years ceased to provide much information of practical utility. Radio broadcasting of time signals along with the nearly universal adoption of the Gregorian calendar – especially in Russia and China – catapulted astronomers into abstract irrelevance. Observatories became the grandest of civilizing institutions. They trafficked almost exclusively in the wonder, revealed by Galileo and celebrated by John Donne, of new objects in the heavens. In antebellum United States of the transcendentalist period, observatories were already 'lighthouses of the sky', illuminating the frontier. By 1900 enormous private fortunes had placed grand observatories at Williams Bay (Yerkes) and Mt Hamilton (Lick) – waqf endowments like those at Dun Echt (by James Ludwig Lindsay, twenty-sixth Earl of Crawford), Athens (by Baron Sina of Vienna, an ethnic Greek) and Nice (by Raphaël Bischoffsheim). None of these nineteenth-century foundations envisaged work related to lucrative gain.

The use of astronomy and geophysics as an instrument of cultural imperialism, so evident in the recent past, derives from a traditional appreciation of celestial observations as a measure of cultural strength. Observatories and the records they produce are objects of general approbation. According to a metaphor sometimes used today, they allow us to peer into the eye of God. During the last third of the twentieth century, when it has become unfashionable for intellectuals to profess belief in the literal truth of religious scripture, astronomers have once more become priests.

We are accustomed to imagine that the science of the celestial sphere is separate from its worship. Children are told about Galileo's condemnation by the Catholic Church for disrupting the Aristotelian cosmos, as well as about Laplace's atheistic retort to Napoleon's query about the place of God in the mechanisms of the stars. They read Walt Whitman's 'When I Heard the Learn'd Astronomer', Wilbur Daniel Steele's 'The Man Who Saw Through Heaven', and Christina Rossetti's feigned indifference to whether the earth circles the sun or vice versa. Yet the Catholic Church especially was a firm patron of astronomy. Late in the eighteenth century, fully one-third of the world's observatories were run by Catholic astronomers. During the nineteenth century and early twentieth century, imposing observatories were erected by religious congregations at places around the world. Jesuit observatories came to Manila, greater Shanghai, Lebanon, Madagascar, eastern Australia, Ebre in Spain, Stonyhurst in England and Georgetown University in Washington, DC.

There is no confusion about why this is so. The stars are unamenable to direct interventions, like vivisection, that have traditionally clouded the spirit of theologians. In modern times they do not speak to disturbing issues like the nature and evolution of consciousness. Telescopes provide direct information about the Creator's cosmical plan. Through photographs the stars reveal themselves precisely in diagram and number.

As church dogma receded in importance among avant-garde intellectuals, great interest emerged in alternatives to traditional faiths and families. Late in the twentieth century, holistic thinking has enjoyed a renascence, especially extra-Galenic medicine, Oriental religious philosophy, and dialectical reasoning in general (from Hegel and Marx to Marcuse and Mao Zhedong). Mechanical ideas about cause and effect are increasingly held in disrepute, and organic notions of physical processes have been proposed for thermodynamics by Ilya Prigogine (b. 1917) and for biomass by James Lovelock (b. 1919). Cosmology and its referent in Einstein's general relativity theory have ridden the crest of this interest, finding an expression in Carl Sagan's (1934–1996) and Stephen Hawking's (b. 1942) reveries for mass readers.

General relativity, with its remarkable astronomical import, broke on the European scene just as astronomers were beginning to share their grand instruments with astrophysicists whose passion was less in the movement of points of light than in the nature of their fires. These men and women focused on the life and death of stars, an interest that naturally related to broad cosmological questions. Understanding the stars as objects required borrowing from the new quantum theory (and, later, quantum mechanics), and the stars – with their massive interplay of gravitation and light – became the testing ground for relativistic ideas. This line of enquiry has been an impractical enterprise. The age of the universe and gravitational lenses are entirely divorced from mercantile fortunes. Astronomy speaks directly to the wonder of the sky seen by people everywhere. Its prosecution has been one of the most inspirational of cultural activities.

5

Showing: Museums

Can you remember your first visit to a museum? Families and schools organize pilgrimages to these temples of nature and art, activities which play a central role in the cultural awakening of the Western child. The memory lasts a lifetime: one never forgets the image of the museum's intimidating steps and vaulted interiors, the hushed tones of reverence appropriate to its galleries, the feel of its cold marble walls and wooden railings, the penetrating smells of aged varnish and paint. The child museum-goer is something of a *hadji*, a visitor to civilization's most holy shrine.

Modern museums present vast storehouses of 'tangible objects' – masterpieces of art, dinosaur skeletons, mechanical gizmos. These are organized according to a system of classification, adopted largely to instruct the visitor. In addition to appealing to the general public, museums also serve the interests of a more limited circle of scholars. These two functions can be either complementary or antagonistic.

In early modern Europe, however, museums fulfilled other purposes. The first cabinets of curiosities, whose name denotes a container or closet, seem distant and remote ancestors of modern museums. These old collections provide a window into a private psyche, reflecting by their extent and design the whim of their proprietor. Because the early curio cabinets manifest one person's unarticulated desires and assumptions, they have been dismissed by historians. They have been seen as hodgepodges of unrelated objects. Many of these objects, claiming to represent imaginary characters (such as unicorns or giants) or to embody magical qualities, have been discounted as hoaxes. One of the responsibilities of the historian, however, is to treat the passions and fashions of times past with sympathy as well as probity. The cultural context of the curio cabinet reveals a great deal about the place of natural science in early modern times.

The development of modern museums

During the sixteenth and seventeenth centuries, Europe could claim hundreds, if not thousands, of curio cabinets. In the view of their owners, they presented a picture of the world, supplying a microcosm of the odd, peculiar, and rare. The personal museum of the Castres physician, Pierre Borel (ca.1620–1671), an early correspondent of the Royal Society's Henry Oldenburg, illustrates the objects sought by Renaissance gentlemen. Borel assembled examples of mechanical and chemical arts, as well as diverse biological specimens. America furnished exotics; Africa provided monsters garnered from the animal, vegetable and mineral kingdoms. Monuments of antiquity occupied a prominent place, with classical vases, urns, statues of gods, medals, engravings, and weapons. Modern artifice also revealed itself in the thermometer, musical instruments, globes, and portraits in oil. Carved corals proved especially interesting because they showed natural materials worked by man. Extremes of nature were prized – the largest, the smallest, the misshapen, and the monstrous. Collectors, no less than scientific virtuosi, feasted on the abnormal and the bizarre.

These collections may be viewed as more than arbitrary miscellanies. Objects were included only if they held value in relation to a larger group. Gold, silver, and jewels represented precious materials; natural history exotica were prized for their rarity. The contrast between these remarkable objects and the commonplace produced a sense of wonder in the observer. Moreover, the ensemble of the objects, due to their heterogeneity, represented 'the variety and plenitude of art and nature', remembering, however, that '99.9 per cent' of the cosmos was ignored in favour of the 'singular and anomalous'. This analysis, articulated by historian Lorraine Daston, also suggests that it is more germane to consider why objects appeared in the cabinet, rather than to search for an understanding of how these materials were organized.

The arrangement of the collections seems to have been highly idiosyncratic. There were probably as many different classification schemes as there were individual collectors like Borel. Sometimes objects took their place for 'aesthetic' reasons – colour or size. A more sophisticated taxonomy related to the four elements, where rocks, plants, and insects represented earth; fish represented water; birds represented air; and artificial productions like glassware or tools represented fire. Alternatively, a tripartite division might be dictated by considering the provenance of objects from the animal, vegetable, and mineral kingdom.

Systems for displaying objects could conflict with explanatory devices, such as the Great Chain of Being. This belief assigned a specific niche to every organism in the universe, from the lowly worm up to humanity and the angels above. The chain of being might have dictated how objects would be displayed. In the view of early collectors, however, all questions of organization were secondary to the knowledge conveyed by startling juxtapositions of objects. These analogies, correspondences, and similarities allowed the viewer to pass from the visible to the invisible universe, from the macrocosm to the microcosm.

Fascination with the *Kunst- und Wunderkammer* proliferated throughout Europe, but interest in these cabinets began to decline on the eve of the Scientific Revolution. At that time, human curiosity began directing itself away from the mystical, hidden world of the microcosm and toward the empirically knowable macrocosm. Passion and desire still drove that curiosity, but the ordinary, natural world became the prime referent for the amateur natural historian or natural philosopher. An infatuation with nature in its greatest extent replaced interest in the rare, the unknowable, and the unseen (that earlier, coveted 0.1 per cent). This was the time when the word 'museum' entered the English language (it was used to denote John Tradescant's [1608– 1662] collection that served as the basis of the Ashmolean Museum at Oxford). Collecting representations of nature became a fad.

By the mid eighteenth century, the 'physiognomy' of collections had changed. The millenarianist zeal to make the invisible visible by means of the curio cabinet had been transformed by a new spirit that held collections up to empirical and experiential investigations. New procedures required living and dead plants, minerals, fossils, shells, animal specimens, machines, and scientific instruments, and these were the objects that came to fill museum shelves. A new ideology of encyclopedism placed stock in displaying the expanse of natural and artificial productions, whose utility derived from their importance to human education, culture, and wellbeing. During the Enlightenment, emphasis shifted to accessing knowledge through the exercise of human intelligence; the realm of spirituality, mysticism, and authority diminished. Treasures of the natural and artificial world were removed from the impermanence and capriciousness of individual exertions, to be displayed on a continuous and permanent basis in the public domain.

This evolution does not mean that virtuosi lost their interest in compilation and collection. Rather, the point of the exercise had shifted. Higher processes and procedures began to inform the assembling of objects. The early Royal Society, for example, expressed disdain towards the cabinets of the virtuosi. It aimed to assemble a collection that would serve as a tool for its collaborative project to reform knowledge. This systematic collection aided the construction of a universal taxonomy and mirrored the order of nature. The execution of the aim, however, fell short, and the Royal

Society's cabinet in fact held little to differentiate it from many others.[1]

Natural objects poured into Europe as a result of the great voyages of discovery, as we shall see in chapter 9. These materials helped to transform the vocabulary of collecting. During the Enlightenment, collectors came to recognize that the usual order of things offered diversity sufficient to delight the observer without recourse to monstrosities or fakes. The Swedish botanist Carl von Linné (1707–1778, known by his Latin name, Linnaeus) recorded this sense of wonder in his epoch-making research on classification. He remarked on the striking colours, extremes of size, and fascinating structures in the *natural* world as he catalogued the treasures of the Royal Swedish Museum in 1754. Enlightened men, in fact, took a sense of astonishment or surprise as an essential stimulus for promoting a deeper knowledge of nature. Awe, however, had to be disciplined. Instruction of the observer – bringing about the refinement of his character – eventually came to overshadow the museum's previously dominant function, amusement.

The dramatic expansion of museums during the nineteenth century is readily documented. In 1910 the eleventh edition of the *Encyclopaedia Britannica* counted no fewer than 2000 remarkably varied scientific museums. Governments, whether municipal, provincial, or national, supported many of them. Museums also belonged to learned societies, universities and colleges, and religious orders. Some traced their ancestry to the whims of an eccentric individual collector; others owed their establishment to the great number of objects amassed by a geological survey or put together for an exposition. On occasion, museums sought comprehensiveness and embraced every known science, both pure and applied. For others, mastering a restricted domain, such as the natural history products of a particular locality, sufficed. Museums prospered in all quarters, some merely having been granted the use of a few showcases or rooms in a library, while others enjoyed a dedicated building.

Paralleling the specialization and fragmentation in science, museums devoted to the splendours of the natural world ceased to welcome examples of human artifice. These artificial products became the preserve of museums of science and technology, which appeared with increasing frequency from the time of the Industrial Revolution. An early museum aiming to present man-made products (construed as the applications of science) was the Conservatoire National des Arts et Métiers, founded in Paris in 1794. More museums emerged to house the scientific and technological instruments brought together by international fairs and exhibitions during the second half of the nineteenth century. These include the Science Museum in London and the Arts and Industries division of the Smithsonian Institution (later, the Museum of History and Technology). Other important museums followed, including the Technical Museum of Vienna, the Deutsches Museum in Munich, and the Museum of Science and Industry in Chicago. Everywhere the function of museums became public instruction.

As museums were recast into educational institutions, questions of purpose, organization, and arrangement became paramount. For the first time, curators felt compelled to develop collections in response to the needs of diverse social groups. Increasingly during the nineteenth century, museums were expected to serve the middle classes experiencing greater leisure, wealth, physical mobility, and educational opportunities. This newly articulate population attached special importance to activities that joined education and amusement, such as visiting a properly organized natural history museum. Their requirements, however, could conflict with those of a professionalized scientific elite. Museum curators, as a result, frequently quibbled over technical issues of museum practice – such as how to display specimens and how many objects to exhibit – but underlying these debates were more fundamental differences of opinion concerning museological function.

Attitudes towards museum work varied according to local

circumstances, institutional loyalties, and national allegiances. The plans and procedures adopted by a government repository located in a major metropolitan centre, as a result, had limited relevance for a small collection developed in the hinterland. Yet, since museums, like other scientific institutions, were seldom started from scratch, successful models served as inspiration and example for less fortunate attempts elsewhere. Personnel were sometimes imported to implement a particular plan or design. With constant international imitation and cross-fertilization, ideas and innovations freely travelled the museum circuit during the decades preceding 1900.

The British Museum and the 'new museum idea'

Having abandoned its parent across town in Bloomsbury, the new British Museum (Natural History) in South Kensington opened its doors to the public in 1881, the jubilee year commemorating Queen Victoria's accession to the throne. At the time, it was the world's most remarkable natural history museum. A gothic 'temple to science', the museum exemplified Victorian architectural taste, which had been seen earlier in the ornate structure built for the New Museum at Oxford. Following the Oxford plan that incorporated biological symbolism into the design of the museum, the walls of the Natural History Museum displayed bas-relief representations of living and extinct species. The eclectic genius of the architect Alfred Waterhouse combined Romanesque arcades and Baroque staircases with cast-iron columns and a glass roof typical of contemporary railway stations. A few observers found the terracotta portico dingy and the high central hall empty, with adjacent side galleries appearing like factory rooms, but most visitors were delighted with the new museum. There were four acres of exhibition space in addition to rooms for laboratories, workshops, and storage.

It was not the dramatic building alone that placed the

Natural History Museum in the first rank, but the overall quality and extent of its collections. Especially in paleontology and mineralogy, the Natural History Museum soon supplied the example to which every other national museum aspired. Specimens were well-displayed, without the crowding that plagued so many government repositories that functioned as 'storehouses'. Exhibits made liberal use of descriptive labels, drawings, photographs, diagrams, and models. Museum publications, too, won accolades from colleagues in sister institutions. Because so many ideas had been tried and the staff were eager to share their experience with others, the Natural History Museum became the most important institution in the world for answering queries concerning museum practice.

The Natural History Museum also became an arena for championing various schemes for classifying and arranging natural history collections. There, as elsewhere, the attractively simple 'new museum idea' – the separation of study from exhibition – had triumphed by 1900. According to its foremost exponent, superintendent William Henry Flower (1831–1899), the concept involved organizing museums around a dual purpose: as research collections and as educational tools. Flower insisted that numerous specimens representing specific and, especially, varietal forms had little significance for the average visitor and belonged in a segregated study section. He compared the practice of exhibiting every museum specimen to the framing and hanging of each page in every library book. The observer who confronted objects row upon row came away if not bewildered, at least bored. Instead the liberal use of storage drawers and cabinets meant that specimens could be well-preserved (away from dust, pests, and light), housed in minimal space, and easily retrieved by researchers. To make the study collections useful, rooms adjacent to the exhibits required adequate tables, reference books, proper lighting, and an accessible museum staff.

The public exhibits, by contrast, were designed to impart

a general understanding of the kingdoms of nature to non-specialists. Only the best specimens were displayed, and these illustrated a particular principle or taxonomic category. There was to be no duplication of materials. Specimens appeared in uncrowded cases at a reasonable height, accompanied by informative labels. The curator might fashion his collection around such specific themes or concepts as geographical distribution or evolution, whose lessons might be enhanced by the distribution of guidebooks and catalogues. His resources could lead him to emphasize one group while excluding others. The function of these illustrative materials was to increase the educational value of the public materials.

Flower's notions ultimately triumphed over the more comprehensive and grandiose schemes of his predecessor, Richard Owen. They also helped to transform museum practice in other parts of the world. By 1900, most museum staff accepted the principle that 'less was more' in museum displays and that the remaining inventory of specimens should be reserved for scientific investigators. Yet the acceptance of the 'new museum idea' posed a quandary for curators for more than two decades. Museums served two distinct audiences simultaneously: a few scholars and the public at large. But it was difficult, if not impossible, to be equally responsive to the needs of both groups. As a result, any concentration on scientific requirements left directors open to the charge of neglecting their responsibilities to the public. Emphasizing the role of popular educator, on the other hand, suggested that museums were not performing their research function. The attempt to fulfil these two conflicting mandates provided a constant source of vexation to museum administrators all over the world.

Museums in Europe and the United States

At the time that the British Museum underwent expansion and reorganization, museums elsewhere in Europe experienced a surge of growth stimulated by the increasing size of their collections. The costly Vienna Naturhistorisches Hofmuseum opened to generous praise from 'new museum' enthusiasts in 1889. In the same year, the zoological collections at the Paris museum were moved to new Romanesque-style quarters replete with animal carvings; anatomy and anthropology found improved accommodations a decade later. The Bohemian Museum opened in Prague in 1894 and the Royal Belgian Museum inaugurated new buildings in Brussels in 1903.

Most continental museums displayed every object in their possession and followed what John Edward Gray (1800–1875), the British Museum's zoology curator during the middle decades of the nineteenth century, had contemptuously labelled the 'French plan'. This called, as Gray put it, for 'attaching each specimen to a separate stand, and marshalling them like soldiers on the shelves of a large open case'. A few museums, though, became converts to the views being articulated across the Channel. The Prague museum, for example, displayed true-to-life exhibits incorporating natural surroundings. It also made liberal use of drawings and models. The new building in Vienna separated exhibits from an internal core of workrooms, storage, and study collections.

By 1900 Germany could claim 150 natural history museums, Britain 250, and France 300, while the United States counted a respectable 250. In America, where museums had grown in quantity and improved in quality during the latter half of the nineteenth century, successful institutions traditionally depended upon individual initiative and philanthropic largesse. Examples include Harvard University's Museum of Comparative Zoology, for which Louis Agassiz raised $300,000, and the museum at Yale University, which benefited from George Peabody's bequests. Public museums, directed by trustees and supported by government, gradually began to supplant the cabinets of natural history typically associated with private colleges, lyceums, and academies. Foremost among these were the American Museum of Natural History in New York City and the National Museum, part of the Smithsonian Institution, in Washington. The American Museum, established in 1869, moved into new quarters in 1877 which spanned thirteen acres. Two years later the National Museum began to construct a new building to accommodate the donations received upon the dismantling of the Philadelphia Exposition of 1876. The museum's functional structure attracted worldwide attention for covering more than two acres while costing taxpayers only $250,000.

Like the British, the Americans had become expert at mounting explanatory and true-to-life displays, incorporating habitats, charts, diagrams, and photographs. The National Museum perfected the use of plaster casts and models to make fragments of bone and fossils comprehensible, permit reconstruction of partial skeletons, or fill the gaps in a series of specimens. Along with the American Museum of Natural History, it established a reputation for portraying animal groups in their natural environments. This method, initiated with birds and mammals, was soon extended to fish, invertebrates, fossils, and finally to plants, through the use of finely worked glass and wax models. Both museums benefitted from techniques pioneered at the taxidermy firm run by Henry Ward (1834–1906) in Rochester, New York, who hired out his trainees as curators.

Colonial museums

Along with advances and expansion of natural history museums in European and American metropolitan centres during the latter half of the nineteenth century came a proliferation in other parts of the world. With the gradual accumulation of capital in the hinterland – coming from the development and exploitation of natural resources, in turn made possible by improved transportation networks – colonial legislatures began to support public museums. As part of a survey conducted under the auspices of the Museums

Association in 1893, British Museum curator Francis Arthur Bather (1863–1934) evaluated museum resources throughout the Queen's empire.

The first characteristic of a colonial museum, argued Bather, was a purview limited to materials garnered from the immediate environment. Visitors to the museum – whether fresh immigrants from abroad or inhabitants of the interior – came to see local products, which were best kept apart from any other collections the museum might possess. In order to encourage and sustain support by residents, Bather reasoned, all donations, no matter how common, should be accepted. The other special interest of most visitors involved applied science. Although Bather pitied the curator who had to spend his days analyzing ores, an emphasis on practical matters had to be tolerated; it was preferable to having no publicly recognized utilitarian role, and hence being under constant threat of closure. If separate technological museums were allowed to be established, support for natural history museums risked dwindling to nothing.

The second characteristic used by Bather to distinguish colonial museums was their dependence upon foreign institutions. Bather argued that external support, particularly from the mother country, enabled colonial museums to weather day-to-day problems, such as periodic financial depressions, invasion by insects, and deleterious climate. Assistance might take the form of specimen exchanges or the advice of specialists whose skills were unavailable in the colony. Without the aid of metropolitan institutions, museums in the hinterland could neither function properly nor ultimately survive.

Bather underestimated and therefore understated the frequency with which fine institutions were almost literally carved out of the wilderness. Images of Victorian science palaces danced in the heads of museum men in the hinterland, and they sought to forge reasonable facsimiles under adverse circumstances. Some of these museums emphasized local objects, including the artefacts of aboriginal peoples, technological implements, and examples of native flora and fauna. But other museums, especially those associated with universities and colleges, sought to display broad collections selected to represent the diversity of animal, vegetable, and mineral kingdoms on a worldwide scale. These were intended principally for the instruction of students destined for careers in mining and medicine.

Like their metropolitan counterparts, colonial museums also set out to educate or at least to 'elevate' the middle and lower classes. As one nineteenth-century writer described the educational power of natural history specimens: 'Let each object represent so much knowledge, to which the very mention of its name will immediately conjure up a crowd of associations, relationships, and intimate acquaintances, and you will then see what a store of real knowledge may be represented in a carefully arranged cabinet.' Advocates repeatedly explained that viewing properly organized natural history museums instilled a sense of order, method, and law. In addition to developing an individual's powers of observation and reflection, museums might stimulate healthy exercise. The resulting interest in some branch of natural history, it was claimed, could lead the eager student into the 'pure air and pleasant scenes' of the countryside, thus offering the 'best antidote to habits of dissipation or immorality'.[2]

Such frequent and enthusiastic endorsements of the multiple intellectual, social, and moral benefits conferred by natural history museums indicate that their promoters were not simply mouthing empty rhetoric. As even museologist Kenneth Hudson – ever sceptical of museums' justificatory rhetoric – admits, science museums rest on foundations that are more democratic than those supporting art museums. Perhaps to a greater extent than for their colleagues elsewhere, colonial curators brought a missionary zeal to their work, which implied, as well, a concern with extending the frontiers of civilization.

Colonial and metropolitan museums: some comparisons

The 'museum movement' travelled around the globe with remarkable speed as museums throughout North and South America, Africa, Asia, and the South Pacific were founded, renovated, or given new quarters during the last decades of the nineteenth century. Some colonial museums copied the general organization of European institutions. The Hobart Museum in Tasmania and the Museu Goeldi at Pará, Brazil, followed the Parisian pattern of supplementing specimen collections with 'living' museums – botanical gardens, zoological parks, or aquariums.

As the *Encyclopaedia Britannica* generalizes the case, the architecture of nineteenth century museums was monumental and imposing, owing to their association with national prestige and civic pride. Architects achieved a grandiose effect by incorporating 'colonnades and arches, high vaulted interiors, [and] vast flights of stairs' into buildings often situated in parks or on broad avenues. On occasion, colonial museums – such as the South African Museum at Cape Town and the National Museum of Victoria at Melbourne – imitated the architectural detail of the cathedral-like British Museum.

More commonly, however, museums in the hinterland followed the design of a classical temple, which carried connotations of dignity, antiquity, and permanence. This architectural convention dictated a symmetrical façade, a commanding entrance, perhaps with a portico, and a multi-storied central hall surrounded by balconies and staircases. The third edition of James Fergusson's *History of the Modern Styles of Architecture* of 1891 singled out Montreal's Peter Redpath Museum as a particularly fine example of neoclassical style emerging from a rather conservative and provincial tradition of British colonial architecture.

Some museums adapted their design to the immediate environment or to local conditions. The Museu Paulista (also called the Museu Ypiranga, after the nearest trolley line) took advantage of São Paulo's mild climate by making its second floor rooms accessible only by an outdoor balcony. The Albert Museum at Jaipur followed Indo-Saracenic architectural style in its new building of 1876. Institutions without the money to erect their own facilities used structures intended for other purposes, such as public libraries and legislative buildings.

The size of the scientific staff and the amount of resources of colonial museums differed widely from one location to another. Budgets ranged from £8000 for the Canterbury Museum in Christchurch, New Zealand, to more than £500,000 (including salaries) for the National Museum at Rio de Janeiro, the showcase of Brazilian museums. The average budget pales beside those of national museums in Paris and London, which each spent more than £800,000 a year around this time. But many first-class British provincial museums spent not even £16,000 a year. Colonial museums, then, failed to attract funding to equal that of perhaps the top five museums in the world, but their budgets rank with those of the better European institutions.

Many colonial museums accommodated scientific staffs comparable to those of major metropolitan institutions. The American Museum of Natural History, for example, housed fourteen scientists among its seventy-one employees in 1877. In South America, where European traditions were especially strong because of French, Italian, and German curators, departmental keepers often doubled as natural science professors in the local university. By the late 1880s, several Argentine museums mustered staffs of eight or nine members.

Statistics on the annual number of visitors suggest how well museums functioned as instruments of popular education. It was not unusual for a colonial institution to attract 100,000 people each year. Even illiterate natives, it was claimed, enjoyed exhibits showing local resources and the history of their people. By way of comparison, the British Museum and the American Museum were receiving about

400,000 visitors annually by 1900; another major institution, Prague's Bohemian Museum, boasted an attendance
of 90,000 in one year. Colonial museums, then, ranked
as popular institutions on a global level, especially given
the more limited size of the local population that they
served.

The size of collections also reflects the significance of colonial museums. Ceylon's Colombo Museum contained
more than 100,000 zoological specimens, the Indian
Museum at Calcutta and the Geological Survey Museum in
Ottawa both more than 350,000 items (emphasizing rocks
and minerals), and the National Museum at Melbourne
more than 500,000 objects. The extent of these collections
may be compared with the world giant at that time, the
National Museum in Washington, DC, which held more
than 3 million items during the early 1890s. By the 1930s
it counted more than 12 million objects and was acquiring
more at the rate of 500,000 a year. Though trailing behind
this leader, many colonial museums attempted to assemble
reasonably large collections to be used for teaching or
research purposes. By the end of the century, most colonial
museums had assembled hundreds of thousands of specimens valued at hundreds of thousands of dollars. Many
of their holdings were irreplaceable and unique, leading
European curators to visit these distant places in order to
glimpse their remarkable treasures.

Moderate-sized colonial towns appear to have been better
endowed than their analogues in England. Ottawa,
Dunedin, and Durban, for example, mustered considerable
resources to support museums. Larger cities, too, such as
Sydney and Cape Town, could be categorized as 'good
museum towns', according to the criterion of providing at
least sixpence a year per capita for museums and art galleries. All these considerations support the view that colonial museums were important loci of scientific activity.

Descriptions of colonial museums

Historians have largely failed to recognize the extent to which non-European museums of the late nineteenth century sought to assemble significant collections that included foreign materials. Partial justification for museum-building came from the desire to edify the local populace, especially younger colonists who had never seen the natural and artificial products that were commonplace in Europe. Apart from this public rationalization, however, different visions inspired curators who found themselves working in unforeseen places and under unanticipated conditions. Using European practices and methods as the measure, these men believed that a proper museum had to include objects of universal value as well as materials of local interest alone. The diversity of the natural world was to be shown at least through representative types, if not by a multiplicity of individual forms. Reputations depended upon the number of specimens amassed, with considerable cachet attached to the acquisition of exotica from abroad.

If a naturalist's fortune had led him to Cape Town or Bombay, the question became, as Melbourne's museum director Frederick McCoy (1823–1899) put it, how to *grow* a museum in the hinterland. Growth required financial resources, which came to most museums at some stage in their development, but rarely at an adequate level or on a regular basis. Certain local objects, however, such as rare natural history specimens or archaeological implements, possessed commercial value and might be sold or exchanged for materials from abroad. International exhibitions, whether the huge fairs held in major metropolitan centres or the more modest colonial affairs, called attention to these treasures from all over the world.

Some colonial museums fared better than others in turning the accidents of nature to good account. In Africa, natural history museums were concentrated in the extreme southern portion of the continent. South African and

Rhodesian museums survived only in centres with large white populations, such as Cape Town, Durban, Pietermaritzburg, and Grahamstown. Although some institutions attempted to attract native inhabitants, more typical was the museum that excluded blacks every day but Thursday, when admittance depended upon wearing boots or shoes. Even the descendants of English and Dutch settlers failed to attend museums in great numbers, because, according to one observer, museums could not compete against the magnificent outdoors. Despite above-average architecture and design, the meagre financial resources of most South African institutions left them impoverished by North American or European standards.

African museums were well endowed, however, in comparison with their neighbours farther east. The impetus for the creation of a number of India's principal museums – in Bombay, Calcutta, and Madras – came from the donation of the collection of a local philosophical or scientific society. Following the initial bequest, though, specimens often fell prey to the forays of insects and extremes of temperature and humidity. These adverse conditions were exacerbated by poor curatorial care. Hindu workers, reluctant to take life, tended to tolerate pests, while the caste system encouraged a rigid adherence to assigned duties. In the words of one observer, there was no other place in the world where museums 'count[ed] for so little, . . . [were] so meagerly supported, or . . . [were] so few and far between'. Many large towns lacked museums altogether; those that existed appeared as 'gingerbread palaces, fantastic and bizarre, or gloomy prisonlike edifices . . . [with] galleries more suited to be mausoleums.' Some still exhibited freaks and monsters to the curious; others showed obsolete maps and charts occasionally displayed upside down. As in Africa, widespread illiteracy, extreme poverty, and rural settlement patterns made museums irrelevant to the vast majority of the populace.[3]

Museums in Canada, South America, and Australasia

Elsewhere the museum movement was more successful. The leading museums of postconfederation Canada, for example, were concentrated within a 150-mile radius of Montreal. The largest collections belonged to the museum of the Geological Survey of Canada (later christened the National Museum), which was transferred to Ottawa after nearly forty years in Montreal, along with Survey headquarters, in 1881. Ottawa's huge reserve of 150,000 paleontological specimens placed it far ahead of any other Canadian museum. Nevertheless, the large number of items displayed in Montreal's Natural History Society museum, together with McGill University's Peter Redpath Museum, made the city a close second to the national capital. Quebec City, which for a time had alternated with Toronto as capital of the young colony of Canada, placed third because of the excellent collections assembled in the museum of Laval University and in Quebec province's Museum of Public Instruction.

Although the balance of power would shift with the creation of the Royal Ontario Museum several decades later, Ontario's best museums, those in Kingston and Toronto, housed only a fraction of the material exhibited in the three main museum centres of Ottawa, Montreal, and Quebec City. The typical Canadian museum of the day – crowded into several rooms and controlled by Catholic educational institutions in Quebec or by some other kind of organization or university in Ontario – usually contained around five thousand natural history specimens. Municipal museums, more closely associated with the needs of the local citizenry than those of a learned society or school, were almost entirely absent. As one authority who surveyed Canadian museums summed up the case in the 1930s, Canada had long been active in collecting, but its educational museums were embryonic and its museum endowments negligible.

South American museums, by contrast, tried to function

both as research institutions and as instruments of popular enlightenment. Supported by national or provincial governments, important museums could be found in every capital city. Rio de Janeiro, Buenos Aires, Santiago de Chile, and Montevideo erected autonomous natural history museums. Bogotá and Caracas combined natural history with art and other subjects, and housed all the collections under one roof. Most of these institutions enjoyed comfortable annual appropriations and spacious quarters that accommodated a large staff. They undertook considerable scientific fieldwork and issued a variety of research publications. According to one observer, the larger museums in South America attracted 100,000 to 150,000 visitors annually, which amounted to about 5 to 10 per cent of the local population.

Museum resources in Australia and New Zealand compared favourably with those of cities of similar size elsewhere. In New Zealand, as a result of the action of public-spirited and energetic citizens, five towns – Nelson, Christchurch, Wellington, Auckland, and Dunedin – possessed museums by 1877. Australian museums ranged from excellent to mediocre, despite the strong hand of individual state governments in funding. Following settlement patterns, museums tended to be concentrated in the southeastern corner of the mainland. By the 1870s about a dozen museums, emphasizing geology or zoology, had been created; another dozen had opened, through the efforts of local learned societies, by 1900.

The success of the 'museum movement' is especially surprising given the adverse circumstances under which colonial institutions, even in these most favourable situations, had to function. Political and financial problems daily tried the patience of curators like McCoy, whose museum fell victim to ministerial whims as internal government reorganizations shuffled it from one department to another. The acute depression of the early 1890s caused the dismissal of the entire scientific staff at the Queensland Museum in Brisbane, and the acquisitions budget for the Australian

Museum in Sydney was reduced more than 80 per cent at that time.

Unanticipated 'acts of God' seemed to strike even well-endowed and long-established colonial museums with distressing frequency. Fires destroyed collections at Sydney and at New Westminster (British Columbia), where the entire town went up in flames. Earthquakes wrecked museums in South America and New Zealand. Less extreme circumstances – fluctuations in humidity and temperature, as well as excessive amounts of sunlight – caused mounted specimens to shrink, crack, and fade. In tropical areas, damage also resulted from the incursions of moths, mites, birds, and monkeys. By contrast, John William Dawson in Montreal had to contend with a snow blockade that delayed a shipment of materials from New York.

The attitude of the local populace was sometimes just as intractable as the environment. One sympathetic Australian politician complained to McCoy that 'it is difficult to indoctrinate people with ideas altogether foreign to those which have already occupied their minds and it is still more difficult to implant a new idea in a mind hitherto fallow'. Several curators worried about vandalism and the theft of coins and revolvers because 'visitors were too indiscriminately admitted'. Indeed, when thieves stole gold from several Australian museums such fears proved well founded.[4]

According to those who surveyed colonial museums, themselves museum professionals, the skill and energy of the curatorial staff spelled the difference between success and failure. In South America, a number of Europeans, such as Hermann von Ihering (1850–1930) at São Paulo and Rudolph Amandus Philippi (1808–1904) at Santiago, transformed the drudgery of museum-building into magnificent testimony to a life's work. In Australia, the sons of prominent metropolitan keepers such as Henry Woodward (1832–1921) and Robert Etheridge (1819–1903) served their scientific apprenticeships as curators at the outposts of the British Empire. In many instances, however, the problems encountered and the low salaries paid made colonial positions somewhat less than attractive to ambitious naturalists.

A handful of active and enthusiastic men directed museums located in the major urban centres of Canada, Argentina, Australia, and New Zealand. Early curators stayed at the helm of their museums for decades, and during this long period they never failed to bring a single-minded dedication to their work. It is not surprising that natural history museums enjoyed particular success in these environments where 'outposts of transplanted European society' were created. Unlike most other colonial or neocolonial situations, these countries had been recently settled by European immigrants. Native peoples, already decimated or otherwise subjugated, posed no threat to establishing young societies that would quickly surpass the living standards of their European progenitors. From around 1870 to 1914, these 'Dominion capitalist societies' became the 'centre pieces of British imperial strategy'. Wool, meat, dairy products, cereals, and lumber were exported from their sparsely settled interiors in return for industrial products, loan capital, and immigrants. Among the European powers engaged in this process, Great Britain was especially active in financing railroads and other mechanisms that helped bring food and raw materials to market.[5] This period of especially rapid economic growth coincided with the time of museum expansion.

Enterprising directors found in Canada, South America, and Australasia particularly fertile environments for museum-building. Most curators initially emphasized local products and resources, subscribing to the theory that residents were most interested in 'the things they can find about home'. Attracting as many visitors as possible helped to justify museums' role as educational institutions. To increase popular appeal, directors often displayed coins, ancient relics, and ethnological materials alongside natural history specimens of every description. It was not unusual to show technological apparatus, such as agricultural machinery. Australian museums developed their geological and mineralogical exhibits, while Christchurch's Canterbury Museum displayed specimens of wool accompanied by practical hints from sheep-breeders.

After a time colonial curators in these favourable circumstances sought prime foreign specimens as well. They aimed at collections to rival those at the great metropolitan museums, to be, as natural historian William Swainson (1789–1855) put it, 'to the naturalist what a dictionary is to the scholar'.[6] Julius Haast (1824–1887) at Christchurch and McCoy at Melbourne mastered the mechanisms of exchanging and purchasing specimens abroad. Museums affiliated with local universities, such as Argentina's La Plata Museum and the Peter Redpath Museum in Montreal, strove to display comprehensive and instructive series of natural objects for visitors seeking the rudiments of systematic biology or mineralogy. Several other institutions, including the public museum of Buenos Aires, established solid reputations for scientific research.

Colonial museums, however, tended to exhibit specimens row upon row and for the most part neglected to incorporate up-to-date techniques, such as explanatory labels and habitat cases. A few institutions did adopt the 'new museum' idea and did separate research from exhibition collections. McCoy set out to develop teaching collections based upon the exhibition of generic types and representative geographical groups of species. This index to organic life required no subsequent rearrangement for new specimens. Making a distinction between local and foreign objects gave other museums some rudimentary arrangement. The remarkable collection of moa skeletons at the Canterbury Museum, for example, strikingly distinguished New Zealand specimens from the other exhibits.

In his survey of 1893, Bather distinguished colonial museums by their adaptation to local circumstances and their dependence upon foreign institutions. Certainly museums in the hinterland did follow metropolitan models and did import European staff. The personnel often stayed on to shape a new kind of scientific career, modifying metropolitan patterns. The very existence of these indigenous museums overseas militated against scientific dependence, because they protected local treasures against plunder by European institutions. They may be seen as encouraging the first steps toward scientific independence, in that they provided a haven for local natural resources and products. They came to offer scientific training and positions to native-born talent.

Colonial museums flourished only so long as the museum movement prospered elsewhere. When the youngsters who flocked to the British Museum in South Kensington during the 1880s and 1890s reached maturity, natural history museums had already begun to lose their vitality. The natural sciences – pursued by specialist geologists, zoologists, and botanists – offered more promising vistas from the microscopic, rather than the macroscopic, level. Even Darwinian evolution, which at first had accelerated the zeal to collect by rationalizing taxonomy and giving new scientific significance to varieties, seemed to offer greater inducements to the geneticist in the laboratory rather than to the ornithologist or mammalogist in the field. Those who remained in the field found that new techniques like photography provided better data than numerous museum specimens.

Developments external to biological discourse, such as the rise of research institutes and the remarkable expansion of universities, also diverted resources and interest away from museums to other scientific endeavours. The intense personalism of many museums sets them apart from dominant trends in the emerging ecology of scientific institutions at this time. With the onset of an era of big science, research began to be advanced through teamwork, using elaborate technologies supported by massive government expenditures. Most museums failed to transform their curators into cooperative coworkers, owing both to the authoritarian dominance of the director and to the atomizing forces exerted by insufficient staff. Resources all too easily went instead to competing scientific institutions. Major natural history museums thus appear as a particular stage in the evolution of scientific institutions. Formations better-adapted to the demands of twentieth-century scientific activity displaced them from an important role.

Until the second or third decade of the twentieth century, the history of natural history – including geology, paleontology, botany, zoology, and mineralogy – was largely caught up with the development of museums. To view them today is to glimpse magnificent monuments to an earlier scientific tradition.

Growing: Botanical Gardens and Zoos

Frances Hodgson Burnett's *Secret Garden*, appearing first in 1911, was one of the great publishing successes of Edwardian England. In the story, the invalid Colin delights in an Indian folk tale which posits that peering down the throat of a python reveals the universe unfolding. Colin's first passage through the gate of an abandoned, secret garden fulfils just this function: He drinks in the verdant foliage, designed to evoke the beauty of the world. Humankind has long experienced Colin's wonder and delight in the scents, colours, and smells of the garden, where horticultural treasures exhibit the vast bounty of nature within a fixed perimeter. Gardens represent our collective attempt to return to Eden, although nature is therein artificially rearranged, ordered, and bounded.

'Living' museums – whether botanical gardens or zoos – possess charm that eludes even the most attractive collections of dried and stuffed specimens, works of art, or technological gadgets. We react instinctively to the presence of other living things, even animals in captivity and topiary trees pruned into artificial forms. Historians, psychologists, and sociologists might trace this affinity to evolutionary processes, to the need to react against rampant urbanization, or to a primitive desire to return to nature. The fact remains that we love gardens – whether botanical or zoological – and we will spare no expense in their maintenance.

The foundation of botanical gardens and zoological parks in modern times relates closely to the same impulse for gathering empirical information that led to the creation of museums. One sees this association in the writings of Francis Bacon, who had emphasized the importance of libraries and workshops in furthering amateur scientific pursuits, but who also argued for the creation of collections representing animal, vegetable, and mineral kingdoms. Bacon called for 'a goodly, huge cabinet, wherein whatsoever the hand of man by exquisite art or engine has made rare in stuff, form, or motion; whatsoever singularity, chance, and the shuffle of things hath produced; whatsoever Nature has wrought in things that want life and may be kept'. That cabinet was to supplement

> a spacious, wonderful garden, wherein whatsoever plant the sun of divers climate, or the earth out of divers moulds, either wild or by the culture of man brought forth, may be . . . set and cherished; this garden to be built about with rooms to stable in all rare beasts and to cage in all rare birds; with two lakes adjoining, the one of fresh water the other of salt, for like variety of fishes.

Bacon concluded that one had 'in small compass a model of the universal nature made private'.[1]

The development of botanical gardens

A different aim – the establishment of gardens as an aesthetic pursuit – predates Bacon's lifetime by millennia. The sumptuous parks of the Egyptian pharaohs or the pleasure gardens of the wealthy Romans celebrated self-indulgence. Even the quiet beauty of the enclosed gardens of the medieval monastery, at first glance markedly dissimilar in function, still appears highly artificial as it served to enhance habits of personal introspection and contemplation. The open spaces of the secular garden of the Middle Ages, in contrast – typified by flowering meadows and bountiful orchards – are associated with the sensual pleasures of daily life. There, wild plants were domesticated over the centuries for agricultural purposes.

152 · *Servants of Nature*

The geometrical formality of the Renaissance garden, with its central fountain and statuary, emphasized artifice. This artistic tradition reached its apotheosis in André Le Nôtre's (1613–1700) design of Versailles, 'the greatest garden the world has seen'. By emphasizing 'the formal subordination of nature to reason and order', it might seem to unite artistic and philosophical traditions. But the attention to the formal conventions of architectural style meant that aesthetic criteria overwhelmed scientific interest in plant diversity. The gardens at Versailles symbolized the Sun King's supremacy, at the same time entertaining and distracting a captive court.[2] The increasingly ambitious plans of Le Nôtre, applauded and supported by Louis XIV, proved nearly as costly in men and matériel as the most grandiose gardens of the Egyptian pharaohs.

The tradition of cultivating plants for their ornamental and architectural effects may be traced into our own times. A more intriguing development, one more closely related to Bacon's vision of capturing the diversity of plant life in microcosm, is shown in the evolution of the botanical garden. The success of these gardens in presenting a large collection of native and foreign plants depended on local conditions. In northern climates, glasshouses offered protection from the cold, but at Buitenzorg, on Java, coolhouses became necessary. These herbaria presented flora for didactic and educational purposes, as much as for the needs of horticulture or agriculture.

Many botanical gardens developed from physic gardens, established to protect doctors and apothecaries against 'ignorant and unscrupulous drug-sellers and root-diggers'. In these gardens, medical men raised their own 'simples', or medicinal herbs, from which apothecaries prepared medicines, the 'remedia composita'. From the beginning, physic gardens were associated with the medical faculties of universities, where one professor usually taught both botany and medicine. The gardens themselves often doubled as classrooms.

Exploration and commerce expanded the physic garden from merely encompassing collections of medicinal herbs to embracing new botanical varieties. In this way, interest in beautiful and unusual plants joined a concern with practicality. Following the discovery of the Far East and the Western Hemisphere in the fifteenth century, astonishingly novel and varied flora flooded into European port cities. The voyages had been undertaken to find spices and agricultural products, but many ornamental plants, including tulips, lilacs, and peonies, were brought back to Europe in the process. This development revolutionized the character and design of gardens, for gardeners paid attention to plant structure, morphology, and physiology. Botany – particularly the collecting of exotic plants – became a fashionable activity for Renaissance princes.

The earliest botanical garden was created in Hamburg in 1316. The great age of the garden is marked by the appearance, in the sixteenth century, of establishments at Padua, Leiden, and Montpellier, and, in the seventeenth century, at Oxford, Uppsala, and Paris. Many of the great names in natural history and botanical illustration are linked with these early botanical gardens. In 1547, Ulisse Aldrovandi (1522–1605) founded the botanical garden at Bologna; in 1560, Conrad Gesner (1516–1565) laid out one in Zurich. Herman Boerhaave became director of Leiden's botanical garden in 1709, where the first greenhouse had been built in 1599 to protect plants brought back to Holland from the Cape of Good Hope. Under Boerhaave, Leiden emerged as the most important centre of medical education in Europe. The botanical garden contributed to this reputation, and as a result itself entered an era of unprecedented expansion.

Outside Europe, as early as 1685, the Dutch East India Company established a botanical garden at the Cape of Good Hope to provide fruit and vegetables to scurvy-ridden sailors in transit to the East Indies. Soon it began to acclimatize plants on their way between four continents: Europe, Asia,

Africa, and South America. In the Northern Hemisphere, America's first botanical garden owed its creation in 1728 to John Bartram (1699–1777) of Philadelphia.[3]

The historical association between botanical gardens and gardens of physic could be real or symbolic. In Padua and Paris, the connection was close. Prospero Alpini (1553–1616), who had been physician to the Venetian consul at Cairo, returned to Italy in 1583 with Egyptian plants and became professor of botany at the University of Padua. Ornamentation in the garden included statues of the ancient physicians Asclepius, Hippocrates, and Galen. In Paris, the king's physician supervised the Jardin du Roi, established by Louis XIII in 1626, which was also called the Jardin Royal des Plantes Médicales. Three doctors from the medical faculty at the University of Paris also served as curators, as they studied the properties of plants and started a cabinet of dried specimens.

In 1729, when the Jardin Royal des Plantes Médicales became simply the Jardin Royal des Plantes, the director and staff no longer had to be physicians. This change inaugurated the period when the garden emerged as a leading institution for research in natural history – the era of curators Antoine-Laurent de Jussieu (1748–1836); Jean-Baptiste de Lamarck (1744–1829); Bernard-Germain-Etienne de la Ville-sur-Illon *comte* de Lacépède (1756–1825); and Georges-Louis Leclerc *comte* de Buffon (1707–1788). As superintendent of the Paris garden, Buffon added greenhouses and a natural history cabinet. In 1793, following the transformations wrought by the French Revolution, the garden was renamed the Muséum d'Histoire Naturelle. It also acquired a director and professors of anthropology, paleontology, comparative anatomy, and zoology.

Convention dictated that all botanical gardens follow a similar plan. The discovery of the New World had transformed the quarters of the square physic garden from representing the four corners of the earth to standing for the four continents: Europe, Asia, Africa, and America. Each quarter divided into individual beds or parterres, the ensemble of which was supposed to exhibit the flora appropriate to each continent. These might follow intricate geometrical designs, as in Padua or Paris, or contain long and narrow beds called *pulvilli* (small cushions), as in Leiden and eventually in Oxford. Each *pulvillus* held a particular family of plant, and was divided again into smaller units to accommodate individual plants. Historian John Prest likens the arrangement to an extended family seated at long dinner tables, and concludes that just as every place-setting was accessible, there was no plant that could not be 'seen, touched, smelled, and sketched' from the gravel or grass walks between the *pulvilli*.

Essentially, the botanical garden functioned as an encyclopaedia. Like the pages of a book, plants of each genus were placed according to a predetermined plan for reference purposes. Unlike the dried specimens in a herbarium, plants in a real garden could be 'read' in the sumptuous diversity of their living form. There observers' senses constantly received the plants' vibrant colours, pungent aromas, and varied textures. Just as with museums, gardens sought to bring the earth's lush flora into a smaller compass, where it could be appreciated, enjoyed, and savoured. It was a feast for the empiricist appetite.

Prest contends that the foundation of these botanical gardens represents an attempt to recreate the Garden of Eden. Humankind could retreat to the quiet of the garden to escape the world or to find remedies for injuries and disease. In the botanical garden, then, 'man could enter into communion with what was green and full of sap, recover his innocence, and shed his fear of decay'. The legacy of the botanical garden was an association of gardening with optimism, hope, and progress. These are also the values that gave rise to Baconian empiricism and the Scientific Revolution.

The cultivators of botanical gardens aimed to increase their knowledge of the natural world and express humanity's power over nature. By bringing all known plants to one place, they could be named, catalogued, and properly classified; their nutritive and medicinal properties could also be assessed and described. Linnaeus's creation of binomial classification greatly assisted this task, because his taxonomic system (laid out in his *Genera Plantarum* of 1737) provided for the unambiguous identification of every animal and plant within a general framework. Linnean nomenclature, using two Latin names, gave a universal, international language to naturalists.

Unlike other systems that depended on external appearances, Linnaeus based his organization on observable characteristics of the plant's sexual organs – its stamens and pistils. Flowering plants, for example, are divided into twenty-four classes according to the number of male organs or stamens. Each of these classes is divided in turn into orders, according to the number of female organs or pistils. The class and order, then, depends simply on counting; the genus and species, however, reflects qualitative decisions based on appearances. Often ancient herbals dictated the genus; assignment to a species came from considering the flower's particular characteristics.

Although New World botanical species could be catalogued by the binomial system, the unexpected quantity of novelty rocked the foundations of the garden. Seventeenth-century Oxford's garden, for example, could scarcely claim to represent the universe when it contained only 2000 species, for naturalist John Ray (1627–1705) estimated that soon around 20,000 different plant species would be known. Botanists competed to grow the greatest number. The Jardin du Roi at Paris counted 1800 species in 1636, and by 1665, 4000.

It was not only an embarrassment of riches that challenged the vision of the directors of botanical gardens. The end of the seventeenth century brought harsher winters than Europe had recently experienced. This change spelled the doom of more delicate plants, because climatic severity made glasshouses prohibitively expensive to construct and maintain. A new importance came to dried specimens, which were kept in a herbarium, the *hortus siccus*. This was also the time of the rise of the travelling collector who went out to view plants in their natural habitat. Abstract defence of botanical gardens became difficult. At the same time, gardening promoters exuded unbounded enthusiasm, something no enclosed garden could ever satisfy.

The design of botanical gardens shifted with their rationale. The British landscape architect Capability Brown (1715–1783) introduced 'the informal dogma of the fluent curve' during the second half of the eighteenth century. This concept swept away the formality of the quartered square garden in favour of an imitation of nature. More remarkably, even, animals were introduced and allowed to roam freely. The rise and fall of the botanical garden, in a certain sense, prepared the groundwork for the zoological garden.[4]

Botanical gardens emerged throughout the eighteenth century, although without the earlier fervour associated with their creation. Linnaeus helped restore the Uppsala garden, which had been damaged by fire in 1702 and neglected until 1741. Strongly influenced by Dutch botany since the time of his studies in Holland, Linnaeus saw gardens as living botanical museums and outdoor schools. Imperial gardens came to Peter the Great in 1713 and Maria Theresa in 1754, both at the urging of Gerhard van Swieten (1700–1772), a Boerhaave student who became professor at the medical faculty of the University of Vienna. In Britain, university gardens started at Cambridge in 1762, Dublin in 1790, and Glasgow in 1817. Private gardens also flourished there, including that of John Tradescant, the elder, whose catalogue of plants had appeared in 1656, in his son's *Museum Tradescantianum*. In a vicarage garden, Stephen Hales (1677–1761) founded the science of plant physiology by experimenting with the movement of sap and recording his conclusions in the *Vegetable Staticks* of 1727. Swiss physician and botanist Jean Gesner (1709–1790) counted 1600 botanical gardens in Europe at the end of the eighteenth century.

Botanical specimens from America and Asia continued to flood into Europe. Expeditions began to include naturalists who returned with all manner of exotic flora. Brilliant scientific collaborations were forged between botanists on different continents. The English Quaker merchant Peter Collinson (1693–1768), also a patron of Benjamin Franklin, worked closely with the Philadelphia farmer John Bartram. Their exchange of specimens over a period of forty years provided 200 plant species from eastern North America for European gardens, with a large number of European cultivated plants travelling west in return. The acceptance and articulation of Linnean nomenclature allowed this tide of specimens to be identified and classified accurately.

Botanical gardens flourished outside Europe and North America, as well. The French created a garden to acclimatize plants in Pamplemousses, Mauritius; British planters established a similar garden on St Vincent, in the West Indies. But the most important was founded in Buitenzorg, Java, in 1817 by Caspar Georg Carl Reinwardt (1773–1854). (Buitenzorg was the residence of the governor general of the Dutch East Indies; the town's name translates as *sans souci*, or without care, also the name of the park established by Frederick the Great.) Dual advantages of heavy rainfall and high altitude made for lush vegetation. Europeans knew it as the most beautiful garden in the world. Like all tropical botanical gardens, Buitenzorg emphasized economic products – cinchona (whose bark yields quinine), rubber, and coffee – under successive directors Melchior Treub (1851–1910) and Rudolph Scheffer (1844–1880). Late in the nineteenth century, Treub made the garden a mecca for botanists all over the world by setting aside one laboratory for the use of visitors. In his herbarium, the sheets of dried plants were kept in tin boxes (as a precaution against insects and humidity) rather than in the customary folios.

The French Revolution and industrialization helped to

change the character of botanical gardens. A new emphasis recognized the needs of the common man. As museums had done, gardens modified their *raison d'être*, aiming to present and classify plants for study and instruction, not simply to display them for ornamental or for utilitarian purposes. Emphasis went to field collection, research, and the demonstration of commercial use. These impulses led to the creation of public gardens, freeing the microcosm of nature from the exclusive preserve of the noble and the learned. In 1802, for example, the merchant and poet William Roscoe (1753–1831) founded the Liverpool Botanic Garden. On the Continent, public gardens were created at Breslau in 1811, Geneva in 1817, and Munich in 1822. Some were newly attached to universities, such as those at Göttingen and Harvard, and occasionally the public could stroll through their grounds. People flocked to gardens as a place for picnics, plays, fireworks displays, and even trysts – finding in botany the pleasure long enjoyed by the nobility.[5]

The fad of gardening during the middle decades of the nineteenth century drew along other institutions in its wake. A host of popular horticultural encyclopaedias, journals, and books were authored and edited by gardener-writers like John Claudius Loudon (1783–1843) and his rival Sir Joseph Paxton (1801–1865). Horticultural societies emerged to suit all levels of society, following the foundation of the London Horticultural Society in 1804 and the Botanical Society in 1839. Some focused on special plants, such as the Ipswich Cucumber Society, but more general interests were also served, as with the York Grand Floricultural and Horticultural Society. Even the fascination with specific flowers, whether hyacinths, tulips, or anemones, followed the dictates of fashion, making one day's rage the next day's weed. There was tremendous interest in hybridization of new varieties for the most popular flowers.

The acclimatization and cultivation of these plants was made possible only by the assistance of new technologies. The development of the Wardian case permitted the transportation of tropical plants to greenhouses. Originally these glass containers had been devised by Nathaniel Ward (1791–1868) for growing plants in London's smoky atmosphere. Another innovation was the indoor garden or 'horticulture under glass', a successor to earlier orangeries. In these huge glasshouses, rare and delicate plants like orchids could be 'scientifically arranged in a long succession of glass-contained avenues'.[6] In North America, greenhouse building was especially advanced in Montreal, due to its harsh climate. Unique in Canada at the time, Montreal possessed an active horticultural society, subsidized by the government. In that city of bitterly cold winters, people were proud to cultivate particularly fine landscape and flower gardens, a tradition that has continued to the present day.

Kew Gardens

The evolution of one of the foremost botanical gardens of all time, Kew Gardens, illustrates these general trends. Kew emerged by partition from the grounds of the country palace of the English royal family at Richmond. From the mid eighteenth century onwards, these monarchs bestowed serious attention on their palace gardens. Frederick, Prince of Wales, and his wife Augusta of Saxe-Gotha appointed the architect William Chambers (1726–1796) and the Scottish gardener William Aiton (1766–1849) to transform the estate. Chambers accurately predicted in 1765 that within a few years the gardens would contain the best collection of plants in Europe, as a result of 'the assiduity with which all curious productions are collected from every part of the globe without any regard to expense'. Over the next seven years, more than 3400 species of plants were grown at Kew. An orangerie, a pagoda, and a 'Great Stove' – a 114-foot long glasshouse – were built. Peter Collinson wrote to John Bartram that Kew was 'the Paradise of our world, where all plants are found, that money or interest can procure'.

George III, the son of Frederick and Augusta, employed the legendary gardener Capability Brown to remodel the grounds of Richmond Lodge. Upon inheriting Kew from his mother in 1772, King George appointed as his horticultural adviser Joseph Banks, who had just returned with Captain Cook from his first voyage to the South Seas on the *Endeavour*, a trip he had undertaken at his own expense as ship's naturalist. Banks, as director of the gardens for the next forty-eight years, persuaded the king to send out plant collectors to all corners of the Empire. As a result of their exertions, exotic plants arrived at Kew from South Africa, Australia, China, and South America.

Under Banks's supervision (by then he had also been elected president of the Royal Society), Kew led the world in botanical exploration and experimentation, with special emphasis on acclimatization and practical botany. A Kew gardener, David Nelson, was aboard the *Bounty* when the infamous mutiny of 1787 occurred. On that occasion, Captain William Bligh had been charged with bringing breadfruit plants (a proposed source of cheap food for slaves) from Tahiti to the West Indies. Nelson survived the mutiny, but he died upon reaching Java after drifting for eleven weeks in an open boat with men loyal to Bligh. Despite this small reversal, a flood of exotic flora continued to wash up on British shores. In 1789, 5500 species of plants grew at Kew; by 1814, there were more than 11,000. In order to help record this botanical wealth, Banks persuaded Franz Bauer (1758–1840) to become artist-in-residence.

In 1820, both George III and Banks died; as a result of the loss of patronage and leadership, Kew began to decline. But a new era opened in 1838, when a parliamentary committee recommended that the gardens become a state-run institution. William Jackson Hooker (1785–1865), former director of the botanical gardens at the University of Glasgow, arrived as director in 1841. His vision of Kew extended Banks's programme for having the gardens serve as an imperial repository of botanical knowledge. Following this ideal, Hooker built glasshouses – most strikingly exemplified by the great Palm House – to protect specimens brought back from the antipodes. The exquisite Palm House became the pivot of the new layout of the grounds, designed by William Andrews Nesfield.

For Hooker, Kew's beautiful appearance, which attracted scores of visitors, was nearly as important as its scientific function. By 1850, the gardens appeared in their modern guise with extended visiting hours for the public. During Hooker's directorship, the number of annual visitors increased from 9000 to half a million. William Hooker died in 1865, to be succeeded by his son, Joseph Hooker (1817–1911), who had been assistant director for a decade. Joseph's special interests involved studies in plant classification and distribution. Together the Hookers directed Kew for almost half a century, giving them the honour of transforming a 'moribund royal estate' into a major botanical garden. Indeed, the Hooker dynasty actually continued at Kew until 1905. Upon Joseph's retirement, his son-in-law, William Thiselton-Dyer (1843–1928), succeeded to the directorship following a ten-year stint as assistant director.

Unlike other botanical gardens, Kew never affiliated with a university or any other educational establishment. With the foundation of a library and herbarium, it emerged as a research centre, giving special attention to economic botany. Both Hookers sent their former staff members to run botanical gardens throughout the Empire, whether in Trinidad, Brisbane, or Singapore. Among the many men who had collected plants for Kew at one time, William Ker became superintendent of the Ceylon botanical gardens in 1812 and Allan Cunningham (1791–1839) directed the Sydney botanical gardens, beginning in 1836. Through these contacts in the colonies, Kew orchestrated the development of profitable plant-based industries, particularly cinchona, rubber, and sisal. Today Kew cultivates around 45,000 of the nearly 250,000 known species of green, flowering plants.

By the middle of the twentieth century, more than five

hundred botanical gardens were scattered across the globe. Some were merely public pleasure parks, such as the large municipal Golden Gate Park in San Francisco and Prospect Park in Brooklyn. Others, in contrast, performed few popular roles, serving almost exclusively to help advance botanical knowledge. At one university garden, the Amsterdam Hortus Botanicus, Hugo de Vries (1848–1935) conducted his important experiments on Dutch transplants from the Texas evening primrose, which culminated in an elaboration of the mutation theory. Among the more scientifically oriented gardens may also be counted acclimatization stations, such as the Carnegie Institution's Desert Botanical Laboratory at Tucson, Arizona, and its Acclimatization Garden in Carmel, California. Those that fall closer to the middle of the spectrum, facilitating research as well as recreation, are Kew Gardens and the Jardin des Plantes in Paris.[7]

The evolution of zoological gardens

During their early days, many royal botanical gardens like Kew or the Jardin du Roi held small menageries of exotic animals. The keeping of animals in captivity, if not traceable back to Noah's Ark, certainly goes back to ancient times. The creation of zoos – restricting wild animals to particular cages, enclosures, or green spaces – may be linked to the presence of cities and their wealthy inhabitants, especially as these captive animals became part of the trappings of royalty. Animals were associated with pomp and circumstance; they played a role in religious pageantry and processions among civilizations as diverse as Indian and Egyptian. Game animals were stocked in large parks to ensure adequate supplies for hunting, in ancient Persia, Assyria, and Carthage.

These animal collections were more menageries than zoos. The word menagerie dates from the end of the sixteenth century, when it was used to describe the government of the family, the care of the household, and everything else relating to such matters. By the second half of the seventeenth century, the word took on its modern connotation; before that date, there was no special word to describe the quarters of wild animals.

The function of keeping animals in cages was to amuse their captors. The possession of exotic specimens conferred power and prestige on the owner; their delivery was an act of homage to a ruler. Keeping animals in captivity symbolized a life of luxury, even nobility. In a few instances, however, zoos took on certain functions that transcended mere entertainment. Aristotle probably observed the menagerie of Alexander the Great; the results appear in his *History of Animals*. Aristotle's benign approach to nature continued during Roman times with Pliny the Elder (ca. AD 23–79) and his *Natural History*. This philosophical or literary orientation contrasts starkly with the brutality of beast shows, in which animals fought men or each other to the death.

Indirectly, ancient menageries came to serve learned purposes. One unforeseen result was the advancement of economic or applied zoology through the domestication of indigenous species and the acclimatization of foreign animals. A well-known case is the domestication of the wild cat in Egypt. In addition, because the courts and estates that supported menageries also patronized *savants* and philosophers, the captive animals served zoological, physiological, and anatomical purposes. During Roman times, for example, Galen used animals in menageries for dissection.

Something more closely allied to the modern perspective on keeping animals in captivity occurs during the Renaissance. Leonardo da Vinci (1452–1519) maintained a small animal collection as a source of models for his art. He carried out detailed dissections on both human and animal corpses, which he subsequently illustrated. In addition, he conducted experiments to elucidate the structure and function of organs in diverse animals, comparing, for example, the movement of wings in bats, birds, and butterflies. He studied vision in a variety of animals, such as the movement of eyelids and pupils among diurnal and nocturnal species.[8]

European menageries or zoos expanded during the sixteenth century as explorers brought back remarkable exotic animals from their voyages. These served as additional materials for descriptive zoology, since they provided the only source for observing wild animals in a living state. Collecting animals also helped to answer the desire to exhibit the diversity of the animal kingdom in microcosm. By gathering together the fauna of four continents, virtually all the links in the Great Chain of Being might be presented. Keeping the animals alive, however, was more of a problem than representing the vegetable kingdom in a botanic garden.

In addition to aiding observation, menageries came to assist experiment. William Harvey used dogs kept at the royal park at Windsor for demonstrating the circulation of the blood and for showing that mammals reproduce by producing eggs, like other vertebrates. In Italy, Francesco Redi took dead animals from the Florence menagerie to show that flesh putrefied by the presence of eggs deposited by flies, rather than by immanent properties. He also explored the physiological relationship between the saliva of poisonous snakes and their venom.

By the seventeenth and eighteenth centuries, French zoos became the most important in the world. This was largely due to the efforts of Louis XIV, who decided in 1663 to embellish his estate at Versailles with a menagerie, or zoological garden. This action allowed him to have his animal collection organized according to architectural principles. He became a pioneer in screening animal enclosures with trees and shrubs. The novel techniques employed by his gardeners and animal handlers inspired imitations elsewhere.

As a result of the court's cultivation of the king's fancies, the menagerie encompassed the most complete and extensive collection of living animals that had ever been brought together in one place. It included fifty-five species of mammals and hundreds of species of birds. By incorporating particularly rare and exotic specimens, Versailles acquainted naturalists for the first time with animals collected from Africa, America, and Asia. It functioned as a huge experiment in animal acclimatization in the broadest sense, and it also served the development of comparative anatomy.

What changed Versailles from being a simple pleasure garden to assisting the advancement of scientific interests was the creation of the Académie des Sciences in 1666, just when the menagerie was beginning to flourish. The Académie resolved to dedicate one of its two weekly sessions to natural history. The medical doctor Claude Perrault (1613–1688) engineered the defeat of an Académie programme for experimentation on live animals, put forward by Christiaan Huygens, arguing instead for the importance of comparative anatomy, which, in his view, included study of both the structure and function of organs. Accordingly, the Académie began dissections of human cadavers and the corpses of indigenous animals. Every year from 1669 until 1690 also saw the dissection of an exotic creature, often furnished by Versailles. After Perrault's death, dissections continued under the physicians Louis Gayant (d. 1673) and Joseph-Guichard Duverney (1648–1730).

The splendours of Versailles, including its extensive gardens and menagerie, became one of the most powerful symbols of the *ancien régime* to the revolutionaries of the 1790s. In the context of this upheaval, a Jacobin mob set out from Paris to liberate the animals, resulting in the senseless slaughter of most of them. The surviving animals went to the Jardin des Plantes in Paris, to serve as the nucleus of a new national collection. Other animals were added that had been seized from private menageries and circuses. The naturalist Georges Cuvier (1769–1832), in collaboration with Louis Daubenton (1716–1800), Jean-Baptiste de Lamarck, and Etienne Geoffroy Saint-Hilaire (1772–1844), reorganized the collections. The Jardin des Plantes provided zoological specimens for their research; at the same time, the collections opened up by order of the Committee of Public Safety.

Debates ensued among the revolutionaries about the scientific value of the animals that had belonged to the court, and over the wisdom of extending the collections. In 1792, Alexandre Brongniart (1770–1844) argued that the royal menagerie had been an expensive and largely useless item of luxury and ceremony, but that a more utilitarian collection could well serve the needs of natural historians and physiologists. Ambitious reports and plans were put forward by Lacépède and Edmé Verniquet (1727–1804). Ultimately, lack of money on the part of the government and lack of consensus among naturalists doomed these proposals. Still, what became the menagerie of the Museum d'Histoire Naturelle – with its small collection of sixty-five mammals and twenty birds – can claim pride of place as the first of the national zoos, founded in 1793.

With the advent of Napoleon, the zoological collections of the Museum expanded further, particularly due to the exertions of travellers on foreign missions. By 1813, the zoo could count 383 animals, with many rare specimens among them. Under the directorship of Frédéric Cuvier (1773–1838), brother to Georges, the land and buildings of the zoo were improved and extended. A chair in comparative physiology came to Cuvier, but he died the following year. Consequently, the directorship of the menagerie reverted to the professor of zoology, in this case, to Etienne Geoffroy Saint-Hilaire, who was succeeded by his son Isidore (1805–1861) in 1841. It became the most important academic zoological garden in Europe, judged by the scientific works that emanated from it. The zoo also encompassed nonliving collections in botany, geology, and comparative anatomy, as well as a library with 28,000 books.

Much work at the Museum concerned the acclimatization

of exotic species and the hybridization of domestic animals. The corpses of deceased animals furnished material for the comparative anatomists. But these advances came to a halt by the outbreak of the Franco-Prussian War in 1870, which inflicted severe damage on the zoo's buildings and grounds. The animals were slaughtered to feed the starving citizens.

A rebuilding programme after the war more than compensated for these losses. With nearly 1700 individual animals in 1910, the zoo could still be called 'the cradle of zoological science, the Mecca of zoologists'. However, as historian Gustave Loisel (b. 1864) points out in his definitive work, the richness of a national zoo consists not in its total quantity of animals, but in an appropriate selection of species, in their healthy appearance, and in the service rendered to science. Although Loisel argues that the Paris zoo met these criteria well, he admits that the buildings were crowded and dilapidated.[9] By 1934, a new zoological park opened in Vincennes, which introduced the habitat design innovations of German zookeeper Carl Hagenbeck (1844–1913). Achille Urbain (b. 1884), an expert in animal nutrition, eventually took joint charge of both the Paris and Vincennes zoos.

When the Jardin des Plantes reached its apogee under the old regime, private menageries still flourished in France. Especially important from a scientific standpoint was the *comte* de Buffon's menagerie at Montbard. There Buffon had purchased the picturesque ruins of an old castle. He reconstructed the castle, planted gardens, established laboratories, and founded a zoological park. He stocked the park with a wide range of exotic animals, which served as a basis for writing his multivolume *Histoire naturelle*. At Montbard, he also undertook varied research in experimental zoology, notably in reproduction and hybridization.

The rise of public zoos

By the nineteenth century, the menagerie began to take new forms, rather than simply reflecting idiosyncratic acquisitiveness. As for museums and botanical gardens, zoos attracted the burgeoning middle classes, who enjoyed a stroll through these parks that brought together exotic animals, inspired buildings, and lush vegetation. The growth of zoological societies gave clearer reason to the keeping of animals. The societies tended to open their gardens to the public for a small admission fee. The word 'zoo', in fact, coincides with the creation of the Zoological Society of London in 1826; by 1847, the word appeared in the *Oxford English Dictionary* as a colloquial abbreviation for that Society. Henceforth 'menagerie' would acquire a pejorative sense by association with outdated and unscientific practices. (Today it refers to the animals in a circus.) According to the letter of its charter, the aim of the London zoo was to advance the sciences of zoology and physiology, as well as to introduce new and curious examples of the animal kingdom.

The distinguished imperialist Sir Thomas Stamford Raffles (1781–1826) presided over the London Zoological Society at its inception. He died within the year, and his successor, Lord Derby, served as president for nearly twenty years. During the early 1830s, the Society attracted more than a quarter of a million visitors to its gardens at Regent's Park. These were the halcyon years when illustrator John Gould (1804–1881) exhibited a collection of mounted humming-birds and when the poet and illustrator Edward Lear (1812–1888) could be spotted sketching parrots.

The Zoological Society's importance continued during the secretariat of Philip Lutley Sclater (1829–1913), from 1859 to 1902, and under Abraham Dee Bartlett (1812–1897), who superintended the gardens during the same period. The Society's *Proceedings* and *Transactions*, established in the 1830s, carried contributions from leading naturalists and lithographic illustrations by the renowned artist Joseph Wolf. The enormous number of rare species gave rise to a range of studies in animal behaviour, descriptive zoology, hybridization, and comparative anatomy. Their museum of mounted specimens eclipsed in importance the zoology department of the British Museum (Natural History). Distinguished foreign fellows included François Magendie (1783–1855), Ernst Haeckel (1834–1919), and Louis Agassiz, whereas the reclusive Charles Darwin seemed to dislike the Society.

Although the London zoo belonged to a private society, it functioned in many respects as a municipal zoological garden. Indeed, the emergence of this kind of zoo may be seen as an extension of the much earlier custom of keeping wild animals in captivity within the walls of a city or town as a symbol of communal power and authority. (Consider the swans at Valencia, the wolves at Rome, the bears at Berne, and the ducks at the Peabody Hotel in Memphis, Tennessee, a tradition that continues in animal mascots for sports teams.) Another variant of this practice involved keeping an animal collection in part of a public park. The oldest example, dating to the mid eighteenth century, is Madrid's *Casa de Fieras* in the Retiro.

Municipal zoos, properly speaking, are creations of the nineteenth century. Three other zoos, in addition to London's, were founded in Britain and Ireland in the 1830s: Dublin, Bristol, and Manchester. On the continent, zoos were established at Amsterdam in 1837 and Hamburg in 1860. The Berlin zoo expanded considerably under the directorship of the zoologist Heinrich Bodinus (1814–1884) during the last quarter of the century. Although these institutions were ostensibly supposed to advance natural sciences, most did little besides exhibiting as many specimens as possible. New importance was attached to the architecture of zoos; impressive façades and ornamentation appeared. Both the number of animals and the splendour of their quarters radiated civic pride.

Associated with the rise and development of large urban zoos during the nineteenth century was the emergence of animal collectors, like Carl Hagenbeck of Hamburg. Hagenbeck employed what seem to us to be cruel and brutal techniques to obtain zoological specimens. He used the profits from his trade in animals to establish a novel zoological park on the outskirts of Hamburg. The Stellingen Tierpark, founded in 1907, aimed to present animals in their natural habitats protected by means of a concealed trench and moat system, without bars and cages, 'the hardest edges of captivity' in the words of historian Jon Luoma. In this way, Hagenbeck succeeded in achieving separation and enclosure, seemingly so central to the concept of zoos, by means virtually invisible to the spectator.[10] For the first time, visitors came face to face with wild animals, encumbered by neither visual obstruction nor locked cages. Hagenbeck also designed the zoo in Rome, which opened in 1911.

Except for the introduction of Hagenbeck's habitat concept, which eventually revolutionized the keeping of animals in captivity, zoos tended to languish early in the twentieth century. Certainly all zoos, like museums and botanical gardens, possess important scientific, educational, aesthetic, and recreational functions, which may be emphasized differently from one institution to another. But zoos experienced difficulty in sustaining the vitality of this range of functions over the years. By 1900, the rationales for their existence became tired; raw numbers ceased to confer prestige; capture and display no longer represented real or metaphorical power. One curator referred to the 'worldwide intellectual mediocrity' that characterized botanical gardens, too, during the first half of the twentieth century.[11]

The solution has been a new emphasis on conservation, research, and popular education. The earlier impulse towards displaying a living catalogue of the plant and animal kingdoms, of exhibiting for the sake of exhibiting, has largely been shunned. Zoos in particular have been developed as refuges or enclaves for the protection of endangered species; successful breeding programmes have allowed them to think of repopulating decimated wild habitats. Animals are now presented as part of interdependent ecosystems, not as 'single trophies – like rare postage stamps from far-off lands'.[12] The successful zoo becomes, in the words of authors Bill Jordan and Stefan Ormrod, 'a sort of Noah's Ark, complete with laboratories and classrooms'.

There are practical results from this policy of factitious survival. Environments denuded of biota can be resculptured with stock from zoos. Breeding programmes can furnish colonies of animals for clinical pharmaceutical research. New disciplines or interdisciplines, such as cognitive science, can draw on zoo populations. Zoos, like gardens, are living libraries, offering a great store of wisdom to those who wish to read the book of nature.

II

ENTERPRISES

Measuring: The Search for Precision

Every year early in December, Tante Inge invites in friends for a holiday party. The invitation is highly prized, especially because of the sumptuous table: goulash and various roasts (served hot and cold), salads of exquisite tastes and textures, and a barrage of Torten, Lebkuchen, and Stollen exuding aromas of nut, chocolate, sugar, and fruit. A prudent and thoughtful woman, Tante Inge prepares for weeks in advance. Recipes are followed precisely, with an occasional eye to the calendar and the stars. (Everyone knows that mayonnaise will not thicken under a full moon.) A peculiar instrument ministers to each dish on the way to the table. It is an ancient tin *Messbecher*, a measuring cup in the form of a truncated cone. The interior face has scales for measures in weight and volume. To add just the right amount by weight of white pepper, vanilla sugar, or crushed almonds, Tante Inge has only to pour the ingredient into the cup up to the appropriate line. The Messbecher has the utility of a slide rule and bears some resemblance to a medieval torque-tum. It would be impossible to bake a Panama-Torte without one.

Cooking is no doubt the most ancient of sciences. It has always been connected with material transformations and therapeutic effects. Cooking relates fundamentally to the working of ceramics, metals, and glass, to dyeing, distilling, and brewing, to preserving life and restoring it. Since Eve, Odysseus, and Socrates, eating or drinking has been the path to wisdom and its droll opposite, oblivion. As Tante

Inge would insist, for the Panama-Torte with its various layers to emerge from the oven optimistically convex, instead of depressingly concave, the cook must have measured weight, temperature, and time. Is it any wonder that when, in the nineteenth century, Oxford University decided to acquire a chemistry laboratory, it borrowed the design of the monks' kitchen at Glastonbury?

When Tante Inge is not cooking she may be found gardening. In her heated 'winter garden', she forces bulbs and cultivates orchids. In spring she plants and prunes. There are scores of rules to follow. Water, sunlight, heat, and humidity are rationed in precise amounts. In her latitude, to avoid a disaster from frost, planting must wait for the first full moon after Easter. Ash from the old-fashioned cooking oven is strewn over the snow around fruit trees; in spring, the trees are surrounded with heaps of compost. There is a perpetual sequence of transplanting and rerooting, grafting and fertilizing. It is all done quantitatively. Seeds are spaced out by finger and foot. Soil is mixed by the shovelful. Water is delivered by pitchers. Sunlight is calculated by hour and by day of year.

Gardening is also an ancient art, supplying many of the basic ingredients for cooking. It is the very sign of civilization, for the word *culture* is sometimes taken to be shorthand for 'agriculture'. Everyone knows that in matters of cooking and gardening, success comes by measurement.

Not everything of importance, you will say, concerns precision. The finest words in any language are a variety of: 'See you for drinks at six.' The sentence is apparently precise. Most people, however, have been in a situation where the meaning of the sentence is entirely contextual. Does one wear pearls or polo shirt, sports jacket or jogging shoes? Does one bring a tooth brush? Is it six o'clock Hamburg time (18 Uhr), six São Paulo (about seven) or six southern California (any time after six)? In many cases, one simply does not know what is expected, and that uncertainty produces delicious anticipation. Everything is at once precise and inexact. The general feeling is familiar to everyone who drives an automobile and interprets pictographic road signs according to his or her lights.

Precision has often been a function of culture. In monumental architecture, that which provides the most striking image of antiquity, the integrity of a structure can be enhanced by its apparent caprice or asymmetry. We admire the Parthenon and the Taj Mahal for their harmony and balance, but admiration is coupled with intrigue when we view the inconsistencies of the great Buddhist temple Borobudur on Java and the broken transept of the Quimper cathedral in Brittany. Why, we wonder, did the medieval Quimpérois decide against a true, right-angle design for their monument, when in all other matters they were more than equal to the precision required? In a similar way, we may ask why the Pre-Columbian Mesoamericans did not think to use right angles in their grand monuments when they had calculated astronomical regularities for cycles of tens of thousands of years. Why, indeed, were medieval mosques oriented so poorly toward Mecca when Islamic civilization possessed the knowledge to line them up precisely?

Precision was just measurement, and measuring was frequently a practical matter. The state established measures of assets, animate and inanimate, the better to control them. From all civilization of antiquity we have long lists of chattels and the means of their maintenance. We know, for instance, how much beer and grain were required to feed the Egyptian pyramid builders. About the size of armies and the civil service we are less well informed, perhaps because no clear distinction existed between state and private service. In Egypt and Mesopotamia we have precise calculations for land area, a precision that led Herodotus (d.ca.425 BC) to claim land tenure and taxes as the basis for mathematics, and notably (as its name indicates) geometry.

Precise measurement, like most mental exertions, has depended on judgment. This is evident in astronomy, the science where, more than any other, one is advised against fakery. Everyone sees the motion of the stars, but how shall the stars be measured? Shall we use equatorial coordinates (the traditional Chinese measure) or ecliptical coordinates (the measure favoured in Europe)? How shall stellar brightness and colour be established? Yet measurement is not merely a matter of opinion. Things constructed from precise measurement – boats, cathedrals, bridges, airplanes – do indeed function everywhere. Measurement is something that crosses cultures. The inflexible demands of precision are evident to Tante Inge, just as they are to every reader of the present lines.

Measurement in antiquity

The notion of precision implies a sense of utility, and utility has been traced to the oldest recorded mathematics, that of Sumer. Sumerian, the most ancient of written languages recorded with a wedge-shaped stylus (producing 'cuneiform' marks), persisted into Hellenistic times. Its principal legacy today is the sexagesimal system of notation used for measuring angles (60 minutes in a degree) and time (60 minutes in an hour).[1] The greatest of astronomical treatises, Ptolemy's *Almagest* (which continued to be taught into the twentieth century), makes calculations sexagesimally, just as we do when organizing our day. Our measurements of the cosmos may be traced back at least four thousand years.

Why sixty? It is a convenient base for allowing halves and thirds of a large measure (sixty) to be expressed in simple units (thirty and twenty), rather than expressing thirds as continuing fractions (as they are in base-ten measures). Sixty is evenly divided by both ten and twenty (the fingers-and-toes mnemonic) and by twelve (the number of complete lunations in a solar year). The Greeks recognized the utility of this system, for they counted sexagesimally in money and weight. The Chinese, too, measured in this way. We must not omit mentioning that sixty is also the sum of digits for man, woman, and child, a convenient explanation for those whose interests turn to symbolism.

Where does practicality enter into measuring? The need for assigning ownership to land arises in many urban civilizations. We have pictures of the world from both Egyptian and Sumerian antiquity. Land was certainly measured precisely 2500 years ago in these and other settings. Such measurements led to diverse representations and styles. In Mesopotamia, the measurements gave rise to complex algebraic problems featuring equations of higher degree, and what we understand as geometry (notably the Pythagorean relation) was known and taught in algebraic guise; in Egypt there is nothing nearly so complex or refined. Maps keyed to a grid are common to both China and the Eastern Mediterranean in the period around 100 AD, but China has nothing like the pictoral geometrical abstractions and canons of proof that we associate with the Hellenic and Hellenistic world. Chinese artistic perspective resembles a Mercator projection of the globe, but even spherical trigonometry entered China from India – and remained for nearly a thousand years an arcane technique known only to court astronomers. In seeking the practical roots of any abstract discourse, we must be wary of naïvely rationalizing what may be, in any particular civilization, an individual variation. The signal innovation in Sumerian precision notation was that of place value. It enabled the Sumerians to deal with fractions. The development is fundamental to higher mathematics – one sees it independently in both Chinese and Mayan calculations – for it allows efficient tallying of assets.

Classical archaeology has revealed that weights and measures had multiples of twelve (a persistent allusion to the Egyptian zodiac) and sixty. In the Byzantine world, units of weight telescoped into each other by multiples of twelve; the system continued with some modification in medieval Islam. Byzantine weights derived from the *litra* (a continuation of the Roman *libra*), a gold coin of about 300 gr. The

Greek *drachm* served as a weight standard throughout the medieval period, finding a reprise in the Islamic *dirhem* (defined, according to one Egyptian commentator, as the weight of sixty normal-sized barley grains without their husk). The Islamic world combined sexagesimal with decimal measure to establish the *miskal* as the weight of 6000 mustard seeds, ten dirhems equalling seven miskals, with dirhems subdivided into quarters, sixteenths, and sixty-fourths – all easy to measure out on a portable balance. Just as we do in time, we keep a memory of the distant past in volume. The French revolutionaries of the late eighteenth century, in attempting to establish a 'natural' system of weights and measures, chose a term for volume, the *litre*, that had been known for more than a thousand years in Eurasia as a standard of weight.

Calculation with Greek numerals was easier than with Roman numerals – that which inspired Europe in the early medieval period. Asia instructed Europe in replacing Roman numerals with Hindu-Arabic digits. These originated with Brahmin script around AD 600, although some researchers have claimed an origin well before then. Initially they were probably used for calculations on a sand table, possibly one engraved with lines for place values. The innovation relates not to the invention of decimal place values – these were known in China – nor to a special representations for zero – present in Babylonian astronomy – but to symbols for easy computation. From the time of their introduction in Islam through a translation by al-Fazarî (fl. ca.760–790), the new ciphers spread in two forms: the 'Indian' numerals (*huruf hindayaah*) preferred by Eastern Arabs and still used in the Arabic world today; and the 'dust' numerals (*huruf al-gubar*) used by Western Arabs on sand tables, which with the reintegration of the Hindu zero became the numerals of Europe.

Much of verifiable precision concerns tabulations of data. The exact sciences in antiquity revolved about calculations based on tables of mathematical functions, like sines, and observations of celestial phenomena. Diverse systems of enumeration did not prevent the achievement of complex calculations. Babylonian mathematics in the period up to about 400 BC recorded heliacal risings of the planets (their periodical appearance in the night sky from behind or in front of the sun) as well as solutions to higher-degree algebraic equations; Ptolemy recorded and calculated sexagesimally, but with Greek notation (where separate digits represent units, tens, hundreds and thousands), which avoided recourse to a mechanical counting board. Just how avidly Europeans of the high medieval period embraced technical innovation is revealed in the increasing use of 'Arabic' numerals by the end of the thirteenth century.

Syncretism and measuring instruments

The European adoption by the thirteenth century of Hindu-Arabic numerals, written on vellum manuscripts and engraved on brass astrolabes, facilitated precise measurement. It is well to emphasize, however, that much of precision computation involves streamlining, or the *avoidance* of actual calculation. Mnemonic habits suffice up to a certain point, but we have always had aids for computing. These took the form of tables for multiplication or division, as well as analogue devices. In classical antiquity the foremost analogue device was the counting board, a device used up to the present day in some civilizations. It was known as an *abakion* (*abacus*, in Latin), on which counters (*psephoi*, or pebbles in Greek; *calculi* in Latin) were placed. In East Asia, there were, from the time of the Warring States (300 BC), counting rods, which by the time of the Thang dynasty, were placed on counting boards. Medieval Europeans have credited Islamic civilization as the source of their own abacus. Evidence suggests that the modern form of the Chinese abacus appeared relatively late, in AD 1593. Here may be a rare instance of European technical innovation filtering back to China during the medieval period. At this time,

Europeans had just formulated several new mechanical counting devices, such as the reduction compass of Fabrizio Mordente (published in 1567) and the sector of Galileo (published in 1606). The field of classical learning depending most on precise calculation and observation – astronomy – benefited greatly from the invention by John Napier (1550–1617) of logarithmic tables (published in 1614) and their mechanical analogue, the sliding number line, devised by Edmund Gunter (1581–1626) in 1623.

Calculating devices are related to another class of object enjoying an independent tradition – simulacra, or mechanical objects that represent natural phenomena. The tradition extends back at least to Hero, whose automaton snail delighted visitors to Alexandria's Museum, and it finds a secure point of reference in the 'Antikythera Machine', a Hellenistic gear-driven device from about 80 BC for indicating positions of the planets numerically. This most remarkable device, recovered in 1900/1 from a shipwreck off Antikythera Island between Peloponessus and Crete, remains the only original mechanism dating from antiquity. The machine makes use of a crank and differential gearing to reveal heliacal risings in a most ungeometrical fashion: they are given as numbers rotating on disks and signaled by dials. In other astronomical simulacra, possibly the one installed in the Tower of Winds of Athens during the Roman period and certainly the colossal Chinese clock erected by Su Sung (1020–1101) at K'ai-fêng in AD 1090, water was the source of power.

Medieval Chinese clocks were simulacra, complete with jackwork to activate bells and moving displays, as well as gears to move an armillary sphere that showed the positions of the planets. This tradition continued in Europe. A notable instance is the great cathedral clock of Strasbourg. Simulacra are represented many centuries later in mechanical orreries – crank-driven representations of the Copernican universe that delighted aristocrats and itinerant lecturers in the eighteenth and nineteenth centuries. Today's planetarium projector, developed by the firm of Zeiss a hundred years ago, provides simulations like the ones of Su Sung.

The essential mechanism in a medieval clock was an escapement, a device that converts continuously applied force into discrete, impulsive motion, as for example in the ticking of a second-hand. Europe's modification to clocks, detailed by Giovanni de'Dondi (1318–1389) in 1364, concerned the verge and foliot escapement, driven by falling weights rather than falling water. How did Europeans convert the slow descent of weights on a string and the irregular uncoiling of a spring to a steady roundabout motion of gears? Cathedrals focused the energies of medieval mechanical adepts, and at least one talented builder must have been struck by the inertia of counterweighted bells rung out by ropes that pulled the ringers up and down. By rotating the bell axis 90° and installing an escapement, the bells could be powered by a dead weight. The association would not have been as apparent to Chinese astronomers, for whom bells and gongs were rung by percussion at ground level. The symbolic legacy of medieval Christianity, the church tower, may be responsible for the weight-driven clock.

Newtonian measurement

The mathematical elaboration of Newtonian physics by Leonhard Euler (1707–1783), Joseph-Louis Lagrange (1736–1813), and the Bernoulli family came just as decisive tests persuaded even staunch traditionalists of Isaac Newton's superiority over his rival René Descartes and his contemporaries Christiaan Huygens and Gottfried Wilhelm Leibniz. Through astute political manoeuvring (he was a vocal supporter of the Glorious Revolution of 1688), Newton became warden and then master of the British mint – a lucrative position allowing him to move from Cambridge to London. There he vigorously defended British monetary standards against debasers and counterfeiters. Central to the philosophy of the new age – for which Newton served as the main prophet – was measurement, especially as it related to establishing the truth or falseness of scientific propositions.

Were Newtonian principles testable? Newtonians contended that the circumference of the earth would be greater than the measure of a great circle through its poles; Cartesian recalcitrants, whose sensibilities were rooted in a distaste for things English, concluded differently. The question engendered outrageously expensive expeditions to the north of Sweden and to Ecuador, where geodesical arcs were surveyed over tundra and mountain. In a related development, pendulums were transported near the equator to see whether they would beat quicker, as Newton predicted. Using the new mathematical tools of differential and integral calculus, natural philosophers recalculated orbits of celestial objects, from planets to comets, and then pointed telescopes skyward to see if the stars were slaves to Newton's prophesies.

The results of these queries, by the middle of the eighteenth century, were unequivocal. The earth was an oblate spheroid, a gravity pendulum beat time quicker at the equator, Halley's comet returned on schedule, and electrical matter (unanticipated by Newton) did not disobey the master's dicta. By the 1780s the foundation had been laid for new sciences of chemistry (by Antoine Laurent Lavoisier [1743–1794]) and electricity (by Charles Augustin Coulomb [1736–1806]), and a seventh primary planet had been discovered (Uranus, by William Herschel [1738–1822]). Precision underlay it all.

Astronomers divide their life between undertaking celestial inventories and calculating stellar motion. Before the telescope, inventory focused on the stars visible to the naked eye, ephemera such as comets and meteors, and anomalies like variable luminosity. Physical astronomy, brought into being by the telescope, revealed new worlds in the heavens. The worlds, there for all to see, proclaimed a new cosmical order. Lenses certainly did magnify distant objects, but aberrations of form and colour rendered telescopic image imprecise; uncertainties of mechanical drive-trains made it all but impossible to have even a moderately sized telescope follow an astronomical object across the sky. For telescopes with a large separation between lenses (combinations of many metres were used at the Paris Observatory in the seventeenth century), precision pointing was practically impossible. Until nearly the end of the eighteenth century, astronomers still practised their science by observing punctual events: the onset of a planetary conjunction or an eclipse, the appearance of a comet, the transit of an inferior planet. Mirrors, used in combination with lenses for reflecting telescopes, added new problems; not only was it difficult to figure the shape required (a parabola) but one also had to keep the mirror's copper or silver surface highly reflective – a thankless and perpetual task. Telescopes were used for marking position, although not by everyone.

The situation changed dramatically at the end of the eighteenth century. Reflectors of enormous proportions, notably the one mounted by John Herschel (1792–1871), made their appearance. The invention of composite lenses, where the two parts had different and mutually compensating refractive indexes, provided high-resolution images. Pendulum clockwork drove these telescopic leviathans, allowing large lenses encased in metal barrels to keep pace with the nightly rotation of the fixed stars. Grand catalogues of new celestial objects were the result.

Instrumental precision behind the giant eyes derived from advances in metalworking, notably in producing flat surfaces and precisely ruled scales. These advances in turn depended on the emergence of professional mechanics who devoted themselves to manufacturing instruments – sextants, spyglasses, compasses, and chronometers, as well as the more lofty astronomical telescopes. The mechanics, self-trained men with unusual power of concentration and persistence, provided sea lords with navigational tools while satisfying a growing demand for devices to illustrate Newtonianism, a demand encouraged by lecturer-demonstrators like the Abbé Jean Antoine Nollet (1700–1770). The secure living provided by selling to forward-looking admirals and country squires provided enough creative leisure for innovation, and competition among various firms sped the desirability of establishing a reputation for *accuracy*.[2] With Jesse Ramsden's (1735–1800) dividing engine at the close of the eighteenth century, unusually precise scales could be turned out in great quantities. These were the scientific equivalent of mass-produced metal pots and pans at the dawn of the First Industrial Revolution.

John Heilbron has analysed the extent of late eighteenth-century interest in measurement. He emphasizes that precision was an adjunct of imaginary quantities, the various ethers and imponderable fluids. Astronomical precision was calculated by measurement of angular arc – the part of a circle separating a star and, say, the horizon. This precision was set at 10 seconds of arc by Jean Picard's (1620–1682) sighting lenses and screw micrometers at the end of the seventeenth century. It increased to perhaps one second of arc by George Graham's (ca. 1674–1751) refractor mounted on a brass-trellis mural quadrant. Then came accelerated mechanical precision in marking the parts of a circle, culminating in Ramsden's dividing engine (and the navigational improvements produced by new sextants). Jean-Charles Borda's (1733–1799) repeating circle brought portable precision down to several seconds of arc. French mathematical physicists domesticated Borda's circle for measuring optical properties of matter in the laboratory, notably refraction. Correlative attention to laboratory measurement resulted in Coulomb's law of electrostatics, countless measurements of specific heats, and attention generally to the quirky attributes of the three stages of matter. The French Revolution canonized the mania for natural precision with a universal measure, the metre, taken naturally enough from French geodesical precedent.

Timepieces

Time and extension are two givens of measurement, and mechanical clocks were the flip side of the revolution in precision. The clockmaking trade sped through the Renaissance and into the seventeenth century with great flourish and irregular improvements in accuracy. Just as in China, in Europe clocks were held to be representations of the universe as well as reminders of mortality. Many could give the day of Easter for a century; few could predict the next dawn to better than several minutes; fewer still, the exceptional instruments with second hands, found use by professional astronomers. Then in 1656 Christiaan Huygens invented the pendulum clock, thereby obtaining two orders of magnitude greater accuracy (his clocks lost only a second per day). From then on, time mastered the terrestrial moment. With cascading improvements, such as George Graham's dead-beat escapement, astronomers could mark the heavens with probity and determine an accurate picture of landforms, longitude differences being calculated directly from different times for a common celestial event.

Not so at sea, where pendulums would not swing freely. What was required, as Umberto Eco has recently elaborated in his fantasy, *The Island of the Day Before* (1994) was an improved table clock or pocket watch. John Mudge invented the detached-lever escapement, contributing to greater pocket-watch accuracy, and he competed for a princely prize offered in 1714 by the British Board of Longitude for an accurate method of determining longitudes (to better than a degree) between Britain and the West Indies. Yorkshire carpenter John Harrison (1693–1776), however, devoted his life to solving the problem by devising an accurate chronometer. He realized that a crucial effect concerned the variation with temperature in the tension of metallic spring balances. He consequently introduced bimetallic strips of steel and brass (which would bend as the temperature varied) acting to displace the distance between the inner and outer end of a watch's balance springs, and in this way adjusting their tension to compensate for shrinkage or expansion. Harrison's chronometer passed the Board's tests, finally persuading them in 1764 that the problem had been solved. Like other mechanical adepts of the time (notably the reclusive experimenter Henry Cavendish [1731–1810] and the polymath Carl Friedrich Gauss), Harrison revealed the details of his practical methods only with great reluctance. By the eighteenth century scientists routinely shared their results, but the general broadcasting of scientific technique in a press conference or pre-print is a twentieth-century development.

By the end of the eighteenth century, pocket watches with balance springs provided reliable time (Harrison's chronometers lost only a second over a month). The instruments of intrepid explorers underwent few improvements between the time of James Cook (1728–1779), who took a prototype of Harrison's successful chronometer on one of his cruises, and Arthur Stanley Eddington (1882–1944), who took along chronometers when in 1918 he verified Einstein's general theory of relativity by observing a solar eclipse on Principe Island off the coast of Africa. The main desideratum was to introduce the least amount of perturbation to the beating pendulum or expanding spring, in the quiet manner of George Graham's dead-beat escapement, in which the mechanical train is arrested at each swing of a pendulum. Innovations such as Clemens Riefler's nineteenth-century mounting of the suspension-spring block on a knife edge (its rocking giving impulses to a pendulum) were ideally suited for laboratories, observatories, and finally radio signals at the time of the First World War.

Portable timepieces, often carried in great number, served to remind Europeans about their precise distance from the seat of empire. Since the eighteenth century, watches have connoted an authority never possessed by pocket sundials and nocturnals. To the modern European mind, time is money. Beyond Europe there was no comparable interest in precision chronometry.

Even after contact with European mechanical clocks, Japanese retained a traditional division of ordinary life which was remarkably like the one of medieval Europe. Japanese night and day separated into six equal periods, each one corresponding to a sign of the Chinese zodiac (medieval Europe had twelve zodiacal hours for night and for day).[3] By the early seventeenth century, the Japanese were manufacturing their own clocks, both weight-driven and spring-driven, mounted as lanterns or on brackets. Early Japanese clocks replicated European ones, which had two independent foliots (one for day and one for night). Later clocks simply used falling weights to indicate hours passed, the length between hours being varied by changing a vertical scale periodically; eventually these falling-weight (or 'pillar') clocks featured universal scales with plotted graphs for taking account of variable hours across the entire year. Horology took strides during what has been called the Japanese 'Industrious Revolution' beginning late in the eighteenth century. At this time, for example, the astronomer-surveyor Ino Tadataka (1745–1818) used a novel clock. It was controlled by a long pendulum, and it featured two crown-wheel escapements. A large dial had an outer circle going from one to ten, with divisions in tenths, and an inner circle divided in the same way. One hand followed each circle in a tandem arrangement that divided each large numeral into tenths, hundredths, and thousandths. The eminently rational decimal clock eventually fell victim along with traditional Japanese clocks to the relentless marketing of European and American wares.

In some cultures today, possessing a multifunctional gold wristwatch makes up for a lack of literacy. This is hardly surprising, since much of European rationality concerns 'making the trains run on time', a watchword for fascist efficiency. The engineer protagonist of John Hersey's *A Single Pebble* (1956) begins to shed his Western

190 · *Servants of Nature*

preconceptions when, tracking up the Yangtze River, he loses his watch. A roman-numeral clockface on the *Bulletin of the Atomic Scientists* anticipates the end of humanity in a nuclear conflagration of European making. Increasingly, in this Digital Age, the image and its symbolism are incomprehensible to a rising generation who cannot read roman numerals and never learn to tell time.

Standardization

With clocks domesticated, the first nation state sought to define its extension for all time. France's Bourbon monarchs employed four generations of astronomers – the Cassini family – to map the kingdom. The project, conducted by theodolite, measuring chain, and astronomical telescope, concluded on the eve of the French Revolution and provided a basic topographical inventory of France, right down to roads and manor houses. The interest of astronomers and physicists in taking the precise measure of nature (the term physicist, in French *physicien*, became current in the eighteenth century for people who measured the properties of matter) led to a new science of measuring matter: chemistry. Precision was everything for Lavoisier and his circle (which included astronomer Pierre Simon *comte* de Laplace [1749–1827]), whose epoch-making discoveries depended on extraordinarily sensitive balances. In this sense, instrument-makers invented the new chemistry of the eighteenth century, just as they had the new astronomy of the seventeenth century.

The beginnings of modern chemistry, like the familiar hexagonal image of France, came in the absence of anything like standard weights and measures. There were special standards for each commercial substance, trade, and region. Because so much of European and North-Atlantic expansion during the nineteenth and twentieth centuries has been concerned with the suppression of indigenous standards of measurement, it is well to emphasize that inventing chemistry and measuring the kingdom did not require a uniform national standard of length, weight, and volume. Indeed, the appendices to Lavoisier's popular introduction to the new chemistry of 1789 concern conversion among various measuring units. Unitary standards are no more necessary for scientific innovation than standardized spelling is necessary for great literature. The fundamentals of European culture were formed without either one.

The elite of *ancien régime* France capitalized on revolutionary fervour for eradicating all traces of monarchial whimsy by insisting on a natural unit for length. They rejected the length of a pendulum that oscillates in exactly one second (advantages of reproduction were compromised by uncertainties of gravitational perturbation due to subterranean mineral deposits) and settled on one ten-millionth of a quarter arc of a great circle through the poles. From length would come area, volume, and eventually mass (determined from the amount of water in a cube of sides one-hundredth of the fundamental length). Remeasuring the arc from Dunkirk to Barcelona (it had been measured twice before) was the most difficult part of the task assigned in 1791 to the most prominent of French scientists by the French government. There was plenty of national glory to pass around. Master mechanic Etienne Lenoir (1744–1827) produced a metrical prototype in 1793, a length calculated from Nicolas-Louis de Lacaille's (1713–1762) meridian of 1740. France then guarded the standard metre for more than 150 years.

The French revolutionaries transmuted precision into ideology. They sought to liberate citizens from the pagan nomenclature attributed to lunations, the primitive superstition implied in a seven-day week, and the sexagesimal Babylonian heritage of minutes and seconds. The First Republic instituted a new calendar, with twelve thirty-day months (each one with three ten-day weeks) named after meteorological and agricultural generalities (and a few days of festival to round out the year); a new ten-hour day followed, along with a 400-degree circle. Just as chemists were not inconvenienced by multiple standards nor astronomers by months of varying days and time measured in a sexagesimal system, so ordinary people saw no reason to uproot nearly two millennia of tradition. The new months disappeared, to remain in our language only as a symbol of disturbing political events, like *thermidor*.

Napoleon, who found little merit in republican institutions and innovations, nevertheless encouraged metrical lengths and weights. But only with the July Monarchy thirty years later did the government finally legislate in favour of the metre, kilogram, and litre. New states conceived in nationalist euphoria generally followed suit. Embued as they were with eliminating old-regime privileges, they wrongly associated Napoleonic *dirigiste* legal reform with republican scientific rationalism. The metric system signalled modernity to new political regimes, – from Latin America and Greece to industrializing Meiji Japan. But nations with political structures conceived in the eighteenth century (the United States), slowly reformed in the nineteenth century (Great Britain and its empire), or assertively unrepentant until the twentieth century (Turkey, Russia, and China) saw little advantage in metrification.

The ideology of precision

Measurement dominated science in the nineteenth century. Europeans flattered themselves that they had conquered the world, and they set about to record its dimensions. Armies of surveyors went out to map North Africa, North America, and Asia. Naval expeditions sounded coastal waters and charted islands and currents. New instruments led to new surveys. The terrestrial magnetic field, which exhibited secular and periodical variations, received attention from hundreds of dedicated measurers. National meteorological networks, where scores of people assiduously measured an array of quantities that could be measured, fed on the hope of predicting the weather.

Nineteenth-century meteorology is a prime illustration of the principle that scientific activity stems as much from instruments permitting measurement as from large theoretical visions. It also shows that expensive activity can continue without obtaining concrete results. As precision devices became readily available on the general market, standard measurements of temperature, humidity, barometric pressure, and wind velocity could be assembled from scores of distant locations. In a central office, subtle mathematical analysts sifted through mountains of data to seek correlations. Early in the twentieth century this Baconian programme, which had been given new life by Alexander von Humboldt, extended to measuring earthquakes with standard seismographs, measuring solar radiation with standard radiographs, and measuring cosmic rays with ionization counters.

Theory played a weak role in the massive state-financed scientific bureaucracies concerned with precise measurement. Precision instruments, however, contributed much to theoretical constructions. Mechanical inventors produced a vast array of engines, and the desideratum of precision became a touchstone of nineteenth-century science. Heat, light, electricity, and their interaction with matter (living as well as inert) were measured with great earnestness. Precise observers provided direct evidence of terrestrial motion (in stellar parallax), the isotropic velocity of light (in the Michelson-Morley experiment, 1887), the reality of electromagnetic waves (by Heinrich Hertz [1857–1894]), the rotation of the galaxy (in radial-velocity studies), and the fundamental insufficiency of classical mechanics (in the ultraviolet catastrophe of black-body radiation). Precise calorimetry enthroned the immanent truths of thermodynamics.

During the nineteenth century, measurement became a signpost of Western consciousness. The French Enlightenment had promoted facility in measurement as a virtue – the art of making measurements, following the inspiration of lecturer-demonstrators like Nollet, provided morally edifying entertainment. By the end of the nineteenth century, measuring the natural world finally displaced mastery of the classical languages as the centre of progressive education. Measurement signalled industry and improvement. Measuring provided chemicals, electricity, and steel – the cornerstones of national might arising from the Second Industrial Revolution.

Measurement and industrial progress

The scientific contribution to modern industry emerged first with chemical precision. In France, especially after Lavoisier's innovations, chemistry slowly moved from the medical curriculum to higher learning in general, and it figured in early industrial procedures for manufacturing acids and bases. The impact of systematic scientific research on industrial growth, however, derived from Germany's preindustrial university system. By the 1850s chemistry had become an established part of university curricula. After decades of effort, chemists like Justus von Liebig taught hundreds of students in university laboratories, where chemistry had separated itself from medical supervision to achieve a pedagogical status equivalent to physics and astronomy. Liebig's students remained for the most part aspiring pharmacists and medical doctors, but in the general climate of economic expansion associated with German unification, university-trained chemists began to turn their talents to synthesizing products of commercial value.

As firearms, artificial dyestuffs, and electrical motors moved across national boundaries, and as independently manufactured parts integrated into larger systems, an inevitable urge arose to standardize measuring units. The last third of the nineteenth century saw a continuing stream of international convocations to establish standards in chemical and electrical nomenclature as well as to disseminate and authenticate metre lengths and kilogram weights. Large railway networks in both the Old World and the New World, requiring new time standards, produced the now-familiar time zones. National standards laboratories emerged in Germany, France, Great Britain, and the United States to certify electrical and mechanical devices. These laboratories provided employment for large numbers of physicists, especially. Some of the early national laboratories, like the National Bureau of Standards in Washington, threw themselves into a frenzy of inventorying objects from the world of technology.

Nations contended for recognition in units by enshrining their savant compatriots in eponymous measuring units (ohms, amperes, watts, volts, angstroms, henrys, hertzes) and chemical elements (Scandium, Germanium, Francium, Hafnium [for Denmark] and the politically subversive Polonium – named by the Curies as a defiant memorial to partitioned Poland). Chauvinist discussions about names must not obscure the extent to which international standards were legislated on a national basis, since only nations could award monopolies or patents to technological entrepreneurs. Establishing patent rights in several national markets nevertheless depended on international standards and generally accepted scientific principles. The endeavour to establish interlocking patents in a number of national markets came to dominate the planning of industries large and small.

In the nineteenth century, France effectively controlled international standards for fundamental physical units. The prototype platinum-iridium metre rested in Paris; symbols for chemical elements came from Latin-French inspiration (iron became 'Fe' for *fer*; lead became 'Pb' for *plomb*; copper became 'Cu' for *cuivre*). Paris was the city of perennial scientific exhibits, where manufacturers showed off their latest wares. But Germany, as we have seen in a previous chapter, was the land of instruction. No one was more sensitive to the opportunities offered by measurement than Hermann von Helmholtz.

Helmholtz began life as a military physician, whom the army educated in medicine at the University of Berlin from 1838 to 1842. Soon after obtaining a medical doctorate and while serving in the army, Helmholtz formulated the principle of conservation of energy in a mathematical way. Then he turned to physiological measurements about the conduction of nerve impulses. He rose rapidly in German academia, coming at last to direct the physics institute at the University of Berlin. In 1888 Helmholtz became the first director of an independent imperial institute for physics and technology, funded by the industrial fortune of Werner von Siemens (1816–1892). It was a research institution; its mandate was measuring things, both to practical ends and in the service of new physical theories.

As befitted an institution funded by one of the pioneers of German electrification, Helmholtz's institute developed a strong group working on electromagnetic radiation. Practically, the group sought a standard for light bulbs. A young researcher there, Wilhelm Wien (1864–1928), developed a theoretical expression for radiation intensity as a function of temperature and wavelength. Wien's expression was held by Max Planck to be as unimpeachable as the second law of thermodynamics. Subsequent work revealed a contradiction in classical electrodynamics and led Planck to propose his quantum theory of radiation in 1900.

Absolute measurement and error analysis

The nineteenth-century ideology of precision had two parts: absolute measurements and error analysis. First, researchers sought direct values for the fundamental properties of nature. It was not considered sufficient to know how measurable quantities related to each other, for example, to know merely that voltage varies directly with electrical current; rather, one sought to establish the elemental quantity of electrical charge. It was not enough to know that the attractive force between two masses varied inversely as the square of the distance between them; rather, one sought the precise proportionality. All physical constants – whether mechanical, electrical, or thermodynamical – had to be related to each other unambiguously and without redundancy.

One of the earliest and most devoted members of the tribe of nineteenth-century measurers, Wilhelm Weber (1804–1891), introduced mechanical measurement into electro-dynamics. His fundamental law of 1846 required a value for the speed of light in a vacuum (in his view, this was the relative velocity separating two charged, noninteracting masses). Physicists especially felt called to establish natural constants as precisely and as independently as possible, and many devoted their life to devising sensitive and ingenious instruments for observing or diffracting rays of light, for measuring electrical current, and for sensing small amounts of heat. This enterprise ultimately depended on mechanical springs, balances, and braces. For the reason of absolute measurements, the first generation of laboratories expressly designed for physics (in the 1870s and 1880s) had as their goal the minimization of distortions, both mechanical and magnetical.

Precision can be transformed into accuracy by error analysis, the great tool of the nineteenth-century measurers. Because the attempt to record changes in nature is a function of instrumental sensitivity, all physical measurements are uncertain. But before the end of the eighteenth century, one finds no systematic use anywhere of significant digits or observational uncertainties. From Babylonian star charts to Lavoisien chemistry, measurers ask us to trust their individual eye and their personal judgment. We see natural regularities presented as subjective guesses.

With the French Enlightenment came the appreciation that probabilities related to data, and by extension that physical data could be analysed by what we now call a bell-shaped curve. The observation is due to Laplace, but its systematic (and independent) use by Gauss for reducing astronomical data (and especially his stunning success in finding the misplaced asteroid Ceres) brought the method of least-squares (with its notions of averages and deviations) to the attention of scientists everywhere – once Gauss revealed his methods.

Error analysis found a firm place as the cornerstone of a new discipline, modern physics, that emerged in a special, multinational setting. The place was Königsberg, the Baltic centre of the eastern Prussia landowners, the Junkers. In the early 1830s physics professor Franz Neumann (1798– 1895), in collaboration with astronomer Friedrich Wilhelm Bessel (1784–1846) and mathematician Carl Gustav Jacob Jacobi (1804–1851), persuaded the Prussian state to finance a physics seminar where students could develop the facility for making and evaluating measurements. These academics sought to measure natural phenomena precisely, on the one hand, and formulate mathematical laws, on the other. They were the first to establish the modern physics syllabus, ranging from mechanics to heat, light, sound, and electricity.

The earliest of modern physicists worked at an institution with a distinguished past (Königsberg was Immanuel Kant's [1724–1804] university). They borrowed from pedagogical innovations in the German university tradition, notably the eighteenth-century philological seminars and a teacher-training seminar in physics at the University of Bonn (located in another German state). They took over the powerful mathematical machinery of Restoration France. The programme turned on mastery of experimental error with a view toward formulating universal, mathematical laws. It succeeded through evangelical fervour, taking root in every German-language university.

The transformation of mechanical precision

The search for absolute mechanical precision lost its clear direction after only two generations. It fell victim to its remarkable success in transforming the urban landscape.

As they careened down city streets on steel tracks, trolleys generated enough vibration to shake measurement piles inside laboratories. With electrification and consequent greater speed, trolley vibration became intolerable. Electric wires for trolleys and street illumination produced electro-magnetic waves that interfered with balances and lenses. Research requiring such measurements – in geophysics, for example – left the city for rural isolation. The effect was immediate and universal. The new geophysical institute of a small town like Göttingen was constructed in 1902 on the Hainberg, a ridge located a number of kilometres to the east. At just this time and for just this reason, the Jesuit geophysicists at Shanghai relocated their geomagnetical instruments to a station seventy kilometres distant.

Around 1900 the measurers of cosmical quantities – in meteorology, seismology, and astronomy – left the urban laboratory for more remote settings. Special institutes blossomed on hill tops and in villages – on mounts Hamilton and Wilson in California, on Puy-de-Dôme in France, on Mont Blanc in the Alps, at Williams Bay in Wisconsin. Measurers of biological quantities relocated to Arcadia – at a handful of European coastal sites like Ostend, Roscoff, Naples, Wimereux, Kristineberg, and Monaco, and in the United States at Cold Spring Harbor, New York, at Wood's Hole, Massachusetts, and at San Diego, California. In these settings, plants and animals could be harvested easily. Just as academic physicians situated themselves in large cities filled with illness of all kinds, scientist-measurers decided to establish themselves near what they wanted to study.

The research programme guiding 'classical' nineteenth-century physics, where mechanically precise measurements were intended to reveal new regularities of nature, also metamorphized from the inside. The ideal scientist was held to be at home equally with experiment and theory, having mastered both the tactile knowledge of the laboratory and the abstract harmonies of mathematics. Neumann's ideal found an incarnation in a succession of brilliant innovators, from Helmholtz to Hertz on the Continent and from Lord Kelvin to Lord Rayleigh in Great Britain, but by century's end exact sciences at least had begun to disaggregate into experimental and theoretical camps.

The symbiosis between precise measurers and precise theorizers weighed most heavily on the theoreticians, who were custodians of generality and arbiters of competing truths. An experimenter was under no obligation to *explain* his data, but explanation was everything for a theoretician. An experimenter could rest secure in his inventive capacity, knowing that an instrument well designed or data well taken – the bright line in the sodium emission spectrum measured to better than eight places by means of a diffraction grating, for example – would remain on the books forever. A theoretician who based his calculations on discredited hypotheses, however, could anticipate superannuation. Profound depression at the irrelevance of their work led the brilliant theoreticians Ludwig Boltzmann and later his student Paul Ehrenfest (1880–1933) to take their own lives.

Old programme, new effects

Since the Scientific Revolution, researchers have aimed at discovering new effects and phenomena. Early in the twentieth century interest focused on the distant, the minute, the large-scale, and the attenuated. Enormous lenses, cast and ground by practical-minded tinkers, went into telescopes that revealed the large-scale structure of the visible universe. Strategically located astronomers devoted a lifetime to studying motions of stars in the so-called fixed celestial sphere (stellar radial velocities, variable and binary stars). Electrons produced by potential drops *in vacua* and diverse subatomic particles emitted in radioactive decay were measured and counted by delicate laboratory apparatus. New particles and new effects, such as artificially induced transmutation and chain-reaction fission, resulted. The distant and the minute became indissociable, just as they had been in the mind of Newton: astrophysics revealed new substances, such as the element helium; nuclear chemistry turned to account for the fire of the sun; new phenomena such as cosmic rays revealed the positron in cloud-chamber photographs. Understanding large-scale geophysical phenomena – atmospheric circulation and seismological disturbances – depended on registering attenuated wave-fronts and analysing a resultant avalanche of data. The ionosphere and the earth's magnetic field united all these classes of investigation, as did the study of plate tectonics. An ambitious (and inconclusive) attempt to measure continental drift proceeded in 1926–33, when global longitudes were measured and remeasured with telescopes and radios.

As the use of radios in measuring intercontinental longitudes reminds us, precision related directly to timekeeping. The Eiffel Tower, converted into a giant radio mast before the First World War, provided accurate time signals at home and abroad; by the early 1920s the world had been ringed by powerful transmitters and receivers. Early radio transmission remained afflicted by vagabond electrical currents and ionospheric disturbances, and to secure the time of day at colonial outposts (as well as to guarantee contact with the imperial seat), the military took a great deal of interest in upper-atmospheric research.

We have seen that the daily time signal came from major astronomical observatories, which had accurate pendulum clocks periodically calibrated by transit observations. The accuracy of mechanical pendulums improved considerably with Charles Edouard Guillaume's (1861–1938) discovery of *invar*, a steel-nickel alloy with an unusually low coefficient of thermal expansion, and then *élinvar*, another alloy with a low coefficient of thermal elasticity. (The practical research netted Guillaume a Nobel prize.) A major improvement in accuracy occurred in the 1920s when W. H. Shortt perfected his free pendulum. Shortt began with the notion of double clocks – two clocks in tandem, a 'slave' doing the work of moving pointers and a more accurate 'master' clock keeping time and periodically correcting the slave's rate, the principle of corrective that had been used for a generation to keep time on the London Underground. In Shortt's innovation, every thirty seconds the slave clock gave a small impulse to the master's free pendulum and then synchronized itself by means of a feedback loop. By doing virtually no work, the master's time varied less than one part in ten million.

Shortt's Hegelian-like dialectic of slave and master set the time standard for a generation. It was surpassed by the quartz oscillator. A ring of quartz could be set to vibrate almost imperturbably by a small electrical current; the vibration, stepped down by electrical and mechanical means, drove a timepiece. First constructed in 1929 by Horton and Marrison in the United States, the Hamilton Watch Company's quartz device came on the market in 1957. At mid century, however, future time belonged to nuclear physicists. They accepted extensive military funding and designed clocks around various molecular and atomic resonances.

In German and in French we merely ask for the 'hour' of the day, and in French we associate time closely with the weather. In English, we ask if someone 'has' the time. The possession of this elusive quantity was of great concern for generals and admirals in the last half of the twentieth century. But the arcane accuracy of atomic clocks is a matter of no account to ordinary people, who worry about time as a thief of youth and wages. People paid monthly know that in leap years they donate an extra day of labour to their boss.

Philosophy and practice

The ultimate limits to precision were discussed during the nineteenth century by statisticians concerned with both human and inanimate norms. Epistemology examined what

might be known. Answers ranged from nothing to everything. The forward march of precision by no means eradicated arguments about sense data and cause and effect. The
end of the nineteenth century in fact saw a rejection of
positivist and progressivist canons, only to give way by the
1930s to depression and pessimism about the goals of
knowledge.

There was no pessimism for instrument-makers, however. Many of the less spectacular parts of twentieth-century
science derived less from the 'demand-pull' of brilliant
theoretical ideas than from the 'supply-push' of commer-
cially available measuring machines. The technology of
vacuum pumps, for example, underlay the machines of
atomic and nuclear physics – from William Crookes's
(1832–1919) tubes late in the nineteenth century to the
colossal particle-beam accelerators of our own time. The
use of these machines also depended on instruments for
measuring what the machines produced, from Charles
Thomson Rees Wilson's (1869–1959) cloud chamber to
Donald Arthur Glaser's (b. 1926) bubble chambers. Analysis
of the data followed behind technical innovations. Platoons
of housewives and bohemians, sitting before microfilm
readers, scanned kilometres of cloud-chamber photographs
for significant 'events', a process that has since become completely removed from the human eye. How far things have
come since the first half of the century, when illustrious
astronomers spent daylight hours on a blink-comparer,
flipping through photographical plates of the heavens for
evidence of double stars, and when distinguished physicists
lugged lead bricks up mountains as they searched for evidence of cosmic rays.

Life sciences have also marched to the tune of instruments. The commercial availability of remote-sensing
machines – from X-ray apparatuses to nuclear-magnetic res-
onance scanners – has revolutionized medical diagnosis.
Research budgets have loomed heavy with lines for electron
microscopes, infra-red spectrometers, centrifuges, and

cryogenic holding-tanks. The machines seem to have func-tioned like the early telescopes, microscopes, and bar-ometers. They were fashionable toys used for making routine measurements. That is how Nobel laureate Sinclair Lewis portrayed a large centrifuge in his novel about medi-cal research, *Arrowsmith* (1925).

Thomas Kuhn placed this kind of routine activity at the centre of scientific disciplines – 'normal science', or science done according to disciplinary norms. In his view, within a particular discipline there is agreement about what is to be measured and how the measuring is to take place. Disagree-ment about these matters heralds a major change in parameters and mind-set – a scientific revolution. Have measuring instruments led to scientific revolutions? Has quantity transformed into quality, to use a notion favoured by Hegel? The telescope promoted Copernicanism, the microscope promoted the notion of blood circulation, and early electromagnetical experiments led to the concept of a field, not by accurate measurement but by revealing qualita-tive phenomena. Mountains on the moon, phases of Venus, the Medici satellites of Jupiter, the existence of capillaries in living fish, and the force produced by an electrical current are in essence qualitative discoveries. Yet Lavoisier's chem-istry depended overwhelmingly on accurate balances, the detailed atomic structure of matter derived from measuring subatomic collisions, and Einstein's formulation of the grand structure of space and time was organized around delicate measurements of both moving electrons and the alignment of stars. These innovations would have been impossible without highly sensitive instruments constructed by tech-nologist specialists.

Precision regnant

In Germany, the ideal of unlimited precision foundered on the reef of cultural pessimism that followed the end of the First World War. The German economic miracle of the

Second Industrial Revolution late in the nineteenth century had stemmed from science. It was based on norms of rationality, criteria of cause-and-effect, and precise information about the unimaginably small and the unfathomably large – and in these matters Germany led the world. That intellectual leadership did not transform itself into a military victory. Germany's defeat produced contempt for the imperial forces that had been in charge of the war. It also became fashionable to denigrate science and technology, the motors of the empire and its war machine. Pessimistic and anti-rationalist prophets, from Oswald Spengler to Hermann Hesse, enjoyed a vogue during the early years of the Weimar Republic. Spirit and sentiment, for these writers and for many millions of Germans, were more important than reason and mechanical explanations. During the course of the 1920s, physicists and mathematicians, no less than ordinary men and women, found themselves caught up in the spirit of their times. In Germany, they also questioned whether causality was rigorously obeyed on the atomic and subatomic level.

In a popular article of 1927 Werner Heisenberg (1901– 1976) announced a principle of uncertainty, a new general law of the same form as the laws of thermodynamics. The principle established limits to what might be known about subatomic particles. All aspects of the motion of matter could not be specified with absolute precision: the more precisely one knew the position of an electron, the less precisely one knew its momentum. German scientists, following the spirit of their cultural milieu, embraced the new proposition without hesitation.

In lands beyond Germany, however, accommodation to cultural norms brought renewed allegiance to the reign of precision. Here are the roots of the astonishing discoveries of Joseph John Thomson and his tradition of experimentation at Cambridge. A classical heritage of laboratory precision – renovated and institutionalized by Louis Pasteur during his term as director of the Ecole Normale Supérieure

– lay at the base of French innovations in radioactivity and the science of matter (notably photography and metallurgy). Precise measurements became a dominant chord in United States physics. Early Nobel laureates like Albert Abraham Michelson and Robert Millikan (1868–1953) were quite innocent of theory, and many of their colleagues saw themselves as glorified engineers. The concern with precise measurement in the United States was connected with the use of machines of grand power and cost – enormous telescopes, expensive diffraction gratings, and colossal particle accelerators. It is characteristic that the great theoretician Josiah Willard Gibbs (1839–1903), employed in a precarious appointment at Yale University, was lionized by Europeans but studiously ignored by Americans.

In this light, then, we may reinterpret the decision of the world's foremost theoretician, Albert Einstein, to take up residence in a country with no tradition of theoretical physics, the United States. For Einstein, measurement and precision were everything. Special relativity is nothing other than a theory providing rules for measurement in the context of classical physics; it systematized and reorganized nineteenth-century mechanics and electrodynamics (Einstein did not hold special relativity to be a revolutionary theory). General relativity continued this synthetic drive by adding in gravitational phenomena, in the process formulating a general and apparently absolute framework for knowing about the physical world. Appearances to the contrary, there was nothing 'relative' about relativity.

Exiled from Germany in 1933, Einstein found a sinecure at Princeton in the new Institute for Advanced Study. Theoretically unapprised American physicists and astronomers resisted Einstein's relativistic physics, largely because they made little effort to understand it. Their mastery of measurement, however, in matters both astronomical and microcosmical, would certainly have appealed to Einstein, who enjoyed laboratory experiments since his student and patent-office years. But above all, Einstein would have been attracted by the American reticence to jettison determinism. Einstein disputed indeterminism from the time of its proclamation. For many years he debated about determinist physics with Niels Bohr (1885–1962), who elevated indeterminism into an epistemological canon. It was in the United States, and with two American coauthors, that Einstein published a stirring indictment of indeterminism, which is now known as the Einstein-Podolsky-Rosen paradox (1935).

American practicality and savoir-faire, financed by enormous private fortunes, set the tone for high-precision research in the twentieth century. The early signs of an overriding concern with building great machines of measurement may be found in the nineteenth century. Americans constructed the world's farthest-seeing telescopes and the finest ruling engines for producing a diffraction grating. American science was based on inventorying naturally occurring phenomena – from the distribution of stars to the spectra of chemical elements to the enumeration of fossil species to the geographical incidence of cosmic rays. When the world's most expensive research programme in physics, directed for twenty-five years by Louis Agricola Bauer at the Department of Terrestrial Magnetism of the Carnegie Institution of Washington, ended in the 1930s, the department abandoned terrestrial magnetism in favour of studying nuclear physics, ostensibly the better to understand magnetism on a microscopic level. It constructed some of the early Van de Graaff electrostatic accelerators used in atom smashing.

The high-precision technology behind particle accelerators obscures the crude nature of the experiments built around them. Atom smashers were precisely what the term implied. It was as if a field gun demolished a pocket watch and then the watch's mechanism was divined from various shards of metal distributed across a football field. The goal was to produce increasingly higher energies of bombardment, for such bombardment broke up the powerful forces that held together the constituent parts of the atom. The quest for power led to machines of increasing size and complexity. The cyclotron, where subatomic particles accelerated in an outward spiral and then hit a target, became the physicist's leviathan. The giant metal monster could not be housed in a conventional physics building, where professors had offices, classrooms, and laboratories. Relegated to outbuildings, financed by extraordinary disbursements, and requiring the full-time ministrations of skilled technicians, atom smashers completed the disintegration of the grand edifice of nineteenth-century physics.

Precision and the human spirit

Has the precision deriving from the domestication of electrons in solid-state physics empowered or emprisoned the human spirit? The question is largely rhetorical, for everything promised by that precision remains beyond the reach of much of humankind. Furthermore, a good deal of scientific enquiry does not depend on proximity to sophisticated particle accelerators or expensive remote-sensing devices. A search for general laws by focusing on particular instances is within the reach of researchers everywhere. A personal computer with 100 megabytes of storage – a calculating device two orders of magnitude more powerful than the machines used by the United States air force to defend North America in the 1960s – is readily available in most bureaucracies, universities, and even many homes; subscribing to the *Science Citation Index* (and purchasing a CD-ROM drive for it) costs much less than one government limousine. These tools, coupled with the creative leisure to use them, can generate new knowledge – if only the users of the tools are freed from the constraint of producing results 'useful' to the political master of the hour.

With the microelectronics revolution, precision has come full circle, back to the humanities. Humanists sometimes contend that precise measurement has overwhelmed other,

1. The 200-inch Hale telescope at Mt Palomar, California, portrayed by Russell W. Porter, 1939. Enormous structures for viewing the stars have been a persistent feature of civilized life.

2. The tube for mounting the 200-inch Hale mirror. Albert Einstein attended the ceremony for completing the mounting in 1937.

3. Igor Sikorsky's *Ilya Muromets* undergoing a
winter test flight near St Petersburg, 1914.
Heavier-than-air flight proceeded rapidly from
a technical curiosity to an armed menace.

4. Jan van Swinden demonstrates Benjamin
Franklin's electrical theory to members of
Amsterdam's Felix Meritis Society. Engraving
by R. Vinkelens after a drawing by P. Barbiers,
ca. 1795. The Felix Meritis Society, founded in
1776, was one of many local companies that
emerged during the latter part of the
eighteenth century to encourage experimental natural philosophy. The society's rooms,
which included a natural-history cabinet and
a laboratory for chemistry, as well as an astronomical observatory and facilities for studying
atmospheric electricity, were more richly
endowed than most university buildings of
the time. See Klaas von Berkel, *In het voetspoor
van Stevin* (Amsterdam: Boom Meppel, 1985),
p. 88, and W. W. Minjhardt, *Tot heil van 't
Menschdom* (Amsterdam: Rodopi, 1988),
pp. 78–123.

5. University Museum, Oxford, and Keeper's House, about the time of the debate on evolution of 1860 between Thomas Henry Huxley and Bishop Wilberforce, in an engraving by J. H. LeKeux. The museum arose despite the protest of of many members of the university.

6. A woodcut of Renaissance surgery, in this case amputation. A tourniquet has been placed both above and below the incision. Hans von Gersdorff (known as Schylhans), *Feldbuch der Wundartzney* (Strasbourg: Joannem Schrott, 1517), a work that appeared subsequently in many editions.

7. Toshiko Yuasa in 1933. A pioneering researcher in nuclear physics, Yuasa studied in Paris during the Second World War, taking a doctorate there in 1943. She worked for the most part in Paris up to her death in 1980. At the peak of her powers, she became the first woman physicist to receive a doctorate in Japan – from Kyoto University in 1962. Seki Shimizu (ed.), *Catalog of Toshiko Yuasa's (1909–1980) Archives* (Tokyo: Institute for Women's Studies, Ochanomizu University, 1993).

8. The Museum of Ferrante Imperato (1550–1625), a natural-history *Wunderkammer*. *Historia natvrale di Ferrante Imperato* (Venice: Combi & La Noù, 1672).

9. The cartographer (literally: 'The Magnetic Polar Stone', seen floating in an enormous basin in the foreground). A number of navigational and geographical instruments are visible. The cartographer works from a flat projection. An engraving from *Nova Reperta*, prints designed by Jan van der Straet, or Stradanus, and published by Philippe Galle, or Gallaeus, in Antwerp in the 1580s.

LAPIS POLARIS, MAGNES.
Lapis reclusit iste Flauio abditum   Poli suum hunc amorem, at ipse nauitæ.

10. An ideal curiosity cabinet, portrayed by the Amsterdam collector Levinus Vincent, *Description abrégée des planches qui représentent les cabinets et quelques-unes des curiosités contenues dans le théâtre des merveilles de la nature* (Haarlem: Levin Vincent, 1719).

IMPRESSIO LIBRORVM.

Poteſt vt vna vox capi aure plurima :   Linunt ita vna ſcripta mille paginas.

11. A sixteenth-century print shop, with compositors, proofreader, paper delivery, inker, pressman, office-boy, and owner. An engraving from *Nova Reperta*, prints designed by Jan van der Straet, or Stradanus, and published by Philippe Galle, or Gallaeus, in Antwerp in the 1580s.

12. Napier power platen press, 1849. Knowledge eventually circulated more widely as a result of the eighteenth-century industrial revolution of coal, iron and steam.

qualitative sensibilities in the twentieth century. Popular interest has remained high for modes of cognition that derive from unmeasurable qualities. Psychoanalysis and literacy deconstruction, for example, owe some of their appeal to their rejection of objective measuring standards. It is well to remember, however, that the sciences have traditionally included the humanities as one of their number. Measurement appears in the traditional attributions of the nine Greek muses, for example, and in the medieval canons of *trivium* and *quadrivium*. From around AD 800 up to the Renaissance, a principal task of educated men was construction of a *computus* – a calendar-almanach fixing moveable feasts, saints' days, legends, folk remedies and, among other things, the age of the earth. A portion of the activity of measurers like Leonardo, Vesalius, Galileo, Descartes, Leibniz, and Newton qualifies as humanistic enquiry, a tradition continued into the nineteenth century with writers like Alexander von Humboldt and Charles Dodgson (Lewis Carroll, 1832–1898). In the twentieth century, measurers like Robert W. Wood (1868–1955) and Leo Szilard (1898–1964) are known for their humanistic publications, especially their respective fables, *How to Tell the Birds from the Flowers* (1917) and *The Voice of the Dolphins* (1961). In modern times, interest in antiquity and medieval Europe led to humanistic disciplines based largely on precision measurement – archaeology and paleography. Today's institutionalization of the humanities, indeed, derives from their claim to precision, whether in dissecting the metre of a poem, revealing the authorship of a text, or establishing the date of an object.

Tante Inge would agree. When she is not baking a Panama-Torte, she chips away at a doctoral dissertation in religious studies. Her topic is astrology. She has a moon garden, laid out in a crescent and planted to take full advantage of planetary alignments. In her view, nothing could be more meaningful than astrology, where precision is joined with verse and rhetoric. For her, this is an essential side to measuring. She knows that most daily newspapers contain astrological readings along with their precise renderings of television programmes, weather maps, and stock-market prices. In an age when quantities are weighed electronically, astrology's use of traditional measures contributes to its enduring appeal.

Reading: Books and the Spread of Ideas

Our family has three heirlooms, four if you count the wooden chest carried by wagon across Pennsylvania to Ohio early in the nineteenth century. The first heirloom is a small willowware plate featuring a traditional Chinese scene in blue on white. It was part of the China trade, presumably shipped from Canton to New England and then on to Philadelphia. It remained intact for nearly two centuries, only to break in a recent, motorized move. The second heirloom is a crib quilt featuring a circle of flowers, stitched by hand about two hundred years ago. It covered many generations of infant and suffered countless ablutions without damage until a picture-framer had the bad judgment to stretch it over pine struts, which resulted in serious stains. The third heirloom is the oldest. It is a seventeenth-century German Bible. The paper is white and the leather binding is remarkably presentable. It has served for nearly four hundred years, as the unbroken genealogy penned in its endpapers attests. This artefact, printed in many copies a mere five generations after Martin Luther's seminal rebellion, still exerts power over a widely scattered and heterogeneous clan of Americans.

From script to print

The German Bible marks a middle point in one of the signal technological developments of modern Europe. More than one hundred years before this family heirloom saw the light of day, a development fundamentally altered our relationship to what we know and how we know it. Before this event, most people learned about the world by word of mouth; in this oral tradition, great value was attached to the art of storytelling, and to mnemonic devices and rhymes for organizing and remembering information. A tiny, literate elite depended upon the idiosyncratic renditions of clerical scribes, who generally sought to record only sacred texts and their glosses.

Impermanence and inaccuracy were the order of the day. Knowledge depended on the collective memory and fell prey to interpretations that tended to amend, omit, and distort. Even the best monastic copyist could scarcely avoid compounding error in transcribing documents that had been rendered indifferently and preserved poorly over the centuries. There existed no mechanism for verification and no means of appeal to the consensus of a group.

All this changed with a revolution in a small town in southern Germany. The revolution transformed knowledge from being idiosyncratic and local to being standardized and universal. In the area around Mainz – a region renowned as a centre of finance and mining – craftsmen who worked in metal became adept at engraving and in forging precision instruments and jewellery. A natural extension of these skills occurred at the hands of Johannes Gutenberg (1397–1468), who cast small letters of the alphabet in metal, which could be put together to construct words. The next step in the process required inked letters to be pressed against paper. Once the impression had been made, the type was disassembled and used again. Gutenberg's invention of movable type fundamentally changed the character of knowledge.[1]

The printing press altered both the inception and dissemination of knowledge. For the first time in the West, identical versions of a text could be circulated, discussed, debated, and preserved. No longer did the transmission of knowledge depend on imperfect memory and on the speaker's individual style. Of even greater significance, the invention of the printing press began to erode the distinction between erudite and vernacular traditions. The monk and the man in the street could turn to one medium for their message – the printed book.

Gutenberg's invention took Europe by storm. By 1500, eight million books had been printed, the output of close to one hundred printing presses. It was an explosion of knowledge without precedent in Western civilization. At issue here was not simply the spur to individual creativity brought about by the new possibility of error-free reproduction, but the fact that multiple copies might be issued with astonishing speed.

Gutenberg's first book was an edition of the Bible. The Catholic church embraced the new technology, for it facilitated the production and replication of indulgences. (These were papal pardons issued for a variety of sins, and they were sold to the transgressor.) What brought the Catholic church short-term gain would, in the end, prove to be its undoing. The printing of indulgences so enraged the clergyman Martin Luther that he posted ninety-five objections on the door of his church in Wittenberg. These 'theses' were printed up and distributed throughout Europe within a month.

Just as the invention of the printing press unleashed the forces of the Protestant Reformation, it transformed the face of Western science. Learning ceased to be the preserve of a clerical elite who had laboriously been able to fashion their versions of ancient texts. The old oral tradition of transmitting village lore from person to person also eroded. The printing press offered the ability to send permanent and standardized information with remarkable rapidity. Individuals far removed in space (and even in time) could examine exact copies of one work. Responses, criticisms, and emendations to these texts could similarly be dispatched quickly and precisely. The revolutionary success of the printing press, then, instantly begat the need for more texts.

Thus were born two of the central characteristics of Western science: propositions in these texts could be verified by a dispassionate reader or distant observer, and knowledge could grow so rapidly as to double in less than a human generation.

This explosion of knowledge associated with the Scientific Revolution created its own audience. The community of scholars expanded over time, as it became possible to build on the work of one's predecessors. Each individual no longer had to return to the beginning; a belief in looking forward and a new sensitivity towards human progress emerged. As knowledge assumed international dimensions, more participants began to engage in the stimulating process of criticism and self-scrutiny. The existence of newly available texts dealing with an ever-growing list of subjects indicated the insatiable appetite of an ever-widening readership.

Which kinds of books were published? Who were their authors? Readers looked for practical, technical information in the form of easily accessible manuals, guidebooks, and reference works on a range of topics, everything from navigation to commerce. For the first time, books provided 'how-to' information for advancement in this world, rather than spiritual prescriptions for redemption in the hereafter. In keeping with the new worldly focus, descriptive works about the immediate environment also achieved great popularity.[2]

Historians have debated the extent to which incunabula – early printed works issued before 1501 – represent a different genre of literature from the manuscripts circulated just before the invention of the printing press. Clearly the printing press did not immediately banish interest in the works of antiquity; many of the incunabula merely reissued classical texts. But the printing press brought about a new attitude towards the *audience* for literature. Editing procedures and conventions changed with a view towards providing greater convenience for potential readers, rather than merely serving the needs of the scriptorium. There is debate about whether literacy increased as a result of the invention of the printing press. But new importance certainly attached to the act of learning by reading, which supplanted older procedures for transmitting knowledge. Early printed works were issued in editions ranging between 200 and 1000 copies. Before the invention of the printing press, in contrast, it is futile to explore the notion of 'readership'; works were not produced for more than a handful of scholars.[3] This 'communications revolution' was if nothing else a revolution in information output.

Facilitating the birth of modern science

The revolution in communications may be seen as a necessary precondition for the Scientific Revolution itself. The enterprise of 'science and rationality', which sought to penetrate the mysteries of the physical and natural world, found its interests particularly well served by the invention of the printing press. For works in the natural and physical sciences, the problem of textual drift characteristic of scribal culture – copies tending to depart from the original text over time – had been particularly acute. The advance from handwriting to print offered tremendous advantages in accuracy when it came to reproducing multiple copies of texts containing numbers, diagrams, and illustrations. New data and observations could be disseminated, displayed, corrected, and compared with more accurate original versions. At any stage, emendations and additions might be incorporated into the works, and these revised editions recirculated. A new public constituency for science defined itself in terms of reading and evaluating printed works. Modern science went beyond the secrecy and limited circulation of manuscript culture and became synonymous with public knowledge. Putting the argument in its strongest form, the birth of modern science as we know it was a direct result of the invention of the printing press. Two areas of the natural and physical sciences illustrate the contention.

Following the invention of the printing press, striking progress occurred in the development of medical sciences, particularly anatomy. A new attitude prevailed when practitioners could actually scrutinize a revised corpus of corrected texts attributed to the Greek physician Galen and review them in the light of new translations and dissections. With the aid of the printing press, recently discovered Galenic teachings could be circulated for the first time; others were cleansed of errors and inaccuracies and then republished. In the hands of the Belgian physician and physiologist Andreas Vesalius, Galen's writings served as a potent stimulus to further discovery. Vesalius's masterpiece, *De Humani Corporis Fabrica*, owed much to the Galenic legacy. But, at the same time, it initiated a new era in medical sciences by disseminating a revolutionary ideology based on comparing inherited wisdom with observation and experience.

De Fabrica shares a publication date of 1543 with *De Revolutionibus*, which itself was issued shortly after the death of its author, the Polish astronomer Nicholas Copernicus. The common publication date suggests the close similarities shared by the revolution in anatomy and in astronomy, both events shaped by the communications revolution of the previous century. Like Vesalius, Copernicus drew upon the records of ancient and Islamic observers of the heavens. Unlike astronomers before him, Copernicus could survey a wide range of printed sources containing information about many observations over past centuries.[4]

The importance of the new medium becomes even more apparent if one examines the astronomical revolution among Copernicus's successors. The Danish astronomer Tycho Brahe, working several decades after Copernicus's death, examined printed and accurate versions of all the major astronomical texts of ancient times, as well as two editions of *De Revolutionibus*. Tycho's new observations were guided and focused by what he had read; because of the communications revolution, his findings circulated readily to the next generation, perhaps best represented by the German astronomer Johannes Kepler. Kepler, in turn, was able to double his computational speed by utilizing Napier's printed tables of logarithms. As a university student, Kepler compared the cosmological systems of Ptolemy, Copernicus, and Tycho – all in print.

By the time of the era of Galileo Galilei and Isaac Newton – whose synthetic theories completed the process initiated by Copernicus – the communications revolution had effectively transformed the face of Western culture and Western science. Intellectual life moved north, away from older, Catholic centres like Padua and Paris, which by then had begun to be associated with suppression and censorship of free thought and political activity. The printing industries flourished in the liberal atmosphere of the Dutch Netherlands, where the new philosophies and world systems articulated by René Descartes, as well as by Galileo, were welcomed, for entrepreneurs recognized an attractive product that could be readily sold to a burgeoning class of avid readers.

The rise of the scientific journal

The career of Isaac Newton illustrates the important role of new printing media in bringing a message to the scientific community. Newton, born in 1642 (the year Galileo died), escaped the prohibitions on free expression that broke the spirit of the Italian astronomer. Newton could flourish in the relatively open intellectual climate of Restoration England. He did not fear that access to any kind of literature would be impeded, and he did not hesitate to ponder the general, philosophical implications of his world system. His milieu offered him innovative tools for receiving information and diffusing his views.

The secretary of the Royal Society of London, Henry Oldenburg, provided young Newton with a window on developments in Europe and the opportunity to engage

Continental scientists. Through Oldenburg's extensive network of correspondents and contacts abroad, Newton learned of Gottfried Wilhelm Leibniz's mathematical contributions and offered his own tentative first versions of the calculus. But as Newton's scientific work matured, a new vehicle provided a more formal and regular avenue for scientific exchanges, one that could serve as a point of reference and a record for subsequent priority claims. (Although erudite letters continued to facilitate learned exposition throughout the eighteenth century and into our own time, they increasingly conveyed a sense of capricious idiosyncracy, even when subsequently issued in printed form.) The new vehicle was a scientific journal, appearing under the auspices of the Royal Society and titled the *Philosophical Transactions*. Newton procrastinated for years before publishing the most important work of his life, the *Principia Mathematica*, but he found it much less daunting to issue shorter versions of his theories in the *Phil. Trans.*

The scientific journal came to offer authors the flexibility and speed that they required for disseminating observations and experiments. It was no longer necessary to wait until one had completely mastered an extensive area of knowledge before publishing; intensive, punctual pursuits became the order of the day. The scientific paper or journal article eventually displaced the definitive and comprehensive book as the appropriate showcase for a scientist's work. This development hastened the process of specialization, whereby scientists strove for mastery of an ever more narrowly circumscribed area of knowledge.

The learned journal became a prime mover of the scientific enterprise. It provided the perfect articulation of the ideology espoused by Francis Bacon, the spiritual voice of the Scientific Revolution. Contributions to natural knowledge consisted of succinct accounts of empirical discoveries instead of long discursive essays. It is no accident that the birth of the scientific journal can be traced to the mid seventeenth century, the time of the Baconian-inspired learned academy. Both academy and journal derive from the remarkable new size of the scientific establishment, which around 1650 had achieved a critical mass. Size produced clear economic benefits. Among these was the ability to divide publication expenses and responsibilities. The scientific journal also offered a convenient format for abstracting or summarizing the torrent of literature that had begun to flow from printing presses all over the world.

The world's first scientific journal, the *Journal des sçavans*, actually owed its creation in 1665 to the exertions of the lawyer Denys de Sallo (1626–1669). Combining the serial periodicity of the newspaper (the Dutch *coranto*) with the content of books devoted to natural philosophy, de Sallo aimed not only to supply extracts from other publications but also to introduce news of experiments, inventions, and university proceedings. That it filled an important niche is indicated by its remarkable longevity; only in 1816 did its name change to the *Journal des savants* and its content become more literary than scientific.[5]

Besides the German and Dutch editions that it spawned, within three months the journal inspired an English imitation, the *Philosophical Transactions*. Unlike the *Journal des sçavans*, however, the *Phil. Trans.* soon functioned as the official organ of a learned society, the Royal Society of London. (Dissension resulted because Henry Oldenburg viewed the journal as his personal undertaking and the Society tended to avoid financial responsibility.) Oldenburg decided to exclude the extrascientific political and theological material included in the *Journal des sçavans*, seeing such matters as potentially divisive. The *Phil. Trans.* devoted its attention to recording experiments conducted by Royal Society members, publishing selections from Oldenburg's correspondence, and reviewing books.

From the outset, the *Phil. Trans.* evoked enormous interest in England and abroad, despite the fact that the plague of 1665 impeded production of the first issue and reduced its early readership. By the following autumn, the journal's new Oxford publisher, Richard Davies, printed more than a thousand copies. Subsequently publication moved back to London, where the bookselling trade had generally suffered devastation in the great fire of 1666. By good fortune, the firm of John Martyn (1699–1768) and James Allestry, official printers to the Royal Society and publishers of the *Phil. Trans.*, escaped relatively unscathed and resumed publication of the monthly journal that cost a modest six pence an issue. Martyn and Allestry also published monographs by leading members which bore the imprimatur of the Society: John Evelyn's *Sylva* (1664), Robert Hooke's *Micrographia* (1665), and John Wallis's *Discourse on Gravity* (1675).

In the intellectual centres of Europe – including Paris, Hamburg, and Florence – the *Phil. Trans.* found an especially enthusiastic audience. Part of this popularity emanated from Oldenburg's circle of correspondents abroad, who had actively been sending him scientific information for years in fulfilment of the Royal Society's aim to compile a Baconian natural history. With the launching of the *Phil. Trans.*, it became especially important to maintain and even enlarge this circle, for Oldenburg needed to print the latest word on developments abroad. Continental philosophers toyed with the idea of issuing a translation in French and Italian, and a Latin version was actually published for a short period of time. Leona Rostenberg calls the journal 'one of the greatest achievements of the Restoration . . . [which] vies [in importance] with the writings of Newton, Hooke, Boyle, Evelyn, Wren, and others – all of whom contributed to its pages'.[6]

The great success of the *Philosophical Transactions* and the *Journal des sçavans* evoked imitators elsewhere in Europe. In Germany, a *Miscellanea Curiosa* began publication in 1670 under the auspices of the Collegium Naturae Curiosorum. It emphasized the medical sciences and closely followed the example of the *Phil. Trans.* An *Acta Eruditorum* was founded in that country, as well, in 1682, which, like the *Journal des sçavans*, included nonscientific information. A *Giornale de litterati d'Italia* also followed de Sallo's example in 1668; in Copenhagen, an *Acta Medica et Philosophia Hafniensa* more closely imitated the *Phil. Trans.* In addition to these multiple clones, the *Phil. Trans.* was, in the words of David Kronick, 'reprinted, abridged, abstracted, reformated and translated in numerous editions'.

But the imitators were neither so successful nor so hardy as the originals. Some perished after the first few issues; the most fortunate managed to survive for a decade or so. The critical determinant seems to be support from a learned society, which provided continued intellectual direction, assured publication over the long term, and gave financial backing. During the first century of the scientific journal's existence, only around 25 per cent were sponsored by academies, which helps to explain why the success rate was so low.

The eighteenth century became the era of the proprietary journal, where an individual assumed responsibility for the publication. Ironically, this occurred just at the time when national scientific societies entered a new era of growth and importance. The Académie des Sciences in Paris finally launched a proceedings imitating the *Phil. Trans.*, its *Histoires et mémoires*, in 1702. (Up until this time, its members had simply used the *Journal des sçavans* as a medium for communicating their work.) But most of the brilliant examples of scientific journal-publishing owe their success to the vision and commitment of individuals, not groups. Individuals were willing to innovate and take risks; scientific societies, in contrast, easily succumbed to inaction through the weight of authority.

The growth of scientific journals across the eighteenth century provides a striking example of adaptability and exponential increase. Numbers grew from around thirty in 1700, to more than seven hundred in 1800. During this period, the scientific journal experienced an era of prosperity and influence that parallels the dominance of the printed book three centuries earlier. The amount of information published in scientific journals was so great that new bibliographical, reviewing, and abstracting publications (themselves usually issued in serial form) emerged to guide, manage, and direct things.

The torrent of literature entailed a set of conventions that characterized the scientific enterprise. The journal completed the shift from Greek and Latin to the vernacular of everyday language, a transition that had been instigated earlier by the appearance of the printed book. National scientific societies sought to establish a permanent record of their collective identity by investing authority in publications written in their native tongue. The journal supplied scientists with a quick, accessible medium for publishing their findings; it also brought them information that otherwise might escape their attention, whether issued in foreign languages or recondite sources. The scientific journal stimulated consensus and criticism as it facilitated the flow of information from one scientist to another.[7]

The very success of the general scientific journal (covering a range of physical and natural sciences) led to new developments. Intense competition for a rather limited audience placed a premium on novelty, a special feature that could provide a competitive edge. One such innovation was the decision to concentrate on a particular area of science or technology. Journal editors chose this course because the growing size of the scientific community could support such specialization.

The earliest specialized scientific journals emerged toward the end of the eighteenth century. Notable examples in chemistry are the *Annales de chimie*, founded in the fateful year of 1789, and (in the nineteenth century) Justus von Liebig's homologous *Annalen der Chemie und Pharmacie*, which disseminated the new science of organic chemistry. Physical sciences were emphasized in the *Journal der Physik*, founded in Germany in 1790, and the equally prestigious and long-lived *Philosophical Magazine*, first issued in 1798.

Natural sciences were represented by the *Botanical Magazine*, created in 1787.

Despite a few success stories, however, most of these proprietary disciplinary journals faced an uncertain future. Specialization might offer an advantage, but it also introduced the possibility of oblivion if insufficient practitioners were found to support the venture. The more erudite the speciality, the more limited the circle of adepts, which means that the editors had to charge their readers more for the luxury of indulging in arcane knowledge.

One example shows how easy it was to overestimate the enthusiasm for a particular scientific speciality, which quickly spelled the doom of even the most promising journal. In 1836, the first issue of the *Magazine of Zoology and Botany* was published in Edinburgh. But the usually frugal editors made the unfortunate decision to pay contributors for their articles, and hence exert total control over content. In this way, the editors believed, they might avoid turning away potential purchasers bored by lists of species or offended by controversial and speculative articles. It was simply unheard of for struggling scientific journals to act in this way, in the manner of powerful literary and political quarterlies of the day, such as the *Edinburgh Review*. A pronounced bias for articles dealing with zoology exclusively, combined with an emphasis on northern England and Scotland, further crippled the commercial prospects of the journal, which languished after fewer than two years of existence. The audience for natural history was too limited and inelastic to tolerate such a high degree of specialization. No more than 400 copies of any issue of the *Magazine* were ever sold, leading its printer to conclude sadly that 'scientific men are far too few in number to pay for such a work'.

In contrast to the short history of the *Magazine of Zoology and Botany*, the *Annals of Natural History* successfully exploited the niche that its predecessor could not find, allowing it to become the leading British journal in biology until around 1900. Much of its success derived from the astute direction of the London printer and publisher Richard Taylor (1781–1858), as well as of his illegitimate son and business partner, William Francis (1817–1904). Since 1800, Taylor had been publishing the *Philosophical Magazine and Annals of Philosophy*, the leading English-language journal for physical sciences.

The *Annals of Natural History* (whose name had been crafted to bring to mind its distinguished relative) was created by merging the *Magazine of Zoology and Botany* with several other natural history journals. It offered Francis the opportunity to edit a biological journal, long a dream of his. Although he obtained a doctorate in organic chemistry under Liebig at Giessen in 1842, he had always maintained a keen interest in entomology and other natural-history sciences. Francis translated articles and proceedings from German originals for the *Annals*. Taylor and Francis exercised strong control over the journal's intellectual content – which included zoology, botany, geology, and paleontology – and also kept down production costs. Expensive illustrations were minimized, and the press run was limited to a realistic 500 copies.

Indicative of Taylor & Francis's emerging status as one of the major science publishers in London was its contract to print the proceedings of the Royal, Astronomical, Geological, Zoological, Chemical, and Linnean societies. Around 1800, the first viable disciplinary societies had been created, many of which chose to issue a journal. These recorded the proceedings of their meetings and supplied a forum for the publications of their members. The Geological Society of London, for example, published a *Transactions* (later superseded by the *Quarterly Journal*), beginning in 1811. In Edinburgh, the Royal Physical Society and the Wernerian Society both issued a *Transactions*. Evidence of the importance of these specialized societies and the disciplinary significance of their journals is not hard to find. Charles Darwin and Alfred Russel Wallace (1823–1913) presented their theory of evolution by natural selection to the Linnean Society of London in 1858, and both subsequently published their contributions in the Society's *Journal*.

In 1858, when his father died after a prolonged struggle with insanity, Francis assumed sole direction of the firm. The profitability of the *Philosophical Magazine* and the *Annals of Natural History* were assured by Francis's position in the British scientific community, and his particularly close business and personal relationships with its leading lights, including John Tyndall (1820–1893), Thomas Henry Huxley, and William Jackson Hooker. Francis held the power of transforming scientific manuscripts into print, which made him the comrade of naturalists who were beginning to fill university chairs and museum keeperships – the 'young and rising race' that periodicals had to attract as contributors and subscribers, in the words of one former editor. Francis's monopoly over the publications of learned societies, however, could not be maintained so easily, especially as new printers and publishers looked to expand in the ever more attractive world of science publishing.

By the middle of the nineteenth century, more than one thousand scientific and technical periodicals were being published. This number includes journals that sought to respond to a heterogeneous readership beyond the confines of the scientific community. Proprietary or commercial science journals possessed a number of advantages over the official journals of learned societies. Enjoying more frequent issue, they offered greater speed to aspiring authors. They presented information to those who lacked foreign languages or extensive libraries. They tended to accept shorter articles and more controversial, unorthodox, and theoretical material. As William Brock puts it, 'they kept the scientific societies on their toes, broke their monopolies, and made them less authoritarian and cliquish than they might have been'.

New forms for new audiences

Just at the time when science publishing became closely tied to special-interest groups, the public embraced general and popular science journals. This new development resulted from an expansion of the coverage of scientific topics that had been included in general literary periodicals, such as the *Gentleman's Magazine*, and in the great quarterly journals, like the *Edinburgh Review*. This initiative also responded to the interests of an increasingly large, literate, and leisured middle class, who had begun to make their presence felt. And by the early nineteenth century, new technologies – including the use of steam power and stereotypic plates – speeded up printing and allowed press runs to exceed the earlier limit of around fifteen hundred copies. Distribution improved by means of new transportation facilities like railways and new popular educational institutions, such as public libraries. As a result, editors could reasonably expect to produce relatively inexpensive works of frequent periodicity for a large audience.

Social reformers and voluntary associations encouraged the publishing of popular-science periodicals that provided 'improving' information to readers who might otherwise purchase potentially dangerous political tracts. The *Penny Mechanic* affirmed that the study of the physical and natural sciences 'by withdrawing the mind from pursuits and amusements that excite the imagination' would tend 'to the improvement of our intellectual and moral habits . . . and to substitute placid trains of feeling for those which are too apt to be awakened by the contending interests of men in society, or the imperfect government of our own passions'.

Besides promoting social stability, popular science periodicals answered the needs of increasingly well-paid social groups. At a time when middle-class families found new pastimes in pianos and parlour games, the *Intellectual Observer* noted 'the striking fact of the great increase in the number of purchasers of microscopes and telescopes, which are becoming necessary portions of the furniture of every well-ordered home'. Better than the simple entertainment afforded by music hall, football match, or seaside holiday was the fusing of amusement with instruction, as in a visit to the great Exhibition or a gallery of science. Reading, a purely intellectual pursuit, was praised for producing 'habits of reflexion' particularly 'favourable to orderly conduct'. Perusing a popular science periodical, accessible by its relatively low price and wide distribution, equally amenable to careful study or desultory page-turning, provided the perfect 'rational' recreation.

Most popular-science periodicals were issued in compact and portable octavo format (approximately 9'' by 6''). They were illustrated, either by woodcuts, line engravings, or colour plates, to increase intelligibility. As one reviewer noted, the popular author 'must teach by illustrations that are a species of representation of what actually occurs, and impress the mind with livelier ideas than the mere abstractions of reason can convey'. Also to aid their readers' comprehension, periodicals usually tried to eliminate technical language. The *Penny Mechanic* associated the use of 'hard names' with a 'pedantic display of learning'.

Popular science periodicals usually contained a mixture of information extracted from other periodical publications and books, as well as articles and book reviews written expressly for the periodical itself. Some readers complained about the obvious signs of 'scissors and paste' work, but editors argued that this variety of articles acted to excite and sustain the reader's curiosity, as well as to provide information from otherwise inaccessible sources.

In nineteenth-century England, the encouragement of amateur scientific activity became a primary goal for these journals. The periodical itself provided a medium where subscribers could communicate scientific observations or technological discoveries. Editors insisted that they would suspend critical judgment and admit any contribution to their pages. In addition, by including copious notes on the proceedings of diverse metropolitan and provincial scientific societies, editors sought to convince unconverted readers to join the bustling ranks of amateur practitioners. Readers were told that the raw materials of science could be found in everyday objects that surrounded them at work or during leisure hours.

An experiential and inductive philosophy of scientific discovery and explanation served to reinforce equality in this Republic of Science. Anyone could derive information from the natural world by the free exercise of the senses. The mere stockpiling of this data gave rise to more general truths. All facts were valued; in contrast, theory – which editors called 'a heap of speculative rubbish' and 'visionary hypotheses' – was viewed with disdain. Scientific progress depended upon the steady accretion of factual building blocks collected by many observers. The 'slow and gradual accumulation' of modest discoveries could ultimately rival the 'proudest moments of genius'. Nor was any extraordinary mental prowess required to derive generalizations from these facts. The amateur scientific practitioner needed neither education nor expertise to manipulate experimental data skilfully and to construct theoretical, abstract chains of reasoning. He simply required an eagerness to participate and communicate his gleanings to others.

The popularization of science in France differed significantly from this vision. There, about a generation later than in England (during the 1850s and 1860s), a close-knit group of professional *vulgarisateurs* undertook to make high scientific discourse intelligible to the layman. Popularizers had to be accomplished in the art of lucid translation, said one periodical, because scientific formulas seemed like the words of a foreign language to the public. Periodical editors interpreted academic science by removing elements unique to specialized and theoretical research. The result would not be an 'elementary science', distinct from high science, but a 'popularized' version of 'advanced science', made accessible 'to all who are eager for progress and capable of a slight effort'.

French popular science periodicals stressed the utilitarian benefits of science. Initially they emphasized household hints and recipes, but by the 1850s, this fare was over-shadowed by descriptions of new technological achievements, particularly railways, submarine telegraphy, and steel-making. Popularizers singled out spectacular or controversial events calculated to catch the fancy of their readers, such as the origin of species and spontaneous generation. In addition, they adopted an encyclopaedic approach by discussing all varieties of science and technology in a single publication, contrasting themselves to the specialization of the scientific elite. Their most uniform feature was a report on the proceedings of the Académie des Sciences, a window on the affairs and achievements of high science.

Despite imitative forms and similar preoccupations, French popularizers fashioned a product strikingly different from that of their English colleagues. French editors were writing for a different demographic group, experiencing an earlier stage of industrialization. They were inspired by a special notion of popularization. They defined their work, their writings, and even their readership by constant reference to the professional *savant*.

Showing science: the art of illustration

Successful popularization of science depended on supplying an attractive product to potential buyers. Science possessed a commercial value; it could be marketed. Indeed, ever since the invention of the printing press, science books had been bought and sold, along with literary texts and how-to manuals. In this marketplace, the presence of striking illustrations gave a competitive edge, a means of luring the reader and potential purchaser away from the competition.

The commercial importance of scientific illustration reached its zenith in the nineteenth century. In the production of fine natural history books – the legendary bird illustrations of John James Audubon (1785–1851) and John Gould, for example – all interest derived from the beautifully hand-coloured plates, which stood as works of art in their own right. Letterpress, in contrast, only appeared to identify the delineated species. Essentially, the value of the scientific word had been completely eclipsed by the scientific image. This new order of merit reversed the centuries-long tradition where the function of scientific illustrations had derived from their utility in supplying anatomical detail, demonstrating mathematical equations, or helping to display the essential taxonomic characteristics of plants.

In the view of Elizabeth Eisenstein, the invention of the printing press transformed 'image-making' as well as letterpress. Mechanical processes brought enormous benefit to the replication of illustrations, where the mistakes of hand-copying could completely obscure the meaning of original diagrams and pictures. Subsequently readers saw for the first time, in the words of Eisenstein, engravings of 'the engines required to destroy a fortress or erect an obelisk'; moreover, anatomical texts revealed ' "veins and vessels" that had been less visible before'. Gutenberg's revolution brought 'fresh images' as well as new words to thousands who did not need to be literate to appreciate their novelty.

What made the introduction of these 'new pictorial statements' possible was the use of woodcuts, which could be inked and passed through the presses alongside the text. In the realm of botanical illustration, however, a new era of naturalistic representation dawned in the early decades of the sixteenth century. At the forefront of this development stand the herbals of the Mainz-born physician Otto Brunfels (ca. 1489–1534) and his fellow medical doctor, Leonhart Fuchs (1501–1566). Pierandrea Mattioli's *Commentarii in Sex Libros Pedacii Dioscoridis* (1554) followed the realistic renderings of these pioneers and circulated in tens of thousands of copies over more than forty editions. During the second half of the sixteenth century, the famous Flemish publishing house of Plantin issued the herbals of leading botanists Rembert Dodoens (1516–1585), Charles de l'Ecluse (1526–1609), and Mathias de l'Obel (1538–1616).

By the beginning of the seventeenth century, etching and engraving on metal plates (where the sunken parts of the plate bear the ink and the surface is wiped clean) had begun to eclipse the use of wood block illustrations (where the raised portion was inked, thus producing a muddier effect). Among the landmark works of the century is Dionys Dodart's (1634–1707) *Mémoires pour servir à l'histoire des plantes*, which portrayed living plants in life-sized versions wherever possible. Louis XIII's wars interrupted the projected work, so that it was not issued in its three-volume entirety until close to the end of the eighteenth century, the age of the great florilegia – sumptuously illustrated books about flowers.

Among the names associated with botanical engraving on the eve of the French Revolution (the era of the father of taxonomy, the Swedish botanist Linnaeus) are George Dionysius Ehret (1708–1770) and François Regnault (1746–1810), author of *La botanique mise à la portée de tout le monde* (1774) (the plates were done by Geneviève de Nangis-Regnault). German-born Ehret came to settle eventually in England, where he became a Fellow of the Royal Society and produced illustrations for the *Philosophical Transactions*. With the Belgian Pierre-Joseph Redouté (1759–1840, official draftsman to Marie-Antoinette), however, botanical illustration reached its apotheosis in his spectacular drawings of roses. The other names that complete the list of outstanding botanical illustrators are those of the two German-born brothers, Francis (1758–1840) and Ferdinand Bauer (1760–1826). Like Ehret before them, the Bauers settled in England, where they found a life's work in sketching the spoils of the expeditions of Sir Joseph Banks and others. According to Wilfrid Blunt, 1840 – the year of the deaths of Redouté and Francis Bauer – 'marks the close of the great artistic cycle which opened four centuries earlier with the first stirrings of the Renaissance'.

In 1828, Redouté, perhaps the finest botanical artist of all time, met John James Audubon, arguably the most outstanding zoological artist of all time, and saw the first plates of his *Birds of America*. Like Redouté's vision of the botanical realm, Audubon's copperplate engravings of birds and mammals effectively held a mirror to the zoological world, as he captured these animals in lifelike poses within their natural environments. So successful were these renditions by the artist fresh from the American frontier, that his portrayals of nature 'red in tooth and claw' appear almost frightening and wholly original still today.

Audubon's detractors denounce his unique and arresting poses as 'contorted', but one need only survey the work of his precursors to appreciate his achievements. Zoology followed the lead of botany (depictions of animals being somewhat more challenging to achieve than those of plants), with the first important illustrated works appearing around the middle of the sixteenth century. These include the encyclopaedic treatments of the animal kingdom by Conrad Gesner, Pierre Belon (1517–1564), and Ulisse Aldrovandi. Their authors strove for representational accuracy, but the coarse muddiness of the woodcut impressions gives them a primitive albeit somewhat charming air by the standards of subsequent centuries.

A new phase in zoological illustration began with the introduction of copperplate engravings in the seventeenth century, which better permitted the portrayal of fine physiological structures and anatomical detail. The English naturalists Francis Willughby (1635–1672) and John Ray used the new technique in their works on birds and fishes. A notable work of the period was Edward Tyson's (ca. 1650–1708) *Orang-Outang* of 1699, whose plates illustrated the dissection of a chimpanzee. Even the microcosm could be captured in the text, thanks to the unanticipated inventory of nature revealed through the newly invented microscope.

Robert Hooke displayed this world in his *Micrographia* of 1665, with detailed plates of the flea and the louse; Antoni van Leeuwenhoek (1632–1723) followed forty years later with pictures of bacteria and protozoa illustrating his papers in the Royal Society's *Philosophical Transactions.*

Regional studies and accounts of extended families of animals dominated zoological illustration during the eighteenth century. Examples of local studies include Mark Catesby's (1683–1749) natural history of Carolina (1731– 43) and Maria Sibylla Merian's (1647–1717) illustrations of the insects of Surinam (1705). Among the most extensive of the day was perhaps Georges-Louis Leclerc *comte* de Buffon's *Histoire naturelle, générale et particulière,* published in forty-four volumes between 1749 and 1788. These works, unlike more systematic monographs, lent themselves to publication in parts, which might be issued as author's supply and subscribers' demand dictated. Indeed, this procedure – where copies were sold in advance of their actual publication – proved well suited to expensive undertakings.

The great voyages of discovery of the late eighteenth and nineteenth centuries ushered in a new era of zoological illustration. Self-interested governments and wealthy patrons alike supported the publication of luxury editions that sported fishes, mammals, and insects never before seen, collected from the farthest reaches of the globe. Among these were Thomas Pennant's (1726–1798) *Indian Zoology,* followed by his *Arctic Zoology* of 1784–85, and William Swainson and Sir John Richardson's *Fauna Boreali-Americana* (1829–37), based upon Sir John Franklin's (1786–1847) early expedition to northern Canada. Perhaps the most beautiful, colourful, and exotic specimens depicted were the birds, especially those from tropical regions.

The nineteenth century became the golden age of the great hand-coloured bird books, inaugurated by Alexander Wilson's (1766–1813) *American Ornithology* in 1808. Wilson's sometimes coarse delineations were no match for Audubon's subsequent *Birds of America,* issued in an unprecedented double-elephant folio size between 1827 and 1838. With pages nearly a metre high, Audubon could capture even the hugest raptors in life-sized portraits. Audubon printed fewer than 200 sets of the work, which he sold at the then considerable price of $1000 the set. (Although even one of the Rothschilds hesitated before purchasing a set, the price commanded at auction today for a single plate of the *Birds of America* places them outside the reach of all but a handful of collectors.) John Gould's birds of Australia and other continents continued this tradition of luxury editions for the well-to-do; among the most famous and sought-after are his plates of hummingbirds. By exploiting the artistic talents of his wife Elizabeth and the poet Edward Lear, Gould managed to publish just under three thousand hand-coloured plates of birds in just over forty folio volumes.

Assisting this final progression in the evolution of the illustrated scientific book was the introduction of lithography. This technique involved the inking of those areas of limestone slabs not made ink-repellent by the use of a wax crayon. Lithography proved to be the process of choice for portraying large animals, mammals and especially birds, providing them with especially vibrant colour. The graphic technique showed itself to best advantage in the consummate works of two European-born émigrés to England, Joseph Wolf (1820–1899) and John Gerrard Keulemans. Especially notable are Wolf's plates for the American Daniel Giraud Elliot's *Monograph of the Felidae*, coloured and lithographed by Joseph Smit (1883), and Keulemans's illustrations for Elliot's monograph on pheasants.

For science, as for other cultural expressions, the message is shaped by the medium. The invention of the printing press provided the perfect tool for advancing the interests of modern science: it offered accurate replication of prose and picture, as well as the capacity for quick and wide dissemination of ideas. The creation of the scientific journal gave an advantage to short communications of recent observations and experiments, rather than to the presentation of complicated theoretical systems articulated over a lifetime. An examination of science printing and publishing shows how science, like literature or art, depends on market demand, and how the introduction of new technologies helps to mould the product to the requirements of its clientele.

9

Travelling: Discovery, Maps and Scientific Exploration

'Have you heard about the latest discovery?' The sentence resonates over coffee or drinks, meeting a friend or casual acquaintance, beginning the working day or ending it. The words refer to more than mere news, which often relates horrifying or unpleasant occurrences. 'The latest discovery' is generally hopeful and uplifting, even when it has no practical relevance. Halley's comet returning as predicted, the fossil representation of a feathered reptile, the gene controlling eye-formation found to be identical in insects and mammals – these are all revelations to inspire both awe and critical reflection. The sum of our knowledge about the world is substantially increased, and we are all the richer for it.

We do not say in ordinary speech that new knowledge has been invented, manufactured, or fabricated; it is *discovered*. Light bulbs are invented, shoes are manufactured, and lies are fabricated. In antiquity, however, discovery, invention, and artifice seem undifferentiated inspirations. Examples include the discovery that $\sqrt{2}$ is an irrational number, the invention of the Archimedean screw, and the fabrication of Hero's mechanical toys. All cultural activity used to belong to knowledge, and people adept at mechanics or astronomy also dedicated themselves to music, history, or poetry – in the style of the Greek philosophers. Before the European Renaissance, new things about nature were found out by talented people in all civilizations, but there was no apparent institutionalization of discovery and innovation.

According to Sinologist Joseph Needham, the formulation of a reliable method for making discoveries is a singular achievement of seventeenth-century Europeans. That method, a collection of precepts and practices, took shape after European assimilation of significant foreign inventions and notions: mechanical clocks, firearms, the compass, the stern-post rudder, Hindu–Arabic numerals, and (between 1200 and 1500) the grand corpus of Greek and Roman authors. Primed with a charge of novelty, Europe then expanded to survey the globe in what is still known as the 'Age of Discoveries'.

Who discovered whom?

Events surrounding the 500th anniversary of the earliest voyage of Christopher Columbus portrayed the first century of European expansion as a great interchange.[1] Europeans subjugated native civilizations, effectively realizing the goal of the crusades in unimagined lands. In Spanish and Portuguese territories, the new, relentless regime reflected Iberian norms. Disease spread in epidemics; infidels were converted or exterminated; and the Inquisition acted to censure creative thinkers. From Santo Domigo to Lima to Manila, institutions of higher learning – both learned academies and universities with professional faculties – slowly disciplined 'immature' minds.

In a complementary part of the exchange, foreign cultivations conquered European tastes. Sweetened espresso with croissant, afternoon tea with chocolate and marmalade, a meal of pepper steak and *pommes frites* followed by a strong cigarette – much of what the tourist today finds quintessentially European is a reflection of extra-European products and practices. The list goes beyond ingestible commodities. The European eye was dazzled by indigo and cochineal dyes, and by cabinetry in exotic woods. European sensibility was heightened by halucinogenics like hemp, opium, and peyote. European morality took sobering counsel from social institutions like the guerrilla armies and participatory democracy of indigeneous North Americans. European manufacturing received stimulation from calico, damask, silk, porcelain, lacquer, and South-Asian steel.

History is about the interpenetration of peoples and cultures. The mixing occurs at various levels, from delicate alloys and careful titrations to more energetic jambalayas or pots au feu. The circumstances and consequences of the mixing have been explored in detail by the past generation of historians.[2] In this chapter we focus on how travel and knowledge about distant lands contributed to notions about the world and humankind's place in it.

Travellers in antiquity

It has been said that book learning in education substitutes for travel. We acquire modern languages in school by imagining ourselves in a foreign culture. We study images of medieval cathedrals and quattrocento paintings, and we strive to recover the ambiance around these works of art which is not relayed by photographs. The writings of two of the great encyclopaedists from classical antiquity, Herodotus and Pliny, constitute just this synthesis.

Herodotus, who lived in the fifth century BC, is usually seen as the father of history. He knew the great Greek literature of his time, but in the model of a true scholar he travelled to see the world first-hand. During his twenties and thirties, he left his native Halicarnassus in today's Turkey to explore much of the Greek-speaking world around the eastern Mediterranean, travelling as far east as Susa and Babylon, north to the Black Sea, and finally to Egypt, where he took up residence. Political turmoil propelled him to the Athens of Pericles, Thucydides, Phidias, Protagoras, Zeno, Euripides, and Sophocles. We know the Greek Golden Age by its writers, but they enjoyed no recognized station in life as scholars or philosophers. As citizens, they were above all active in affairs of state. Socrates was a member of the army, Sophocles was an admiral, and Thucydides a general. Herodotus himself became a founding member of the Athenian colony of Thurii.

Herodotus has received apotheosis through his story of the struggles between the Greeks and the Persian 'barbarians'. Fully two-thirds of the account consists of general background. We learn about the rise of Persia as a great empire; displayed before us are the physical and human geographies of nations from Thrace to Assyria. The Greek states are situated in their geographical and political context. These parts of the account reflect Herodotus the traveller. He begins with what he has seen and goes on to create a compelling literary picture. There is no long disquisition on the causes of events and no attempt to show relationships among various events. It is tempting to read Herodotus as a travel commentary, a companion for Greeks abroad, much in the way that Heinrich Wölfflin promoted art history late in the nineteenth century as an entertaining guide for German readers vacationing in Italy.

The second great traveller of classical antiquity, the first-century Roman noble Pliny the Elder, began his career as a rhetorician and writer of various historical accounts while commanding cavalry in Gaul, Germany, and Spain. In Rome, under the reigns of Nero and Vespasian, Pliny completed his only surviving work, the *Naturalis Historia* or *Natural History*. A dedicated observer of nature, Pliny died when he came too close to erupting Mt Vesuvius, which had devastated the coastal cities of Pompeii and Herculaneum.

The *Natural History* is the finest encyclopaedic account of classical antiquity. To provide the contents with modern titles, the work begins with a mathematical and physical account of the world, then covers geography, ethnography, anthropology, zoology, botany, agriculture, pharmacology, veterinary science, mineralogy and metallurgy, and techniques of the fine arts. Pliny draws on more than four hundred authorities (although he states in the preface that he has used only a hundred select authors). As with anyone who undertakes a general work, Pliny depends on available sources, not all of which are equal in sophistication or reliability. From these he extracts a pessimistic vision, Stoically-inspired, which discounts mechanical invention and control over future events. It has consistently been remarked that Pliny sacrificed simplicity and candour for poetic norms of his time. Pliny's accounts of mythical peoples, circulating in print during the fifteenth century, provided a *vade mecum* for Christopher Columbus when he toured the Caribbean.

Maps

Pliny's accounts of physical and human geography, for all their credulity and exaggeration, furnish an inventory of antiquity. We look to his second-century successor Ptolemy, however, for a quantitative account.

Ptolemy's *Geography* (ca. AD150) provides a picture of the world known to him, but it does so in words and numbers. Most of the work consists of places identified by latitude and longitude, whose principles were to be found in Ptolemy's astronomical writings. Since he (like all mathematically sophisticated Mediterranean scholars) knew the world to be a sphere, the prime question of practical utility concerned projecting part of a spherical surface on a flat surface to make a map. He had his known world stretch east 180° from the Canary Islands (this error is at the root of Columbus's overly optimistic prediction of a short Atlantic crossing) and extend in latitude between 16° south and 63° north. Longitude differences were determined by observing one astronomical phenomenon at two locations (the local times differ), and Ptolemy knew about only one such measurement – the lunar eclipse of 331 BC observed (inaccurately) at Arbela in Assyria and at Carthage (here is a reason for the overly long dimensions given to the Mediterranean Sea). There were distortions, and after the slow decline of the Roman world, Ptolemaïc knowledge widely degenerated into primitive, gridless, disk-shaped *mappemundi*. But for 1400 years the picture remained authoritative.

It was authoritative, at least, for Mediterranean peoples. Chinese civilization developed roughly comparable accounts and constructions, even to the point of recounting myths similar to those of Pliny. The early travel accounts, from the third century through the Thang, are rich and detailed, giving rise to the ninth-century illustrated geographical encyclopedia of Li Chi-Fu. Many innovations occurred in Chinese astronomical observation over the next twelve hundred years, but there was little interest in relating astronomical events to the terrestrial grid of cartography.

Before European ships docked in Canton in the sixteenth century, the Chinese knew a great deal about Mediterranean lands, as evidenced by their sailing maps. These were in the general nature of *portulans*. That is, the maps were schematic representations of the rim of the Indian Ocean, with notions about the Mediterranean and indications of compass readings and sailing instructions; European portulans, indeed, derived directly from the importation of the Chinese magnetical compass. The schematic nature of Chinese nautical maps is surprising in view of embassies and naval expeditions sent west beginning in the Thang; by the middle of the ninth century, Chinese trade missions were navigating the east coast of Africa.

Among the grand naval expeditions, the most storied is the one led by Admiral Chêng Ho, the Muslim Three-Jewel Eunuch of the Imperial Palace. In 1405 he took command of 37,000 men in 62 ships and sailed west around Indochina. It was the first of seven missions, whose functions were at once diplomatic, commercial, and exploratory. Over the next thirty years Chinese fleets asserted the hegemony of the Middle Kingdom from Java in the east to India, Somalia, and Madagascar in the west. Among the prizes taken home were live giraffes for the edification of Chinese naturalists. There were also crates of rare birds and the perennially satisfying caches of gems. Natural products for medicines were prized acquisitions, and the voyages of discovery provided new items for Chinese pharmacopeia.

And then, with Chêng Ho poised to become the Chinese Vasco da Gama and discover a new trading route to the western Mediterranean, imperial politics intervened. Antinaval elements put an end to the magnificent explorations; by the sixteenth century, regulations actively suppressed maritime commerce under the Chinese flag.[3] The expense of constructing and maintaining such large fleets must have been considerable – archaeological discovery of a gigantic stern-post rudder confirms that the largest vessels were more than 100 metres long. We must not be surprised at the evaporation of support for promising exploration. We have only to consider the experience of travel to the moon in the 1970s, an adventure deemed unworthy of sustained effort by the world's richest nation.

A great torrent of knowledge and manufactures flowed outward from China, but the Chinese saw very little in the West that merited acquisition, except gold and precious metals. Joseph Needham has reminded us: 'The Mediterranean region acted for two millennia as a kind of monstrous centrifugal pump continually piping off toward the East all the gold and silver which entered into it.'[4] When we reflect on Japanese foreign trade at the end of the twentieth century, we sense something of a continuing tradition.

Progression of people and ideas in the Malay Archipelago

Through the fourteenth century there were persistent trade and diplomatic connections between Indonesian and Chinese regents, just as there were between Indonesia and South Asia. The China trade continued after Kublai Khan's unsuccessful invasion of 1294, the outcome of which saw establishment of the Majapahit empire. For all the traders – Chinese, Indian, Arab, and European – Indonesia was the source of highly valued raw materials. Indian treatises were translated into Javanese, and words of Sanskrit origin were used by Indonesian physicians. The traders brought highly refined manufactures, such as Chinese ceramics. Foreigners lived permanently in large settlements at market towns (the tenth-century Arab quarter at Canton is a typical Asian pattern), but trade was by no means entirely in their hands. Indonesian embassies went as far away as Madagascar, a land in fact colonized at some point by Indonesians.

Traders and missionaries (or, as historian Jacob Cornelis van Leur insisted, traders who were necessarily missionaries), not conquereors, brought Islam to Indonesia. Evidence suggests that the new doctrine did not find strong state adherents until the end of the fourteenth century, in the final century of the Majapahit empire, and Islam spread, ironically, through the diaspora of Muslims following the Christian Portuguese conquest of large Indonesian cities. This late conversion helps explain the absence of traces left by typically Islamic institutions and the late emergence of the religious college, the madrasa.

European expansion

One of the remarkable patterns in history is the way young nations manifest an appetite for geographical expansion. We see this in the first European power, Norway, to undertake systematic maritime exploration. The earliest European colonies, rimming a frozen desert, studded an arc in the frigid North Atlantic, from the Faeroe Islands through Iceland and Greenland to Newfoundland and beyond. The transoceanic expansion occurred without astrolabe, compass, stern-post rudder, or sophisticated rigging, and broadly speaking it may be compared with the peopling of the Pacific islands. It is unreasonable to imagine navigating the open sea without a basic sense of the stars, and regular transits suggest maps, or at least mnemonic guides. Newfoundland settlements date from AD 1000. They expired, but colonization succeeded in closer mid-Atlantic islands. Around the fifteenth century, overseas rule from Norway strangled the first transatlantic civilization, though not before it had provided an organizational model for the English parliament and a literary model for the early Anglo-Saxon epics.

Action shifted south. Several generations after they expelled the Castilians, the Portuguese expanded down the African coast. Henry of Avis, known as the Navigator, encouraged exploration from his position as governor of Algarve at Lagos. He sought to continue a holy war against Islamic civilization, which in his lifetime (1394–1460) still flourished on the Iberian peninsula. He craved objects of commerce, notably slaves and gold. In 1415 the Portuguese conquered Ceuta, across the Straits of Gibraltar; Henry became its governor. By the time of Henry's death, his explorers had attained Sierra Leone. We might think that the fall of Byzantium to the Ottomans in 1453 would have added impetus to the African explorations as an alternative route to the East. But Portugal and Spain contested the spoils of the West African coast, and expeditions proceeded in fits and starts over the next generation until in 1488 Bartolomeu Diaz rounded the Cape of Good Hope. In 1498 Vasco da Gama put in at Malindi, just north of Mombasa on the East African coast, some fifty years after the Ming navy under Chêng Ho had anchored there.

In short order the Portuguese set up strongholds in India and pushed on to the Moluccas, the Spice Islands that were the commercial object of their hearts' desire. The way east passed through the Sultanate of Malacca, an Islamic empire that over the course of the fifteenth century had risen to control much of the Malay Peninsula and Sumatra. The city of Malacca, situated at the narrowest part of the straits separating Sumatra from mainland Asia, fell to a Portuguese fleet in 1511. The Portuguese monopolized European trade until the end of the sixteenth century, when the Dutch supplanted them. Then over the next three centuries, the Dutch progressively relieved local regents of their authority.

Portugal's early maritime rival was Venice, which controlled commerce to Europe in the eastern Mediterranean and which, in the fifteenth century, had also made forays down the African coast. In the western Mediterranean, trade was something of a free-for-all, with Genoa, Catalonia, and England, among others, as players. But it was Castile, having abandoned its struggle with Christian Portugal to focus successfully on cleansing Iberia of Semitic civilization, that took the initiative to seek new realms for plunder. Asia beckoned, but Castile had no real experience in maritime matters. So Madrid purchased expertise – from Italy, from Portugal, from anyone available. Enter Columbus, with his exaggerated, Ptolemaïc claim for a short sailing time west to China.

The Portuguese exploratory imperative began without the benefit of Ptolemy's *Geography*, a version of which had been recovered and printed, with maps, in a succession of editions beginning in Bologna only in 1477. Lands unmentioned by Ptolemy found rapid inclusion in an avalanche of updated pictures. About 1500, Columbus's pilot Juan de la Cosa (ca. 1460–1510) provided the first map of the New World. Less striking on his parchment than the Greater Antilles and the adjacent continental coastline is a rather good presentation of Africa. When Johann Schott (1477–ca. 1550) printed a world geography at Strasbourg in 1513, he separated Ptolemy's *Cosmographia* in its twenty-seven maps from a modern presentation of the world; and in the modern part, prepared by Martin Waldseemüller (1470–ca. 1518), is found the attribution 'America', after Amerigo Vespucci's apocryphal discovery. This was not the first time the name appeared. Vespucci was a famous pilot, navigator, and diplomat who made several voyages to the New World around 1500, and 'America' appears on a map in Bavaria dating from 1507. But Waldseemüller is generally seen as the man who arranged for the name of half the world.

A century of wonders

The remarkable fact of the sixteenth century is that the world's peoples experienced an unusual and extraordinary intermingling without being guided by a new worldview. To account for new phenomena – from customs to biota – people fell back on tradition. Civilizations in the New World, when they encountered Europeans, relied on traditional prophecy, and as a result they were almost completely demolished. South and Southeast Asians, already familiar with travellers from many civilizations, established a *modus vivendi* with the newcomers. East Asia and Africa were more circumspect, and as a result they preserved their political and cultural integrity longer. Europe – for the most part Iberia – tallied up riches in precious metals and spices.

When dealing with quantities of gold or silver, the chances of miscounting are small: too many people have a stake in exactitude. What about reporting on everything else? May not prejudices overwhelm the senses, instructing the eye in what to see and the hand in what to draw? The answer, in sixteenth-century cartography, is negative. The sea and the coastline were accessible to all people with a ship, just as a century later the moon and planets were accessible to all people with a telescope. Overly imaginative land forms invited maritime disaster, or at least censure and ridicule.

Verisimilitude notwithstanding, many maps from the Age of Wonders are in the nature of either mnemonic devices or objects of contemplation. Let us first consider the mnemonic devices, the portulans with their fabulous, mid-ocean compass roses and bearings. For the practical end of transatlantic navigation and of finding one's way to a New World garrison town, personal experience was the key. A state authority would frown on publishing details of wind, water, and shoreline; a clever pilot would keep much to himself. In any case, the business at hand was conquest and plunder. Sixteenth-century ships did not take latitudes frequently, and their longitudes (determined on dry land from infrequent astronomical events, like eclipses) provided few solid coordinates for large-scale maps. The flat projection of Gerhardus Mercator (1512–1594), where the polar regions are stretched to have meridians and parallels intersect at right angles, promised navigational advantages – a straight line traced a great circle on the globe. But real gains were minimal when one's position away from land was indeterminate.

As to objects of contemplation, what is more arresting than a coloured world map, complete with allegories, allusions, and unknowns? Ornamental maps, printed in broadsides, are conversation pieces. They invite councillors of state to evaluate their standing among nations, merchants to imagine the progress of their returning ship, and literate citizens to consider their place in the cosmos. The limits of European civilization, in Abraham Ortelius's (1527–1598) *Theatrum Orbis Terrarum* of 1570, is apprehended directly in thirty-five pages of text and fifty-three copperplate maps. Humankind's place in the cosmos is apparent only to specialists reading Copernicus's *De Revolutionibus* of 1543, but most people could appreciate the world's revolutionary new configuration in maps.

Coastlines and rivers progressively assumed today's familiar shapes – all without the benefit of telescopes, chronometers, differential calculus, and 'scientific method'. The century culminated in Petrus Plancius's (1552–1622) ten-sheet publication of Pietro Petrucci's world map of 1592, which included secrets of Portuguese navigational charts filched by Cornelis (d.1599) and Frederick de Houtman (ca. 1570–1627), and Plancius's *Orbis Terrarum* of 1594. The latter appeared, ornately embellished, in Jan Huygen van Linschoten's (1563–1611) book about the Portuguese overseas, *Itinerario*, which provoked the first Dutch fleet to penetrate the East Indies in 1597–8.

The new encyclopaedia

The Spanish arrived in the New World as plunderers. Across the sixteenth century, they took the reins of power, exterminated indigeneous peoples, and enslaved entire societies. The Portuguese in Asia confronted civilizations already adept at accommodating foreigners; they were unable themselves to extract primary or agricultural resources. To an extent, surveys of the new lands were episodic manuals for material exploitation. The spiritual side of Spanish domination, however, contested with the colonial administration for control over lands and people. The Roman Catholic authorities sought to infuse New World society with Old World norms by building a religious utopia. The church in Europe having evolved diverse national styles (as well as devastating schismatic wars), religious planners understood the importance of assembling a coherent picture of the status quo in America. Two writers, Bernardino de Sahagún and José de Acosta, indicate the extent of interest in the new natural world.

Bernardino de Sahagún (1500–1590), a Franciscan educated at the University of Salamanca, went to Mexico as a missionary in 1529. He became a professor at the Colegio de Santa Cruz de Tlatelolco, one of the first European-inspired institutions of higher learning in the New World. His order's mandate was the care of indigenous peoples, devastated by a holocaust. De Sahagún and his college focused on materia medica, attempting to inventory local cures for endemic scourges. They assiduously assembled information on botanical simples, producing manuscripts, notably in 1556 the *Libellus de Medicinalibus Indorum Herbis*, written in Nahuatl by the indigene Martín de la Cruz and translated into Latin by Juan Badiano. The manuscript collected and illustrated medicines for illness from the head to the feet; it was offered to Philip II with the hope of eliciting funds for the college. Like a number of first-hand accounts of nature in the New World, de

Sahagún's manuscript seems not to have circulated widely.

José de Acosta (1539–1600) is exemplary of the European philosophical approach to inventorying the New World which characterized two centuries of Jesuit activity. At the age of fifteen, he joined the Jesuits in his native city of Medina del Campo, Spain. Brilliant progress led to a call to a chair of theology in Rome, but de Acosta asked instead to be sent to America. In 1570 he left for Lima, where he spent fourteen years, living in the Jesuit college on Lake Titicaca. In 1583 he produced a catechism in the Quichua and Aymara languages, the first book printed in Peru, after which he spent three years in Mexico. Returning to Spain in 1587, he summarized his New World experiences first in *De Natura Novi Orbis Libri Duo* (1588–9) and then in his *Historia natural y moral de las Indias* of 1590. The second work, in seven books, both defends Spanish rule and describes the terrain and natural world of Peru and Mexico. In it, de Acosta devotes much space to Inca and Aztec history and customs; he describes Aztec glyphs and Inca quipus (the sheaves of knotted cord that recorded affairs of state); he discusses altitude sickness, the use of the narcotic coca leaves, and local farming practices. The *Historia* produced a European sensation, being translated rapidly into Italian, French, Dutch, German, Latin, and English. Its success helped de Acosta survive political imbroglios to become the Jesuit superior at Valladolid and then rector of the Jesuit college at Salamanca.

New World biota, no less than lands undreamed of by Ptolemy, encouraged a fresh approach to the observation of nature, but old analytical traditions persisted for generations after the first voyages of discovery.[5] The sixteenth century sits halfway between ancient authority and modern, empirical accounts of natural phenomena. The Segovian savant Andreas de Laguna (1499–1560) is a typical transition figure. He published a rigorously classical anatomy in Paris in 1535, a book studied by Vesalius. Then de Laguna surveyed pharmaceutical weights and measures for a general pharmacopeia ordered by Philip II. In his Spanish translation of Dioscorides (fl.AD 50), de Laguna introduced New World species, notably tomatoes and corn. The slow penetration of new botanicals is also registered in the publications of Nicolás Bautista Monardes (1493–1588), a native of Seville. His first materia medica, in 1536, contained only Old World plants; but in 1569–71, at the age of about seventy-five, he published a work on American drugs – the result of decades spent inspecting samples on the Seville docks. His book (including a discussion of tobacco) appeared rapidly in numerous translations and editions.

The Age of Wonder reached a denouement in Charles de L'Ecluse, known as Clusius. The son of a French Protestant noble from Artois, Clusius became interested in botany in 1551 when he undertook a collecting expedition in Provence for his teacher at the University of Montpellier. He then issued translations and editions: a work of his teacher's on fishes, a translation of Rembert Dodoens's herbal. In 1574 he produced a Latin version of Monardes's book on American drugs. In 1582 came an appendix to Monardes, a translation of the materia medica of Cristóbal Acosta (ca. 1525–ca. 1594), and a reworking of Francis Drake's notes on the western coast of America, all the while producing herbals for Europe. In 1593 this extraordinary activity transported Clusius to Dodoens's chair at Leiden (carrying with it curatorship of Dodoens's botanical garden), which he held until his death in 1609. As much as any mapmaker, Clusius heralded the Dutch overseas empire. He was also an active experimenter whose labours contributed to the Dutch bulb culture. During his tenure at Leiden appeared the early printed discussions of epoch-making imports, notably coffee and the potato.

Classifying nature

A cautious inventory of the natural world, based on over-seas exploration, helped inspire the writings of Francis Bacon. In Bacon's mind and in the opinion of his devoted followers, dispassionate inventory revealed more than new objects – lands, peoples, plants, whatever. Inventory informed the discovery of true principles of nature.

The aim of Baconian inventory was domination over nature. In one of his writings given the title *The Masculine Birth of Time*, Bacon discusses the marriage of masculine mind with feminine nature. The result of this marriage would be a race of heroes with the power to inaugurate an era of peace, happiness, and prosperity. Bacon's language is redolent of sexual mastery over the natural world.[6] Science was no mere abstract enterprise – idly setting ducks in a row or collating the work of ancient writers. It was active, interventionist, and enterprising. Discovering new knowledge was the robust domain of the early transatlantic navigators, not the sedentary reflection of Renaissance humanists.

Elsewhere in the present volume we indicate that the Baconian urge, inspired by more than a generation of natural inventory, led to collective scientific endeavour aiming to understand the heavens newly revealed by Galileo and elaborated by Tycho and Kepler; the airs and vacua uncovered by Evangelista Torricelli (1608–1647) and Robert Boyle; and the unsuspected workings of people and animals, examined by William Harvey and Antoni van Leeuwenhoek. But the seventeenth century saw no extraordinary innovations in collecting, the technique that Bacon advocated. In cartography, a slow, steady refinement charts the path from Plancius's world map of 1594 through Henricus Hondius's (1597–1651) of 1630 to Frederick de Wit's (1610–1698) of 1670 to perhaps the first of modern world maps (by virtue of its clean lines and uncompromising scepticism), that of Guillaume Delisle (1675–1726) appearing in 1730. If he had been required to choose a trade, Bacon would surely have become a mapmaker.

The Baconian cast of the Royal Society of London is nevertheless best illustrated not in maps but in the extensive publications of John Ray, one of the Society's early luminaries and England's premier botanist. Ray sought nothing less than a complete picture of Europe's natural realm and a systematic account of exotic flora. His research was sped by the edition of, among other botanical works, Francisco Hernández's (1517–1587) generations-old manuscript on American plants. Ray preferred to classify plants by genetic affinity. Most of his descriptions are verbal, made without the aid of illustration.

The prestige of Baconian botany, we have seen, is illustrated in the career of Herman Boerhaave, known today principally for his chemical research and teaching. The possessor of a philosophy doctorate from Leiden (1690) and a medical doctorate from Harderwijk (1693), Boerhaave devoted himself to chemistry and medicine, teaching both topics as a *lector*, or reader, at his first alma mater. In 1709, rising to the zenith of his career, he received the lucrative chair of botany and directorship of the botanical garden at the Leiden medical faculty, an offering designed by the curators to keep him in town.[7] Boerhaave had little botanical experience, but he assiduously worked in the new field (to which he added the duties and salary of the chair of practical medicine in 1714 and the chair of chemistry in 1718). He continued to acquire exotic Asian and American plants for the university garden, and he conducted extensive botanical correspondence; indeed, the death of one of his beloved botanical colleagues and suppliers, William Sherard (1659–1728), precipitated his retirement in 1729. Keeping in touch with foreigners was one of Boerhaave's mandates: a grant of 300 guilders per year, on a salary of 1000 guilders, went to support the cost of his correspondence.

As Boerhaave's work at Leiden suggests, early in the eighteenth century, knowledge of botany and the art of transplanting were fashionable in academic circles. Just as chemists sought to make sense of procedures for producing compounds, so organizing the explosion of particular botanical instances became an important item on the agenda of natural philosophers. Then systematics took Europe by storm.

We have also seen that the new apostle of classification was Carl von Linné, or Linnaeus. During his university studies Linnaeus botanized in his native Sweden (travelling to Lapland) and set down principles of taxonomy based on plant sexuality and binomial identification (genus and species). He then spent three years in Holland, obtaining a medical doctorate, circulating among Dutch and English naturalists, and issuing a stream of botanical publications. He passed through Paris, returned to Sweden, and found his way to a professorship at the University of Uppsala. He was a founder and guiding light of the Royal Swedish Academy of Sciences. He worked tirelessly as a teacher and self-promoter.

In the glow of the Newtonian synthesis, and with the advent of the rationalist Enlightenment, systematics enjoyed a vogue in many branches of learning, from mineralogy to linguistics. Linnaeus's rigid rules did not find acceptance everywhere. Significant opponents were Michel Adanson (1727–1806) and Antoine-Laurent de Jussieu, who proposed an essentialist system of classification based on so-called natural affinities among species. But Linnaeus's self-esteem, as well as the appeal of his firm *a priori* principles, stimulated his followers to organize botanical knowledge around the world according to a universal framework. The Linneans transformed Baconianism and became missionaries of a new rationalism for the natural world.

The scientific expeditions

By the second decade of the eighteenth century, mapmakers were confident enough in their abilities to survey any piece of real estate. The Newtonian-Cartesian controversy over the shape of the globe soon applied the confidence to detailed topographical traverses – by rod-and-chain and by theodolite – of Lapland and Ecuador. The practical advantages of this kind of detailed surveying became clear during the wars of the mid eighteenth century, fought both in Europe and in the New World colonies. The consolidation of European colonial realms, as well as knowledge for their systematic exploitation through rational principles, resulted in a significant institutional innovation during the Enlightenment's waning years: the scientific expedition.

Scientific surveys have antecedents in enterprises of lucrative gain, for example, the circumnavigations of George Anson in 1740–4 and John Byron in 1764–6, but they really begin with the three voyages of James Cook, between 1769 and 1778. In the 1750s and 1760s, Cook rose through the ranks of the British navy, undertaking hydrological surveys in Quebec and Newfoundland. The excuse for his first scientific voyage was the 1769 transit of the planet Venus across the sun, a rare celestial event that allowed astronomers to determine fundamental astronomical quantities as well as to verify terrestrial longitudes. Cook took the *Endeavour* to Tahiti for the transit and continued with secret instructions to dawdle around the South Pacific in search of new land masses. Accompanying him as natural-history observers were Daniel Carl Solander (1733–1782), Linnaeus's apostle to England, and Joseph Banks, both fellows of the Royal Society. The wealth of information acquired in Cook's voyage led to a second tour, in two vessels. Circumnavigating the globe from 1772 to 1775, Cook definitively disproved the hypothesis of a large southern continent, and he brilliantly proved the practical, navigational utility (for determining longitudes at sea) of John

Harrison's portable marine chronometer. In a third voyage, beginning in 1776, he sought a northwest passage around North America from the Pacific Ocean.

Cook's tragic death in 1779 in Hawaii (which he was the first European to see), and the publication of the accounts of his voyages, led to a mania for scientific exploring. Joseph Banks, who served as Royal Society president from 1778 until his death in 1820, orchestrated the explorations. Banks supervised two generations of naturalists in foreign parts. Much of his interest concerned economic botany – transplanting and adapting cultivations from one climate to another. Under his patronage the Polish emigré Anton Pantaleon Hove (fl. 1787) roamed India in search of useful plants, and following his instructions the East India Company established agricultural gardens in India and on St Helena. It was Banks who chose William Bligh for the ill-fated experiment of the *Bounty* (the ship's name signalled its mission of transplanting breadfruit from Tahiti to the West Indies). Banks's commitment to moving plants and animals around the world finds graphic illustration in his promotion of Australia's Botany Bay as a convict colony.

Britain's longstanding political antagonist France, the land of Linnaeus's rivals Jussieu and Adanson, was slow to pick up on the rational reorganization of the world. There were some French voyages of discovery, notably Louis-Antoine de Bougainville's (1729–1811) circumnavigation in the 1760s and Yves-Joseph de Kerguélen-Trémarec's (1734–1797) southern expedition early in the 1770s, but Paris ordered nothing in the line of what Cook had undertaken. In 1785 Louis XVI finally appointed Jean-François de Galaup *comte* de Lapérouse (1741–1788) to head up a tour of the South Pacific. It was to be the grandest scientific expedition of all time. The *marquis* de Condorcet (1743–1794) at the Académie des Sciences supervised the plans, which ranged over all fields of knowledge. The explorers, setting out in two ships named *Boussole* (Compass) and *Astrolabe* to signal the expedition's peaceful and scientific purposes, included astronomers, physicists, and naturalists. Their instruments and books, including everything from editions of Linnaeus and Adanson to a complete set of the *Journal de physique* (the journal's editor Abbé Jean André Mongez [1751–1788] also shipped aboard) and the latest, thirty-six-volume Lausanne-Berne edition of the *Encyclopédie*, were valued at 17,034 livres. Lapérouse effectively directed a floating university.

The French exploring urge foundered and failed. Lapérouse sailed around the Pacific basin, sending home reports from various ports of call. In 1788 the expedition attempted a circumnavigation of Australia and dropped from sight. Two years later, the Paris Académie urged a rescue under Bougainville, but the vanguard of the French Revolution, the Constituent Assembly, delayed until finally authorizing one million livres for a search under Joseph-Antoine Bruny d'Entrecasteaux (1737–1793). The searchers blundered about the Pacific until, reduced by disease, they docked at Batavia on Java in the middle of 1793. To their great chagrin they learned about the French Republic, the trial of Louis XVI, and the state of war between Holland and France. As to Lapérouse, his ships apparently went down at Vanikoro in the Solomon Islands. It was a watery end for the count's crew who, in the course of their wanderings, had become thoroughly disabused of the notion of 'noble savages' proclaimed by naturalistic *philosophes* like Jean-Jacques Rousseau.

Aggressive expeditionary rivals of the English were, in the first place, the Russians, who beginning in the seventeenth century had progressively expanded their control over the northern reaches of East Asia. In 1721 Vitus Bering (1681–1741), a Dane sailing under Russian colours, navigated the straits that now bear his name. The St Petersburg Academy of Sciences published Pacific maps in 1758 and 1773. By 1784 the Russians had a permanent fur-trading post on Kodiak Island, and they pushed south to California in search of animal skins; on his last voyage, Cook in fact met with

Russian traders in the Pacific Northwest. Then from 1803 to 1806, Adam Johann Krusenstern (1770–1846) circumnavigated the globe in the *Nadezhda* (Hope), revisiting many locations charted by other explorers; the voyage was followed by the Antarctic expedition of 1819–21 of one of Krusenstern's officers, Fabian von Bellingshausen (1779–1852).

More sustained effort came from the Spanish Enlightenment, the *Ilustración*, under the reign of Carlos III. During the last half of the eighteenth century, the British were much concerned about Spanish presence in the Pacific, notably on the northwest coast of America, and British explorers from Cook to George Vancouver were sent to assert British control there. The voyage of Alessandro Malaspina (1754–1810), an Italian commissioned to explore Alaska, was a response to British and Russian voyagers and to the so-called Nootka Sound crisis of the late 1780s; Malaspina returned to the reactionary Spain of Carlos IV with an enormous collection of notes and analyses, which have remained unpublished to this day.

The Spanish also financed a series of ambitious botanical expeditions in their more secure possessions. One, led by Hipólito Ruiz López (1754–1815) and José Pavón (fl. 1800) under the supervision of Casimiro Gómez Ortega (1740–1818), the first director of the Royal Botanical Gardens in Madrid, surveyed useful plants in Peru and Chile between 1777 and 1788. A second, undertaken by Juan José Tafalla also from the Madrid gardens, surveyed the region around Guayaquil from 1799 to 1808. A third botanical mission of the Spanish Enlightenment ran in New Granada (the Panama, Colombia, and Ecuador of today) from 1787 to 1806 under the polymath José Celestino Mutis (1732–1808); its herbarium and library at Bogotá, assembled by a staff including forty painters, was according to Alexander von Humboldt one of the world's finest. And a fourth mission extended from 1787 to 1803 under Martín Sessé (1751–1803), appointed from Madrid to found a botanical garden in Mexico City. The Spanish initiatives of decades-long expeditions led, during the nineteenth and twentieth centuries, into the primary form of colonial scientific institutions: establishments dependent on the home country for staff and instructions.

Into the 1790s there was still nothing of the *vocation* of scientific traveller, for expeditions proceeded by favourable political or commercial conjunction rather than by a guiding strategy or overarching vision. Alexander von Humboldt provided the latter. Humboldt, whose father was a major in the Prussian army and whose brother Wilhelm founded the University of Berlin, received his early education with private tutors; then he went to a number of German universities and related institutions of higher learning. When he was twenty-one he travelled to France and England with Georg Forster (1754–1794), a disciple of Linnaeus who had been on Cook's second voyage. He worked as a Prussian mining inspector, touring mines and mountains throughout Central Europe and looking into a series of scientific questions, from plants in mines to galvanism. The death of his mother in 1796 provided him with the means to resign from the civil service and plan a 'great journey beyond Europe'. As preparation, he studied astronomy and surveying as well as exotic plants available in Europe.

In 1798, buoyed by a growing scientific reputation, he left Paris with Aimé Bonpland (1773–1858) for Marseilles, Madrid, and at last Venezuela. Until 1804 he and Bonpland travelled over Hispanic America, and Humboldt concluded with a short tour of Philadelphia and Washington. Humboldt noted everything and everyone he saw and collected 6300 new plant specimens. The Prussian native remained in Paris from 1807 to 1827, and he returned there for much of the 1830s and 1840s, producing his expensive and informative travel journal as well as a torrent of related writings on physical and human geography. He eventually occupied a sinecure in Berlin and travelled to Siberia. The enormous range of his scientific and political contacts, and his mastery of everything from higher mathematics to botany, provided him with a wide audience for his advocacy of geography and climatology. Only by amassing an enormous amount of information about the natural world, he believed, would humanity be able to divine its regularities and laws. His scientific programme is epitomized in his *Cosmos*, a popular scientific work appearing in five volumes at the end of his life, which presented everything from star clusters to nonflowering plants.

The Humboldtian ethos offered a clear justification for dozens of long-term Baconian research programmes, from establishing a global chain of observatories for recording terrestrial magnetism to maintaining zoological and botanical gardens. Humboldt's writings revived the Enlightenment scheme of Lapérouse (Humboldt was as much French as he was Prussian), where scientific expeditions measured everything they encountered; *Cosmos* became a touchstone for science in the last half of the nineteenth century. Humboldt had a direct impact on a wider range of people than any scientist since Newton. Those whose career he helped include the chemist Justus von Liebig, the physiologist Emil Heinrich DuBois-Reymond (1818–1896), and the naturalist Louis Agassiz. The English edition of Humboldt's travels fired the imagination of Charles Darwin when he was a student at Cambridge (it contributed to Darwin's desire to sail on the *Beagle*), and no less a figure than the young Albert Einstein was drawn to Humboldt's German prose.

Humboldt provided a standard and a centre of gravity for the proliferation of scientific exploring expeditions beginning in the middle third of the century. The very quantity of expeditions has led to a classification by name. The naval expeditions are known largely by their ships: the Russian *Predpriatine*, the French *Uranie* and *Astrolabe*, the British *Beagle* and the *Challenger* (the latter in deep-sea oceanography), the Dutch *Siboga* (in marine biology). Polar explorations, arriving by ship and crossing or surveying ice, go by their feckless leaders (the American 'Wilkes', the Canadian

'Franklin', the Norwegian 'Amundsen', the British 'Scott'). Tropical expeditions – smaller and less costly – follow leader and region (Alfred Russel Wallace's 'Malay Archipelago', Elie van Rijckevorsel's [1845–1928] 'Brazil'). Scientific expeditions had a political connotation, for they established priority of ownership and they accumulated intelligence useful to future invaders or administrators. The anthropological and geographical societies of London (and their counterparts in other nations) played a central role in nineteenth-century imperialist expansion. 'Permanent' scientific expeditions in colonies like Algeria and Indochina sought out exploitable resources. But the idea of a dispassionate scientific expedition nevertheless enjoyed such favour that it became a Trojan horse for political designs. Ethnologist Auguste Pavie (1847–1925) collected information essential for pacifying Indochina, and geographer Louis Gentil (b. 1868) performed the same service for the French army in Morocco. Superficially dispassionate science led to an apparently independent nation in Africa. In 1876 King Leopold II of Belgium created and funded a scientific exploring society, the International Association for the Exploration and Civilization of Africa. The association engaged the services of the widely travelled Sir Henry Morton Stanley (1841–1904). It then obtained diplomatic recognition in 1884–5 for the Congo Free State, which was absorbed as a Belgian colony in 1908.

Science in the nineteenth century is a story of accelerating and relentless specialization. By 1900, the days of the universally learned Humboldtian traveller had disappeared. Centrifugal forces pulled Humboldt's cosmos into specialty disciplines. The analytical earth sciences separated out into seismology, gravimetry, atmospheric electricity, and meteorology. Disciplines arose for special climates, for example tropical medicine and high-altitude physiology. Even large-scale expeditions increasingly limited themselves to particular tasks. Jean Abraham Chrétien Oudemans (1827–1906), for example, constructed geographical maps of the Dutch East Indies, while magnetical maps of the same lands went to a succession of colleagues working independently; the Dutch navy undertook its own soundings of coastal waters. Franz Boas (1858–1942) and Vilhjalmur Stefansson (1879–1962) focused on ethnology in the Canadian north, leaving geophysics to other explorers. Louis Bauer from the Carnegie Institution of Washington set out in a special naval vessel to survey terrestrial magnetism. Felix Andries Vening Meinesz (1887–1966) surveyed marine gravity on world-ranging submarines. Jacob Clay (1882–1955) and Robert Millikan circled the globe measuring cosmic rays at various latitudes, altitudes, and depths.

All this activity led to a proliferation of maps. Already at the end of the eighteenth century there were charts of magnetic deviations from true north and ocean currents (notably the Gulf Stream in the Atlantic Ocean, identified by Benjamin Franklin). The climatology advocated by Humboldt contributed further to this interest. Captain Ahab of *Moby Dick* (1851), we recall, compiled maps charting the natural history of whales. Daily newspapers produced maps of battlefields. Public-health officials followed the geographical spread of epidemics. Topographical maps early in the twentieth century featured language groups, climatic zones, and the picture of the world most familiar to us today – the daily weather patterns of isobars, cloud cover, and precipitation. (The metaphor of weather 'fronts', where cold and warm air masses battle for supremacy, drew inspiration from the experience of neutral Norwegian meteorologists observing the First World War.) By the middle of the nineteenth century, councillors of state sought precise, detailed inventories of their realm. National geographical services, often supervised by the military, crossed the length and breadth of nations and continents; hydrographical services charted coastal waters. Into the twentieth century, these surveys constituted a great reservoir of talent in the exact sciences.

The empire of mapmakers declined after the First World

War. In part they had accomplished their task – a topographical map, once constructed well, remains a monument to posterity. The finest maps of Indonesia, for example, are the ones produced by Oudemans in the nineteenth century. In part the mapmakers succumbed to technological innovation. Radio beacons broadcasting time signals obviated the necessity for mappers to carry a large collection of chronometers for field determination of standard time, and aerial photography produced instantaneous pictures of inaccessible terrain.

The twentieth century has seen a transformation of the general aspect of travelling. Use of coal and oil led to regular and unremarkable sea voyages; air transport dramatically shortened distances and opened up the parts of continental interiors that had not been served by railway. The passing of the American frontier, especially, gave rise to the notion of preserving wilderness. Eco-tourism, promoted by Theodore Roosevelt in his scheme for national parks, revived the notion of natural goodness, touted by the *philosophes* and bitterly rejected by Lapérouse and his fellow explorers. Automata now handle routine data-collecting – from self-recording weather stations and seismometers to satellite photography. There is increasingly little *scientific* justification for a researcher to spend a lifetime surrounded by the raw source of his observations.

Much essential travel information today derives from journalist-raconteurs. They paint a picture at once broad and intimate of people living with the consequences of unusual choices. In the academy of information to which we all belong, they are our foreign corresponding members. There is an entertaining side to the exotica they reveal. The erotic commentaries and translations of the diplomat and ethnologist Sir Richard Francis Burton (1821–1890) created a sensation in Victorian Britain. Great writers of fiction, from Robert Louis Stevenson to John Steinbeck, Antoine de Saint-Exupéry, and Graham Greene, have recounted trips to far-off places. Over the past century travelling has produced a hazy line between journalism and science. One traveller, anthropologist Paul Rivet (1876–1958), founded the Musée de l'Homme in Paris. Another traveller, the future senator from Connecticut Hiram Bingham (1875–1956), presented Machu Picchu to the world. Thor Heyerdahl (b. 1914), an amateur archaeologist, stimulated interest in South Pacific antiquity.

Extraordinary films of culture and nature, from *Nanook of the North* to *Dead Birds* to the natural histories of David Attenborough (b. 1926) and Gerald Durrell (1925–1997), have expanded our horizons. Yet what we find most arresting are the still photographs. The temper of an age is found less in pictures of atrocity (these have been with us from the beginning of civilization) than in moments of epiphany. Edmund Hillary (b. 1919) pictured atop Mt Everest in 1953 epitomizes a sense of conquest over nature. Less than a generation later, earthrise photographed from the moon radiates a strikingly different message of environmental fragility. Travellers reveal as much about where they are from as about what they see.

Counting: Statistics

You drive to work on a highway that crosses a river. It is spring. Will the bridge hold through the thaw? It is designed to withstand a fifty-year flood, determined by hydrologists who have measured sediment flow, velocity, and volume. The bridge is thirty years old. There is a good chance – an exact probability – that it will give way. You accelerate across the span, travelling 15 per cent over the speed limit. You think that this is as fast as you can go without being stopped for speeding – unless it is the end of the month, when traffic police need to boost their revenue.

The chances are that nothing will happen. The sentence almost defines the faith of people in industrial societies. The chances are that you will not contract a sexually transmitted disease; the chances are that the government tax office will not detect a bit of legerdemain; the chances are that you will not die from eating raw hamburger; the chances are that an asteroid will not impact on the Earth and extinguish human civilization. Modernity deals in probabilities. We apply cause and effect when all the boundary conditions are known; we hope that the boundary conditions will produce an outcome in our favour. The result of this statistical haziness is a central theme of the greatest twentieth-century novelists. Robert Musil's tale of Vienna on the eve of the First World War, *The Man Without Qualities* (1930–2), and Thomas Pynchon's vision of the California counterculture, *Vineland* (1990), teem with probable encounters and improbable occurrences.

Numbers descend from etherial abstraction when we

count objects and events, and when we use the sums to guide our life. Registers of births and deaths, numbers of religious converts, revenue from investments, manpower in state enterprises – these are some of the accounts that have informed our understanding of history and that continue to guide our actions in the present. In this chapter we look into the role of counting in shaping our society.

The odds

Calculating the odds, which each one of us does every day of our life, finds a referent in ancient games of chance, but it has always been made real by the presence of wagers. While it is possible to wager a cow against a knife (or, as modern lending institutions prefer, one enterprise against another enterprise), the exchange normally occurs on the level of money. Wagers are settled promptly (a gentle-man honours gambling debts first). Because the outcome of gambling tempts retribution from fate, the winner seeks to take possession of his good fortune immediately. Unlucky gamblers may lose the family estate, but so far as we have records, the usual wagers have been made in currency.

Mercantile undertakings remain at the mercy of war, weather, and disease, not to mention the market economy. Prudent investors seek high payoff from risky enterprises where the spectre of total loss is omnipresent; saner invest-ments carry lower yield. Just as the careful shopper scrutin-izes apples in terms of price and quality, so the investor ranks investments by their *probable* rate of return.

For an investor to undertake a calculated risk is not unlike divining, a signal feature of many civilizations. Chinese oracle bones, dating from before the Han, were thrown to divine the future – a practice finding elaborate codification in the permuted symbols of the *I Ching*, or Book of Changes. A toss of the dice to intimate fate persists as a theme in Western literature. Secondary-school students now study

Queequeg's oracle bones in Herman Melville's novel, *Moby Dick*. Clairvoyants, using tarot cards instead of dice, make a good living in many towns.

Civilization is an affirmation of faith against the odds. A familiar part of classical antiquity, for example, relates to drawing by lot – whether for sending a soldier into a dangerous mission, dividing a sacrifice among various gods, or assigning an inheritance; the Etruscan Museum of the Vatican has preserved for us an image of Achilles and Ajax throwing dice. Athenian democracy late in the sixth century BC was in fact predicated on choosing magistrates randomly. The executive government, the Council of Five Hundred, at any single time comprised fifty citizens chosen by lot (by means of white and black markers) from each of the ten enfranchised tribes; by lot, each group of fifty directed affairs of state for a tenth of the year, and lot determined a daily president. In the third century BC, Athenian mechanical ingenuity developed an analogue device for choosing randomly among alternatives. The practice seems rooted not in egalitarian impulse but rather in elitist preference for divine guidance. The gods, not the people, would determine the outcome, a principle that enjoined enquiry into the natural laws of chance. In his *Republic*, Plato defines democracy in part by random selection of rulers; in the *Laws* he insists on the divine character of random selection.

Nothing is more characteristic of Plato's affirmation than the uncertainty of seaborne commerce. It follows that marine insurance has a long pedigree. Demosthenes described the form that it took in Athens, since known as bottomry or respondentia bonds. A merchant received money in anticipation of his ship's safe return. The money would be repaid with interest in the event of a successful voyage; nothing would be repaid in the event of failure. Bottomry received no formal encouragement in medieval Islam, where moneylending entailed censure. Lending contracts could be upheld, nevertheless, by non-Muslims, and documents signed with a Believer in infidel lands carried

authority in the *Dar-al-Islam*, the world of Islamic civilization. The practice of bottomry never left the Mediterranean.

The fourteenth century saw marine insurance flourishing
in the Low Countries, and it engendered a sophisticated,
European-wide system of credit and commerce. Evidence
comes to us in legal actions undertaken by merchants in
Flanders and in profit and loss accounts of insurance policies
from the archives of the Strozzi Bank in Naples. As historian
Alfonso Leone has shown, Naples provided maritime
insurance for the trade in cloth and wine that Florence
organized between Venice, Galicia, and England. The
insurance premiums, debited to the insured on current
account, ranged between 5 per cent and 16 per cent, in
addition to duties and fees. Florentine merchants used the
debit to cover risks involved with northern trading. To dilute
the risk, typically the contracts were shared among individuals, or individuals shared in a number of contracts.
More than half of the insurers on the Naples market were
Catalans, eager to invest trade surpluses in a profitable
enterprise. Insurance provided the mechanism of long-
distance trade. In Leone's words: 'The Perpignan cloths
which were the principal Neapolitan import from Catalonia
financed the cost of insuring the transport of English cloths
to Pisa.' The system of insurance underwriting, later to
assume large proportions in Lloyd's, is evident here, just as
historian Charles Farley Trenerry showed that it was
accepted in Flanders about 1400.[1]

Accommodation to temporal uncertainty expanded during the formative stages of European capitalism. Provision
against adversity built upon insurance practices in marine
commerce and revenue-generating mechanisms in civil
annuities. It expanded to cover destruction by conflagration
– although here the early insurers were often a fire-fighting
service. Beginning with the seventeenth century, there was
sufficient statistical evidence – notably in mortality tables
– to allow for a first-principles construction of insurance

premiums and payoffs. The insurance industry, however, made no use of actuarial science, even as the elements of probability took form during the eighteenth century by such authorities as Gottfried Wilhelm Leibniz, Abraham De Moivre (1667–1754), Marie-Jean-Antoine-Nicholas Caritat *marquis* de Condorcet, and Jean le Ronde d'Alembert (1717–1783). Insurance remained a gamble, and clever insurers bet on the basis of diverse, accumulated experience, using the tacit knowledge displayed by judges today in civil proceedings.

The uncertainty of life itself has produced a long history of attempts to achieve income stability in the form of annuities. The third-century Roman jurist Ulpian (d. 228), for example, described annuities sold by the state. For a flat fee, one could purchase a lifetime Roman pension – a practice that revived under strong European states a thousand years later. Beginning early in the thirteenth century, for example, Amsterdam sold annuities to raise money. The naïve Amsterdam annuity rate, like Ulpian's, took no account of the age of the purchaser.

Annuities were seen as investments, to be repaid with interest, but true gambling with lives – for example, betting on the life of a king or pope – repelled civilized Europeans. For this reason, except in England and Naples, life insurance was generally outlawed by the seventeenth century. Life insurance clearly distinguished itself from social assistance offered by guild or sect, which provided restitution for misfortune (including death of a wage-earner) from a members' fund of regular assessments.

The limited effect of high statistical learning on mundane affairs is seen in the history of income-security associations, a form of bourgeois investment that, beginning in the last decades of the seventeenth century, mushroomed far beyond its origins in late medieval mutual-aid associations and guilds. The new associations normally required investors to deposit a substantial entry sum in addition to yearly premiums; survivors received handsome annuities. From the Netherlands and England, the societies spread rapidly throughout Europe. For nearly a century the societies attracted subscribers. Their solvency generally depended on the uncertainty of investing premiums in government bonds. Their premise was quite innocent of statistical evidence, notably mortality tables.

Historian J. C. Riley has observed that the income-security boom correlated with the popularity of the tontine, an invention of Lorenzo Tonti whereby subscribers each pledged a sum to go to one or several survivors. The rise of these associations related, Riley contends, to new middle-class attitudes toward young and old people – a primitive attempt to rationalize security. The unsophisticated associations generally paid out overly generous annuities, and for this reason they came to a bad end. But as in any pyramid scheme, a lucrative payoff attracted investors. The manifest insolvency of the unmathematically grounded associations, pitilessly exposed in the 1770s by Dutch and English writers, contributed to their decline. Demographic trends also had a role to play. Beginning late in the eighteenth century, people were living longer – a trend that would wreak havoc with annuity schemes. But just this trend was a boon for the life-insurance industry. Richard Price's (1723–1791) mortality tables, constructed from the records of mid eighteenth-century Northampton, were used well into the nineteenth century and guaranteed the fortunes of British life assurers.[2]

The sequestration of high learning is not an unusual circumstance. We are led to imagine, from Max Weber's (1864–1920) enunciation of the Protestant Ethic and its instantiation in Benjamin Franklin's scheme for fire insurance, that eighteenth-century savants ransacked the storehouse of rationality for items of practical utility. But in fact, practical innovations generally derived from practical people: the steam engine did more for science than science did for the steam engine. The First Industrial Revolution – the steam engine, iron and coal, mechanized weaving, canals, and ultimately railways – was the province of auto-didact tinkers, and (as in the story of probability) the tinkers did not turn their talents to every domain. The hardware used by the military, whether sabres, muzzle loaders, ordnance, ships of the line, or earthenwork fortresses, changed hardly at all between the Glorious Revolution of 1688 and the Congress of Vienna of 1815. Napoleon's *chasseurs* would not have been out of place in the time of Charles II. Nor did the restorative powers of medicine and surgery differ much between Blenheim and Waterloo. Weather was predicted no more accurately during the French Restoration than it had been during the early days of the Royal Society. The vaunted rationality of the Enlightenment, like the long dominance of the Baroque style, contained a great deal of unreasonable prejudice, and nowhere more so than when dealing with probabilities.

Precision and numbers

Enlightenment reason was a creative mix of ideal prescriptions and tacit prejudices; this is certainly how differential and integral calculus found a place in real-world calculations at the hands of 'mixed' (we would say today, 'applied') mathematicians. The promoters of rational mechanics – Leonhard Euler, the Bernoullis, and ultimately Pierre Simon *comte* de Laplace – provided a handy form for Newtonian mechanics, and they showed how Newtonian principles could derive solutions for many empirically verifiable situations, from the motion of planets in the ether to the properties of liquids in jars. The verification required familiarity with a vaster complement of mathematics than seventeenth-century natural philosophers had known.

Numeracy, the counterpart in mathematics to literacy in language, arrived on the European agenda during the middle of the sixteenth century, in the generation after Copernicus. It took nearly four centuries after Leonardo of Pisa (known as Fibonacci, ca. 1170–ca. 1240) brought

Arabic numerals to his colleagues, in 1202, before merchants, sailors, and surveyors calculated preferentially with
the 'rules' of arithmetic taught to schoolchildren today.
Merchants and tax collectors calculated with counters or
mnemonics – much in the way that the Mediterranean
world used the abacus and Central Europe (until recently)
used knuckles – as they navigated in an extraordinarily
diverse sea of weights and measures. This diversity in reck-
oning impeded neither commerce nor straight thinking.
There is no internal contradiction in thinking about birth-
weight in pounds and ounces, distance in kilometres, land
in acres, soil in cubic yards, wallpaper in double rolls, liquor
in fifths, wine in litres, and temperature in Celsius. Simultaneous use of independent systems of measurement
enriches language and supplies metaphor, just as the possibilities of expression are expanded by daily exchanges in a
bilingual environment.[3]

Even after the rise of Baconian inventory and the triumph
of the exact sciences during the seventeenth century,
numeracy spread slowly. Samuel Pepys (1633–1703),
appointed clerk to the Admiralty, woke up early in the
morning to learn multiplication tables. Benjamin Franklin,
a well-read autodidact and promoter of practical rationality,
esteemed mathematics but never mastered calculations; his
electrical science was quite entirely descriptive. Where
exactness mattered was in the parcelling out of land. George
Washington and Thomas Jefferson had first-hand experience with surveying. This lucrative occupation extended the
possibility of employing new techniques and calculations
from astronomy, although even in enlightened America
astronomical surveyors were in short supply. The Mason-
Dixon Line, it is significant to note, takes its name from the
labours of two Englishmen during the 1760s.

Patricia Cline Cohen has contended that numeracy spread
in eighteenth-century America less as a correlate of commercial activists (and none were more industrious than the
readers of Franklin's 'Poor Richard' maxims) than as a

decline of fatalism and uncertainty.[4] Enlightened thinkers loosened the grip of prescriptive theology as the natural world, in all the stunning profusion revealed by overseas exploring expeditions, opened new vistas for enumeration. Inventory – the accumulation of scientific capital – was necessary for synthesis, but nature could also be mastered through mathematical reason. Mathematics, from slipshod differential and integral calculus through rule-of-thumb arithmetic, demonstrated control and prediction.

Surveying and statistics

The spirit of seventeenth-century natural philosophy became modern physics largely through the search for accuracy in triangulation, especially the measurement of arcs down to a second or less. This passion for precision was for the most part innocent of what carries precise meaning today: error bars and significant figures. Reading through hundreds of pages of measurements and calculations from the late eighteenth century, we are struck by the long tail of digits unreasonably ascribed to physical measurements. The ten-thousandth and hundred-thousandth part of a measurement seem included to punctuate the precision-fetish of the writer. In the hands of mathematical adepts, differential and integral calculus provided an apparently precise structure for derivations from Newtonian physics. Terrestrial measurement came into line with the ideal harmonies of mathematics, just as astronomers had attributed imaginary precision to their calculations for a thousand years.

Belief in the power of measurement produced confidence in the explanatory power of statistics, from mortality tables (for the purposes of insurance and medical inoculation) to temperature and barometric pressure (for weather) to the value of imports and exports (to support one or another perspective in political economy). The prescriptive elaboration of statistics in the work of polymath geniuses like Laplace and Gauss may be seen as evolving naturally from this tradition. What strikes a modern eye are the special areas of enquiry that absorbed heavy statistical machinery, to which we now turn.

First to the skies. We have seen that mathematical precision originated with stellar measurement, not with terrestrial surveying or assaying, so statistical means and norms found their clearest expression in the stars. As high-quality lenses without chromatic aberration found their way into the hands of hundreds of astronomical enthusiasts across Europe, a cascade of celestial discoveries rolled off academic presses. Interest focused, in the afterglow of Newton and Descartes, on the incidental and ephemeral, Charles Messier's (1730–1817) youthful search for comets being typical of the genre. Sharp images from powerful lenses produced chronicles of heaven's structure, from nebulae and stellar clusters to the small planets called asteroids. The immediate result is found in the mature Messier's and William Herschel's star catalogues.

Inventory of the routine, not discovery of the spectacular or particular, is the cleverest dodge to secure long-term funding for a scientific establishment. At the end of the eighteenth century the world had more than a hundred observatories, the overwhelming majority of which were small, precariously funded operations run by an astronomer and his assistant or wife. These intimate teams directed much attention to things beneath heaven: surveying, navigation, meteorology, and the time of day. By the nineteenth century, star-catalogue inventories, the great *Durchmusterungen* that continue into our own time, supplied rationale for maintaining large staff at state-funded observatories. Even the Paris Observatory, for more than a century a family guild specializing in terrestrial surveying, metrology, and experimental physics, eventually turned much of its attention to the stars under Urbain Jean Joseph Le Verrier. Grand observatories housing telescopes unsuited for practical tasks like time-keeping became a sign of high civilization. Copper-sheathed domes, mirroring the vaulted heights of the Hagia

Sophia or St Peter's basilica, sprouted from Pulkovo (near St Petersburg) to Berlin, Leiden, and Washington. New states – Mexico, Ecuador, Argentina - paraded their noble intentions by mounting large telescopes to scan the stars.

Statistics gave reason to these enterprises. Just as the published position of a star was a mean value, so sundry mean values could be averaged for a best probable position. Rather than have the worth of an observation depend on the character of the observer, good and bad observations alike could contribute to scientific progress. Discussion of systematic error, whether slippage in micrometer gears and pendulum-clock trains, the aberration of lenses, or the physiological time-delay between observing and signalling a celestial transit (by voice or by hand), came to dominate the introductory pages of the observations published by both major and minor observatories.

Terrestrial means

The incorruptible stars have always appealed to bright young minds. They seduced Lambert-Adolphe-Jacques Quetelet (1796–1874), a poet, writer, and mathematician who in the early 1820s was one of the leading scientific lights of Brussels. He persuaded the United Kingdom of the Netherlands to set him up in an astronomical observatory. He travelled to Paris for intercourse with great savants and to England for purchase of equipment. Catholic, industrial Belgium gained bloody independence from the Protestant, agrarian north in 1830. Astronomy was so potent a symbol of civilization that the new Belgian monarch, Leopold of Saxe-Coburg (widower of Princess Charlotte of England), retained the commitment of his rival, King Willem of the Netherlands, to install Quetelet as the stargazer of Brussels. Quetelet, like Leopold married to a subject of France, took up residence at his observatory in 1832.

In astronomy Quetelet perceived the power of statistics. From his youth he had experienced the destructive wrath visited by hundreds of thousands of 'ordinary' soldiers. From his early professional career he saw the injustice of autocratic rule. And from his romance with things French he fell under the spell of social reformers like François-Marie-Charles Fourier and Claude Henri de Rouvroy *comte* de Saint-Simon. As his new state's most prominent academic pensioner, he devoted his life to promoting statistics in the service of human rule. Inspired by the notion of 'social physics' proposed by French philosopher Auguste Comte, Quetelet sought to discover the natural laws that governed the social life of the mean, normal, or ordinary specimen of humanity. His instrument was data on commerce and crime, births and deaths, imports and exports, tons of coal, and kilometres of canal.

From having considered the metric system, we see that the reception accorded a new doctrine depends on the slowly turning wheels of commerce and custom. There is striking self-reflexivity in the success of Quetelet's social statistics, for what he wished to discover were just the trends that guaranteed his fortunes. New imaginative literature, from Jane Austen to Gustave Flaubert focused on the large mass of bourgeois citizens – their aspirations, doubts, and pleasures. New theology, from the American John Smith to the Englishman William Wilberforce, urged normal people to take charge of their life and devote themselves to good works and pious thoughts. New wealth, generated by men like Werner von Siemens and Andrew Carnegie, extended the promise of social prominence and ease to millions of normal labourers. Norms were in, exemplars out. The society of mass action had emerged, and like the heavens it was waiting to be enumerated.

Social urgency brought a new spin to statistics. In Belgium and in the British midlands, the First Industrial Revolution climaxed in coal pits, iron foundries, and textile mills. Canals and eventually railways facilitated the inland circulation of bulk goods. European society, already stirred up by the convulsions of revolutionary France, was in the process of forming itself into new bodies politic. The twilight of the Baroque, with its persistent focus on timeless harmony and balance, faded into the Romantic, a plaintive cry in favour of the ego and decisive, individual action. The equiponderate tool of Enlightenment science, the chemical balance capable of differentiating one part in one-hundred-thousand, was eclipsed by the irreversible machine of power and pollution, the steam engine.

Statistics physical and social

Auguste Comte's social physics, as elaborated by Quetelet, extended to the entire living world through Charles Darwin's notion of the normal member of a species. It seized the imagination of mathematicians at Cambridge and Glasgow who awakened British physics from its long, post-Newtonian slumber. And it inspired James Clerk Maxwell's understanding of probability and statistics in the kinetic theory of gases.

The reflection of social physics in a framework for understanding all living creatures is not difficult to identify. It is Darwin's elaboration of natural variation from the norm in the traits of a species. Darwin, however, was far less sophisticated mathematically than Comte, and the force of his evolutionary argument proceeded by prodigious example and masterful prose. Maxwell shared Darwin's mastery of language, but to this he added an incomparable facility in expressing new concepts in symbolic logic.

The core of Maxwell's theory of gas molecules in motion is the bell-shaped curve that indicates, for a given temperature, how many molecules have various velocities: molecules cluster around an average velocity, although they also exhibit extreme departures from it. Maxwell appropriated a model of critical importance for insurance firms. By doing so, he at first gave no thought to formulating a fundamental uncertainty or indeterminism for natural processes. According to his early view of things, although many laws of nature took statistical form, individuals were not governed by them. Later, however, Maxwell came to believe that predicting certain kinds of result might be entirely beyond human capability. If so, there could be no acceptable proof that what happened on the molecular level derived from the laws of mechanics; hence, statistics – sophisticated counting – provided the only possible scientific explanation.

Doctrine of certainty

Maxwell's Austrian counterpart Ludwig Boltzmann, a resolute seeker after observations of the atomic structure of matter, worried about paradoxes arising from his elaboration of Maxwell's velocity-distribution law for gases. In particular, he wondered how the Second Law of Thermodynamics could allow a large quantity of gas molecules – whose individual collisions were entirely reversible – to evolve in an irreversible way, say, by losing energy overall. It has been suggested that rejection of his statistical research contributed to Boltzmann's suicide in 1906, almost on the eve of direct confirmation through the observation of Brownian movement by Jean Perrin (1870–1942). But Boltzmann and the generation of his younger admirers – from Albert Einstein to Paul Ehrenfest and David Enskog (1884–1947) – were resolutely classical determinists.

The search for deterministic laws extended to the study of life. It is apparent in nineteenth-century epidemiology, where the incidence of disease was tracked by statistical correlations. Statistical surveys became an essential tool in understanding the dynamics of disease and the laws of contagion, from William Farr's (1807–1883) analysis of cholera in 1840 to Ronald Ross's (1857–1932) investigation of malaria in the 1890s. Edwin Bidwell Wilson (1879–1964), a physicist-turned-epidemiologist, became an apostle of statistics in the study of population. A wide-ranging student of Yale physicist Josiah Willard Gibbs's (and one of the first Americans to publish on relativity theory), forty-three-year-old Wilson became professor of vital statistics at Harvard in 1922. From this post he applied classical, deterministic statistics to many areas of biology, notably the matter of inferring true values from averages and establishing criteria for the effect of therapeutic agents.[5]

The deterministic rigidity of nineteenth-century statistical innovators is apparent in the early biometricians. Francis Galton (1822–1911), Karl Pearson (1857–1936), and Ronald Aylmer Fisher (1890–1962), all members of the English intellectual aristocracy, passionately counted and correlated everything, whether the frequency of coughs in a concert audience, physical attributes of diverse species, or, most importantly, intellectual capacity of people in various English social classes. Through unflagging labour, the three counters and their disciples developed sophisticated statistical machinery to transform and thoroughly mathematize the biological sciences. They called their enterprise *biometrics*. Multivariate and regression analyses and tests for correlational significance found ready takers among psychologists, sociologists, and naturalists, all eager to establish causes or origins for observed phenomena in a way that transcended obvious prejudice and ideology. Their work played a major role in the rise of population genetics during the 1920s and 1930s and the correlative rise of a genetically informed Darwinian renascence. Notwithstanding the racist current in the thought of these men, the second half of the twentieth century has come to apply their formulas almost uncritically to demonstrate such propositions as 'smoking causes cancer' and 'eating animal fat leads to heart disease'.

Twentieth-century uncertainty

The biometricians *worried* about social trends. This sense of the word worry, meaning to be anxious about or to fret over future events, emerges clearly in the nineteenth century. It derives from an older notion of worry that meant physical harassment or vexation over importunate behaviour.

Writers like Charles Dickens, Louisa May Alcott, and Thomas Hardy portrayed people worrying about what will happen. Just at the time that security seemed at hand due to life insurance, the mass production of consumer goods, and rising standards of living, people began to manifest introspective concern over life's uncertainties. When the future was rationally provided for, the European mind began to focus on possible departures from the probable norm, whether variations in interest rates or disease. This rising sense of worry about the future correlates with a dramatic decline in European fecundity. For the first time people seem to be worried about providing for their children.

Uncertainty is fundamental to worry, and both have been central to the twentieth century, which W. H. Auden called, in a collection of poems published in 1947, *The Age of Anxiety.* In literary and philosophical circles, the years leading up to the First World War saw increasing doubts about classical determinism and attendant notions of social progress and order. These doubts are apparent in the writings of Rudyard Kipling and Anatole France. They are a leitmotif of the antirationalist German *Jugendbewegung,* or Youth Movement, as well as the energetic avant-garde movement of Paris. The desire to infuse spontaneous feeling into cold reason figures in the writing of Henri Bergson, Rudolf Eucken, Maurice Maeterlinck, Rabindranath Tagore, and William Butler Yeats, to mention only Nobel laureates in literature in the years up to 1927. And it is in 1927 that Werner Heisenberg formulated the uncertainty relations. These were essential restrictions on knowing that led into a new probabilist epistemology, the so-called Copenhagen interpretation of quantum mechanics developed by Niels Bohr.

We have previously seen that the uncertainty relations, which put an end to classical determinism, occasioned almost no opposition in Weimar Germany, where they were formulated and published. Physicists and mathematicians embraced them as an acceptable outcome of serious research. This reception is remarkable in the light of other, apparently revolutionary developments like Copernican astronomy and Darwinian evolution, against which people with traditional views mounted sustained criticism. The explanation, Paul Forman has contended, relates to the nature of Weimar society. The end of the First World War produced intense feeling against linear modes of thought and mechanical models, the spirits that had guided Germany's extraordinary rise during the Second Industrial Revolution. Determinism fell casualty to Germany's defeat in the war. Pessimistic and spiritualist sentiments suffused popular thought, from Oswald Spengler's 'decline of the West' to Hermann Hesse's celebration of romantic irrationalism in works like *Siddhartha* (1922) and *Steppenwolf* (1927). Physicists and mathematicians responded to the general spirit of the times by embracing it. In effect they claimed, 'We, too, are indeterminist.' And they added, 'Now give us money.'

Since their formulation in 1927, the uncertainty relations have functioned as something of a general interdiction. Like Newton's Third Law of Motion and the Second Law of Thermodynamics, uncertainty has been invoked to deny an outcome in principle. An engineer may say that a design for a train of gears and hammers does not take into account action-and-reaction; a patent examiner may reject a proposal out of hand because it purports to generate more energy than it expends; and a physicist may caution that an experiment cannot succeed because it proposes to establish cognate variables too precisely. The probabilistic uncertainty relations, however, gave rise to an indeterminist epistemology. Mind re-entered the physical universe through chance. God became the croupier at a great gaming table.

With the advent of quantum mechanics and their indeterminist epistemology, physicists set off to explain everything in sight though quantized lenses. The task became a Herculean labour. Exact solutions to probabilistic equations of motion (either in the matrix calculus of Max Born [1882–

1970] and Werner Heisenberg or in the more traditional wave-mechanics format of Erwin Schrödinger [1887–1961]) could not be obtained for any except the most elementary situations, certainly for nothing resembling the complex nature of nearly all atoms. Approximations and limiting conditions were the order of the day, and it was said that a clever physicist was someone who arranged parameters so that a complicated equation never had to be solved. Candidates for quantization after the laws of motion were the great system of electrodynamics, gravitation, and statistical mechanics. Quantization came to electrodynamics only at the cost of monstrous, *ad-hoc* assumptions, notably the requirement of neglecting diverse, infinite quantities for no apparent reason. Gravitation, as represented until the last third of the twentieth century by the continuous field equations of Einstein's general relativity, persistently resisted becoming discontinuous (as a quantization would require).

Statistical mechanics was apparently the easiest system to quantize – one was, after all, concerned with nothing more than a large number of particles in motion – but the enterprise entailed substantial problems. Since the time of Newton exact solutions to the so-called many-body problem required felicitous approximations. Furthermore, twentieth-century physicists continued to debate the root concepts of statistical thermodynamics: entropy, state functions, and irreversibility. Indeed, statistical mechanics has been unique among branches of the exact sciences in the way that modern practitioners continually appeal to the notions of their nineteenth-century predecessors. Just as the thorny problem of thermodynamical irreversibility was coming under control, basic statistical notions drew fire from writers trying to understand living processes.

Average lives

By the middle third of the twentieth century, life was held to possess a fundamental secret that could be discovered by scientific analysis. The search owes much to materialism – whether in its Marxist or its positivistic guise – which had all but driven spiritualists outside establishment laboratories. One avenue of attack was microcosmical and reductionist, applying techniques from physics (such as X-ray diffraction photography) and biochemistry to elucidate the chemical structure of genetic material. Twenty years of varied enterprise led to the DNA helix, published by Francis Crick (b. 1916) and James Watson (b. 1928) in 1953, a discovery that took nearly two generations to produce practical results in the treatment of disease. A second avenue of attack on the secret of life was macrocosmical and synthetic. It was population dynamics.

The statistical behaviour of large populations finds a nineteenth-century origin in both the biometrics of Karl Pearson's English biometrical school and, more generally, the statistical forensics of epidemics. By the first decade of the twentieth century, in response to thermodynamical worldviews like the one offered by the German chemist Wilhelm Ostwald (1853–1932) and called energeticism, naturalists and sociologists alike talked about population 'energies' in an attempt to bring mathematical precision to Auguste Comte's dream of a 'social physics'. Around the outbreak of the First World War, Charles Christopher Adams (1873–1955) sought to elaborate ecological analogues to Henri Le Chatelier's (1850–1936) criteria for equilibrium in chemical reactions. Vilfredo Pareto (1848–1923), whose lectures so inspired the young Benito Mussolini, infused complex mathematical notation into his thermodynamics of society, where people were simply molecules of various racial sorts possessing innate drives leading to ideologies.

It is difficult to underestimate the popularity of this kind of analogical transference – the construction of a worldview (*Weltanschauung*) from a specialist picture of natural processes (*Weltbild*), against which both Max Planck and Max Weber polemicized. Impressionable minds, like Mussolini's, were overwhelmed by the analogy. Consider the reflection among the Brahmins of Boston: Henry Adams's dynamical laws of history; historian of science George Sarton's flirtation with eugenics; and biologist Lawrence J. Henderson's (1878–1942) thermodynamically inspired ecology.

By the 1920s, the students and followers of Karl Pearson were seeking universal laws for population growth. Raymond Pearl (1879–1940), for example, a prolific populationist at Johns Hopkins University, imagined that fruit flies, Plymouth Rock chickens, and people all obeyed a law whereby fecundity decreased with increasing population density. Pearl's belief in the truth of correlation statistics led him to conclude that people with tuberculosis had a lower than average probability of developing cancer, and he began a programme to infect cancer patients with the reputed palliative disease. The physicist-turned-demographer E. B. Wilson fingered Pearl as a charlatan. Alfred Lotka, an associate of Pearl's and an actuary at Metropolitan Life Insurance Company, was more eclectic in his searching and less certain about codifying the correlations that he found as laws. He was both gratified and chagrined when the Italian mathematician Vito Volterra (1860–1940) formulated differential equations to describe simple ecological systems.

Thermodynamical analogies, and the differential equations used to describe idealized worlds of predators, prey, and parasites, assailed the laboratories of researchers at work to control insects and vermin that damaged crops. By the 1950s, economic entomologists and botanists had come to appreciate the promise of the Lotka-Volterra equations for ideal ecological systems. Ecologists struggled to turn Ronald Aylmer Fisher's bible for bean-counters – *Statistical Methods for Research Workers* (1925) – to serve their enquiries into adaptation and speciation. Within twenty years the cutting edge featured appeals to nonlinear, stochastic mathematics, as well as to the statistical mechanics of irreversibility. Information theory, the natural child of early twentieth-century eugenics, returned to inform the study of ecology and evolution.

The popular triumph of averages

Probability and statistics today underpin nearly all sciences of the field and the laboratory. Early in the century they invaded both macrocosm and microcosm. Is a source of light in the sky a double star? Is a radioactive atomic nucleus prone to disintegrate? Statistics provided the answer. The statistical approach has extended to meteorology and genetics. Will it rain tomorrow? Will your unborn child develop cystic fibrosis? The answers are given probabilistically. Predictions are made in percentages. When you visit a physician and ask, 'Doctor, how long can I expect to live?' the answer comes: 'For someone of your age, sex, and weight, and with your vital signs, one reads the years and months from this curve.'

We see here the popular triumph of Niels Bohr's indeterminist epistemology, advanced in the aftermath of Werner Heisenberg's uncertainty principle. At the centre of Bohr's worldview was the Principle of Complementarity – the notion that parts of the subatomic physical world could behave simultaneously like particles and waves. It made no sense to ask what the world was really like; whatever probabilistic quantum mechanics revealed was reality. Bohr admonished that we cannot hope to know all parts of the world with an ideal precision, like the one behind Julien Offray de La Mettrie's (1709–1751) Enlightenment notion of life as a collection of pumps, sluices, and alembics. Knowing, in Bohr's view, could not be established from fantastical representations of the world – and classical mechanics counted among these fantasies. Knowing did not prefigure doing. Rather, it derived from doing. The essence of knowledge related to the practical question 'how', not to the spiritual interrogative 'why'. How did the experiment work? How many times was the experiment run? How many hits and misses did one obtain? The answer, 'Nine times out of ten the electron will appear here,' has become, following this view, the finality of science.

Bohr promulgated his new epistemology through a committed band of acolytes, who spread the doctrine of the 'pope', as Bohr was sometimes known among younger physicists. Bohr's complementarian missionaries included Pascual Jordan (1902–1980), who applied indeterminism to biology and arrived at an affirmation of the Soviet doctrine of Lysenkoism, according to which acquired characteristics could be inherited from one generation to the next. Jordan also contended that a vital principle, or teleology, overrode mere statistical probability in living organisms, and he used probabilistic quantum mechanics to suggest how an adaptation might be transmitted to an offspring. Then Jordan and Bohr took on psychology, proposing an indeterminist solution to the problem of free will.

To many people – notably nearly all British and American physicists, some Continental physicists (Erwin Schrödinger, Louis de Broglie [1892–1987], and Albert Einstein), and psychologists and biologists generally – the Copenhagen foray into philosophy seemed vacuous or even, in Einstein's words, 'reprehensible'. Jordan, in fact, used the Principle of Complementarity for more than deriding traditional canons of reason: he became a Nazi and wrote admiringly about Adolf Hitler's plan for national renewal. Philosopher Bernhard Bavink also saw complementarity as an essential foundation stone for Nazi race science.[6] Jordan's excursus into quantum biology and human destiny found willing admirers, for example the Italian mathematician Luigi Fantappiè (b. 1901).[7]

Bohr laboured mightily to repair the damage, insisting that his epistemological approach was a research programme proper to physics, but it continually spilled over into other realms, and mind was one of them. For about a century beginning around 1850, following the rage of reductionist materialism, teleological explanations were considered indecent among professional scientists. Bohr's new epistemology allowed volition, a demon that had been exorcized by the brilliant nineteenth-century mechanists, to re-enter laboratory and academy.

British mathematical physicist Arthur S. Eddington wrote in 1928 that 'religion first became possible for a reasonable scientific man about the year 1927', the year of Heisenberg's uncertainty principle.[8] There was no shortage of people who agreed. In France, where Darwinian evolution waited a hundred years to find a firm place, many naturalists and philosophers embraced Lamarck's teleological, vitalist notion of speciation; mind re-entered the natural world in Henri Bergson's creative evolution and Pierre Teilhard de Chardin's religious destiny of humanity. Science was never an obstacle to devout believers of religious doctrine, and in the wake of statistical uncertainty, revealed knowledge once more entered the court of scientific opinion. If correlations could reveal general laws, as population geneticists were claiming by the middle third of the twentieth century, there was no contradiction in asserting the existence of intangible forces – whether or not these originated in Deity and doctrine – to equilibrate the world of living things.

One need only survey the political success enjoyed by various 'Green' movements over the past generation, many of which infuse the Earth with a consciousness, to appreciate the gap separating the existential angst of the 1950s from the goal-directed spiritualism of the 1990s. Whereas organized religion suffered stunning reversals at mid century, knowledge based on divine revelation is today resurgent everywhere. Anthropologist Gregory Bateson (1904–1980), one of the prophets of the new deism, did not avoid seeking allies among devout religious believers. That many of the scientific spiritualists took inspiration from

Oriental religion lends support to Arnold Toynbee's observation that the twentieth century may be remembered as the time when Buddhism entered Western civilization.

It is a synergism that would not have been excluded by John Theodore Merz, advocate of synthesis and promoter of a pan-European approach to science. Early in the twentieth century, Merz concluded his treatise on general approaches to knowing nature with a discussion of the 'statistical view of nature'. He wrote:

> In spite of the wonderful increase of scientific knowledge and the general diffusion of scientific thought in the course of the century, uncertainty is still the main and dominant characteristic of our life in nature and society; the atmosphere and climate of each are as fickle and changeable, as incalculable and unreliable, as ever.

Statistics attempted to dispel uncertainty, but in the end only philosophy and religion would do the trick.

> Those thinkers who in the nineteenth century, as well as in former ages, have dealt exhaustively with these the most abstract and highest conceptions of which human thought is capable, have not been, or have only very rarely been, led to their inquiries from the side of purely scientific interests; they have approached them with a full appreciation of the great moral and religious interests which lie hidden in the deeper significance which we attach to the words.[9]

The law of averages has given rise to moral debates about questions of life and death. How shall public money be invested in education and social welfare? Who shall be tested for one or another 'genetic' disease? Who shall be carried on a group insurance policy? Statistics, deriving from a passion for counting objects and events, may well have precipitated a late twentieth-century revival of religious sentiment.

Killing: Science and the Military

In the modern military-scientific complex, we see the twin incarnations of Athena – goddess of war and avatar of wisdom. Among the cultures of the West today, the advancement of knowledge has become inseparable from the destruction of ways and means of knowing. Whence the affinity between research scientists and professional killers?

Let us consider the matter of killing. A dispassionate reader may ask whether soldiering is not an honourable profession, whether generals and admirals, resplendent in their costumes under television lights, are not simply one form of specialized manager. There is much precedence for military conquerors acting as regents or functionaries, from the times of Alexander the Macedonian and Wang Mang to Francisco Franco, Henri-Philippe Pétain, and Dwight David Eisenhower. Indeed, the training for military command has often included a large dose of polite culture. The *epheboi* of Hellenic antiquity and the great schools of Han China both produced gentlemen who became, in due course, generals. For such commanders, ordering an army to execute a flanking manoeuvre was of a piece with undertaking an irrigation project or hunting wild animals. Military leaders have been schooled as gentlemen down to the middle of the twentieth century. Today's soldier-administrator is certainly heir to the ruling-class customs of diverse societies, but his authority derives from more than familiarization with the classics or drawing-room etiquette. Military authority derives from a decisive ability to kill. In the modern world, the ability is manifest most clearly in the firearm, to which we now turn.

Gunpowder

In his *New Organon*, Francis Bacon signalled the three inventions that distinguished European civilization: printing, gunpowder, and the magnet. Unknown to Bacon, all three of these epoch-making innovations appeared first in China. The middle innovation alone concerns physical power, although its applications went beyond destruction. Gunpowder found use in quarrying and mining, and in the seventeenth century Christiaan Huygens and Denis Papin (1647–ca. 1712) experimented with gunpowder engines. The internal combustion engine, fuelled by petrol at the suggestion of Luigi de Cristoforis in 1830, is nothing other than a domesticated cannon, with the projectile transformed into a piston.

Petroleum flame-throwers, developed in Byzantium around 670 by Callinicos of Heliopolis, found a place in Mediterranean warfare throughout the Middle Ages. By 900 they were also part of Chinese ordnance, complete with a projector-pump, and we infer that the Chinese army manufactured their own distilled petroleum. At about this time gunpowder, known in China for centuries, powered various incendiary devices and propulsion systems before it found a place in what Joseph Needham and his colleagues call a 'proto-cannon', or a barrel discharging small projectiles by an explosion. The barrel may have been inspired by a petroleum-discharging nozzle, or (especially in view of its early bulbous explosion chamber) it may have taken shape by analogy with a penile erection.

By the fourteenth century, the Chinese military depended extensively on firearms. One account indicates that a typical batallion had 700 cannon of various sizes and intentions, more than 600 hand guns, nearly seven tons of gunpowder, and over one million bullets. By this time Europe had received the Chinese gifts. Roger Bacon (ca. 1219–ca. 1292) reported on firecrackers in 1265; metal-barrelled gunpowder weapons – both handguns and bombards – made their presence between 1290 and 1310. The precise path of transmission is not yet known, whether through Islamic cultures or through Russia, but independent European invention is quite improbable.

The late fourteenth century is when Europe arrived at the decisive improvement of breech-loading – a modular, removable explosion chamber (in the form of a beer mug packed with powder) that dramatically increased the rate of firing. In the fifteenth century came the matchlock, an improved mechanism for firing. The musket emerged in Europe by the middle of the sixteenth century, with its wheel lock and then flintlock, and it remained a standard military weapon until the beginning of the nineteenth century.[1]

China received the European innovations early in the sixteenth century. By the seventeenth century, Western embassies, notably the Jesuits, were providing the Chinese with accounts of the latest European guns. In the last decade of the Ming dynasty, in 1642, Adam Schall von Bell (1591–1666), the Jesuit head of the Astronomical Bureau in Beijing, built a bronze cannon foundry; he produced 20 large cannon and 500 smaller pieces. Schall's successor Ferdinand Verbiest (1623–1688) undertook a similar commission for the new Manchu regime in 1675, and two of his cannon survive in the Tower of London collection.

Ferdinand Verbiest's labours in promoting Chinese firepower fit squarely in the tradition of the European Scientific Revolution. Military needs were present from the beginning. Galileo sought to promote the telescope as a military instrument, in Italy and in Spain. Robert Merton emphasized more than fifty years ago that much of the early activity of the Royal Society of London concerned military questions. Soviet historian Boris Hessen even contended, in 1930, that Newtonian mechanics emerged from interest in three practical questions: projectile trajectories; vacuum pumps, notably for mining ore; and the difficult matter of determining longitude at sea. The thesis has been debated since then, but there can be no doubt that each question related directly to military tactics or strategy – the first two focusing on manufacturing and aiming firearms.

The story of science and explosives may be taken up to the present. Revolutionary France created the Ecole Poly-technique to train military engineers, and revolutionary French savants conducted extensive research on explosives and ordnance. From the domesticated descendants of fire-arms – the steam engine and the internal-combustion engine – came not only the early industrialization of Europe but also the science of thermodynamics. Benjamin Thomp-son, Count Rumford (1753–1814), conceived of the mech-anical equivalent of heat while supervising cannon-boring in Central Europe, and the military controlled the early deployment of steam engines in nineteenth-century France. Explosives exerted an attraction generally on men of learn-ing, and they stimulated interests in metallurgy, especially to sink or to shield naval vessels. Alfred Nobel (1833–1896), the great patron of scientific genius, made his fortune with dynamite. Explosives were the object of the Kiangnan Arsenal near Shanghai, which from its foundation in 1865 served as a key conduit for Western scientific ideas. In this way gunpowder made a full circle, returning to its place of origin.

The First World War saw scientists constructing enormous guns, like Germany's railcar-mounted 'Big Bertha', and experimenting with new explosives. Aircraft technology – pioneered by physicists like Alexander Alexandrovich Fried-mann (1888–1925) in Russia and Theodore von Kármán (1881–1963) in Germany and the United States – came to focus on delivering ever larger quantities of explosives. The largest explosive of all, the atomic bomb, occupied the atten-tion of thousands of scientists and engineers during the Second World War. Ballistic-missile technology – including questions of guidance, propulsion, and payload efficiency – established a balance of military terror during the last half of the twentieth century and absorbed the talents of tens of thousands of scientific personnel. The heritage of gunpowder thoroughly pervades modern science.

The vocabulary of military science

The West has invented a vocabulary to insulate soldiers from the havoc of war. Soldiers are enjoined against becoming vandals, thugs, assassins, or hooligans – denying to Germanic, South-Asian, Arabic, and Gaelic peoples the dignity of honourable murder. Soldiers are warned not to go berserk or run amok. They are encouraged to imagine that they serve a nation, not merely a warlord commander. In attaining a goal by the invocation of death, however, soldiers wreak indiscriminate horror. Despite injunctions against torture or rape when enslaving or 'liberating' a subject population, soldiers inevitably inflict violence of this kind. Notwithstanding their ability to extinguish life – immediately and without compromise – military personnel have become sanitized as 'specialists on violence'.[2]

The association between killers and creators apparently begins with practicality. Military officers must lead men-at-arms to control society at large. The respect leading to victory in battle required, according to the nineteenth-century military strategist Carl Philipp Gottlieb von Clausewitz, both power and knowledge. In his view, the knowledge was of a special kind. It depended more on the *doing* that 'cannot properly stand in any book' than on the *knowing* that can be displayed in print. For this reason, war was unlike physics. 'Where the object is creation and production, there is the province of Art; where the object is investigation and knowledge Science holds sway. – After all this it results of itself that it is more fitting to say Art of War than Science of War.' Clausewitz went further, seeing war as an extension of state policy or business activity with 'competition on a great scale'.[3] The modern army in fact emerged from the leading business community of the seventeenth century, the Netherlands. Forged by Mauritz of Nassau, the Dutch army emphasized continual drill to make the mechanics and choreography of armed assault routine, constant exertion in the way of constructing fortifications to lessen the possibility of mutiny, and modular articulation in the vertical file of battalion, company, and platoon to insure coordinated action. 'The creation of such a New Leviathan,' historian William H. McNeill has emphasized, 'was certainly one of the major achievements of the seventeenth century, as remarkable in its way as the birth of modern science.' The 'military-commercial complex' that underlay the modern army, in McNeill's view, was a signal characteristic of post-Renaissance Europe.[4]

Permanent military institutions dispensed knowledge and radiated power. Engineering (the Dutch uncle of physics) was, until scarcely six generations ago, the province of military and naval garrisons; we retain this memory in our designation of 'civil engineer', a nineteenth-century neologism for a nonmilitary bridge-and-road builder. Surgery, the Western skill that astonished the Tokugawa Shogunate in Japan (1603–1867), was a military specialty requiring technical dexterity and speed as well as detailed anatomical knowledge; professional standing armies depended on such expertise, as it was easier to patch up a seasoned soldier than to train a new one.

By the eighteenth century, the military controlled engineering and surgical schools where the standards of instruction equalled or exceeded those in the universities. In France, for example, there was a prestigious engineering school at Mezières, founded in 1748. Among its professors was the Abbé Charles Bossut (1730–1814), who was succeeded by Gaspard Monge (1746–1818) in 1769, Abbé Jean Antoine Nollet, and Etienne Bézout (1739–1783). Professors in the artillery schools at La Fère and Baupaume included Nollet, Bézout, and Pierre Simon *comte* de Laplace. The navy also ran its own schools, notably at the Garde du Pavillon de la Marine. In early eighteenth-century Berlin, Surgeon-General Ernst von Heltzendorff founded a medical-surgical college that eventually affiliated with the medical faculty of the local university; in Vienna, a similar military medical institution opened in 1785. Through the nineteenth century the finest of scientists – Urbain Jean Joseph Le Verrier, Henri Poincaré (1854–1912), Hermann von Helmholtz, Albert Abraham Michelson – received years of systematic instruction courtesy of military authorities in Paris, Berlin, and Annapolis.

North-American academic writers, schooled in the virtues of disinterested science, have expressed surprise at the military regency over scientific research during the past half century. This domination seems unusual only because it followed the cometary passage of the research university. The ascendency of research universities, we have seen, is a short-term historical phenomenon, existing in Germany for about one hundred years (roughly, 1830–1930) and in the United States for half that span (1890–1940). The research university never quite succeeded in Russia, France, or Spain, nor in much of Latin America, Africa, and Asia. By the second third of the twentieth century, major research installations everywhere had quit the groves of academia for the *campus martius*.

The 'military-industrial complex', against whose influence United States President Dwight D. Eisenhower cautioned in a valedictory address, depended on cooperation from the academic world. Although universities in Eisenhower's time were nurseries for industrial employers – in the natural sciences as well as the humanities – industry required its own, mission-driven laboratories. Furthermore, much of industrial research concerned proprietary formulas and processes, from the brewing of Coca-Cola to the manufacturing of Corning glass. This was entrepreneurial activity that assumed the eponym research-and-development (R&D) because it did not take place in the transparent, public fashion preferred by university scientists.

R&D was well suited to the requirements of military general staff. The great military research efforts in the middle of the twentieth century were really an extension of industrial enterprise. The atomic-bomb laboratories at Oak Ridge, Los Alamos, and Hanford operated under the command of a state monopoly. Most industrial nations erected similar institutions. From Siberia to Patagonia, and at many points in between, more scientists laboured for the military on secret projects than sat at university laboratory benches.

To understand the connection between science and the military, it is useful to focus on the nation that established a model for the modern military-scientific relationship, France. In so doing, we shall see that nothing could be farther from reality than political scientist Robert Gilpin's statement, appearing a generation ago, that 'a great cleavage has long existed between the academic world and the military establishment in France'.[5] We shall see that the strong military component of French science eventually brought about its attenuation. The conclusion can be stated at the outset: although the military mind may excel at commanding grand scientific projects, it slowly but relentlessly stifles creative work.

The perspective offered here is not easy to reconcile with the image of France as an open and tolerant nation, the habitus of free-thinking literary and artistic innovators. For this reason, it is well to emphasize that various artistic disciplines, no less than scientific ones, follow the letter of special social contracts. The notion of an artistic avant-garde resembles the picture of researchers pushing back the frontiers of learning, and there have certainly been golden ages characterized by literary, artistic, and scientific achievement in a national setting. It is nevertheless presumptuous to imagine that all the arts and all the sciences necessarily flourish or fall together. Art can certainly flourish while science sleeps.

National needs are articulated, and often invented, by the state, which possesses the means of punishing dissidence. Nations have indeed been given life by the state through force of arms. In such circumstances, the military (and its manifestation in state security services) will naturally seek to discipline and direct the nation's cultural productions. Among national monuments celebrating military conquest are memorials to learning, and monstrous destroyers have included savants in their retinue. Napoleonic France, and especially its Egyptian Institute, is instructive in this regard, even though Napoleon's patronage of science continued a tradition of the *ancien régime*. Napoleonic precedent extended to distant realms of French influence. There is a bonapartist cast to General Gabriel García Moreno, the theocratic architect of the Quito Institute of Technology (one of nineteenth-century America's most distinguished scientific schools), just as there is to General Louis-Hubert-Gonzalve Lyautey, the early twentieth-century French conqueror of Morocco who patronized literature and science.

French military builders

The military builds things. Roads, railways, canals, dams, ships, planes, castles, prisons, and sporting palaces have all arisen under military direction and through the toil of conscripted labour. These constructions are at the centre of civilized society, but it would be a mistake to infer benign intent from the use-value of the object. The structures and machines reinforce the institution that called them into being.

Military construction serves two interrelated ends. First, it provides tools to help the military do what the military does best: wreak large-scale destruction. The glorification of this destructive capability plays a role in military cohesion: vulgar army marching songs, childish designs painted on airplanes and boats, and macabre allusions to blood, gore, and death. We have a great deal of evidence about the extent to which this side of military culture may dominate the life of a tribe, nation, or state. There is nevertheless another side to what the military builds: the design of monuments to enhance its prestige. An arch of triumph, for example, serves no practical function in battle; it is, rather, an object of general awe. The military has used science for both practical and inspirational ends. And people of learning have profited from military patronage. In comfortable surroundings they have built hellish devices, and they have also elaborated notions of no conceivable destructive effect.

The two sides to the interaction between science and the military are apparent from antiquity. The death of Archimedes, intent on solving a geometrical problem on his sandtable and oblivious to the command of a soldier who had invaded his lodging, is revealing, for Archimedes designed mechanical devices to forestall the intrusion. The pattern of considering problems of both military and non-military import is found in the activity of penetrating scientific intellects, like Leonardo da Vinci and Enrico Fermi (1901–1954). Both sides of military sciences are manifest in the first modern nation state, Revolutionary France. Let us consider the abstract side first.

We have seen that during the *ancien régime* in France, astronomy was tied to practical surveys of the earth's surface. The matter related to the figure of the globe and not less importantly to the dimensions of the Bourbon kingdom. The early Republicans identified mapmaking at the centre of their new reign of reason. Associated with this identification was a strong military presence in the exact sciences, which has persisted to the present day.

We have also seen that the metre, called into being by the Constituent Assembly in 1790–1, required for its instantiation a detailed survey of the terrestrial meridian passing from Dunkirk to Barcelona. The task was accomplished by surveyors associated with the Ecole Polytechnique, the military school placed at the summit of expertise in the exact sciences. The surveying work proceeded intermittently, finally propelling Jean-Baptiste Biot (1774–1862) and Dominique François Jean Arago (1786–1853) to the Island of Formentera in 1808. The arc they surveyed gave greater precision to the standard metre, and the surveying brought lustre to the Bureau of Longitudes, the independent astronomical office set up in 1795 to collect and publish observations and generally to oversee the French observatories. The bureau came to reside at the Paris Observatory, its highly paid staff – an assembly of astronomers, mathematicians, hydrographers, and geographers – charged with organizing astronomical missions and publishing an ephemeris. By the value of its emoluments and an accompanying freedom to undertake research of a general nature, a senior appointment to the bureau was highly prized; and the nature of French learning guaranteed that the bureau's senior staff would be members of the Académie des Sciences in Paris.

The bureau functioned as the astronomical research institute of the Paris Académie from its inception, and its hand figured in every major French astronomical expedition over the succeeding two centuries. Yet it operated no major observatories of its own – except for the portion of the Paris Observatory that it came to control by sufferance. And although it projected a strong presence in distant parts during eclipse or planetary-transit expeditions, it spent very little, if anything at all, on permanent overseas installations. Furthermore, the bureau was by its nature an interdisciplinary fellowship that transcended the mandate of its administrative superior, the Ministry of Public Instruction. Soldiers and sailors freely entered its ranks. Military and naval assistance was indeed crucial in mounting astronomical expeditions, but multiple allegiances discouraged the emergence of a scientific school that could claim to be attached to the bureau. The bureau's *modus operandi* facilitated the appointment of Admiral Amadeo Ernesto Bartolomé Mouchez (1821–1892) as director of the Paris Observatory in 1878. Mouchez then set the course for French astronomy over the next two generations with an ambitious project to compile a photographic chart of the sky's two million brightest stars.

Naval stars

A military regimen was precisely what the Paris Observatory required. Its director at mid century had been Urbain Jean Joseph Le Verrier, perhaps the greatest of French astronomers and the worst of French administrators. He sought to cut off the observatory from its quasi-feudal past by transforming it into a state bureaucracy (the Cassini family and eventually Arago had run the observatory as a private fiefdom, and when Le Verrier inherited it, the staff were bound together by blood and marriage). The annals of early nineteenth-century French science, however, were written in individual accomplishment and discovery. Le Verrier's subordinates would not suffer anonymous submersion in his bureaucracy, and his tenure as director marked the most tempestuous years of French astronomy.

Mouchez's claim to Le Verrier's mantle has nothing to do with the Ministry of Public Instruction (the observatory's source of funding) and everything to do with the navy's scientific ambitions. Naval engineers were assimilated into a separate and well-paid corps in 1814, an elite numbering seventy-three over the next century. Scientific work – whether on astronomical expeditions, with the Bureau of Longitude's training school at Montsouris (organized by Mouchez), or in the navy's own observatory at Toulon – was also confided to career naval officers outside the corps. Mouchez himself was in the latter category.

Naval engineers and regular officers constituted the single largest reservoir of astronomical talent in nineteenth-century France, a circumstance freely exploited by the Ministry of Public Instruction. In 1868, for example, the naval officer and mathematician Jean-Philippe-Ernest de Fauque de Jonquières (1820–1901) urged the Ministry of Public Instruction to lend a spectroscope to hydrologist Philippe Hatt (1840–1915), then charting the coast of Indochina, so that Hatt could observe the upcoming solar eclipse at the nearby island of Poulo Condore. When France constituted expeditions to observe the transit of Venus in 1874, Mouchez and hydrologist Jean Jacques Anatole Bouquet de la Grye (1827–1909) were seconded from the navy to the Ministry of Public Instruction; and the navy sent out four teams of its own to observe the 1882 transit. This general pattern is found in all French overseas astronomical expeditions during the nineteenth century.

The star chart

Admiral Mouchez was the man of the astronomical hour. After graduating from the Ecole Navale, he quickly distinguished himself as an astronomical calculator. He rose through the ranks to become a member of the Bureau of Longitudes in 1873, a member of the Académie des Sciences in 1875, and a vice admiral in 1878. The veteran of numerous foreign tours of duty (Mouchez spoke Spanish fluently), he was the navy's leading hydrographer and expeditionary. Installed as director, he spent nearly a decade renovating practical astronomy in Paris. Then he committed resources to an ambitious and far-reaching project.

The project turned on advances in photographical astronomy. By the early 1880s, David Gill (1843–1914) at Cape Town had demonstrated the feasibility of using celestial photography to inventory the stars. In 1885 he began a map of the southern skies; he took the photographical plates, and Jacobus Cornelis Kapteyn (1851–1922), professor of astronomy at Groningen in the Netherlands, measured them. Mouchez had inherited a star-mapping project of his own, an inventory of the ecliptic started in the 1860s by the brothers Paul (1848–1905) and Prosper Henry (1849–1903) at the Paris Observatory. The Henrys abandoned their catalogue when they reached the profusion of the Milky Way. They began experimenting with photography. By the middle 1880s, with a photographic refractor of their own design, they discovered celestial objects that had escaped astronomers using traditional resources. Their achievements led Mouchez to correspond with overseas colleagues about the possibility of a photographical map of the heavens. Gill urged an international congress to discuss the proposal. Mouchez was only too happy to offer Paris as its site. Here was a way to supersede all the star catalogues that had been produced or were under production by foreign rivals like Friedrich Wilhelm August Argelander (1799–1875) at Bonn in Germany, Benjamin Apthorp Gould (1824–1896) at Córdoba in Argentina, and of course Gill himself.

The organizing conference of 1887 assembled more than fifty astronomers from nearly a score of countries. With military precision, they divided the sky into portions assigned to various observatories; they planned for photographical illustrations (in 22,000 sheets showing stars down to fourteenth magnitude) and tables of some two million brighter stars. Uniformity would be insured by using identical telescopes and ruled photographical plates. As the project was underwritten by the Paris Observatory, the Académie des Sciences covered the cost of publishing its memoirs and bulletins (although not, of course, all parts of the catalogue). A permanent commission came into being to supervise the project, assign work schedules, direct special studies, and generally handle problems as they arose.

Within several years, the commission established standards and conventions – even for the thorny issue of stellar magnitude. The authorized telescope would conform to the 34-cm refractor designed by the Henrys and built by the firm of Gautier in Paris. Photographical plates were to be had from the appropriately named company Lumière Frères in Lyons. Stellar zones were divided among eighteen participating observatories. The commission had no money of its own to distribute, however. Each participating observatory had to find independent funding. United States astronomers abstained from committing any of their impressive resources to the project (and the Germans committed theirs reluctantly), but Admiral Mouchez nevertheless believed that naval discipline would be able to set the enterprise on a firm path. French presence overseas – in which the navy played a large role – confirmed his faith.

Among the eighteen observatories taking slices of the sky, three were on French soil: Paris, Bordeaux, and Toulouse. Another, Algiers, was in France's major overseas territory. France effectively controlled two observatories in the Southern Hemisphere, Santiago de Chile and La Plata in Argentina, as French astronomers had recently arrived to take charge of each operation. A third South American observatory, that at Rio de Janeiro, had been managed earlier by a refugee from Le Verrier's tyranny and then had passed into the hands of a native of Belgium. And the Mexican national observatory at Tacubaya was also Francophile. Each of the five overseas observatories would be harnessed to a long-term project, using French astronomical technology, for the greater glory of the imperial seat. To the constellation of eight observatories owing greater or lesser allegiance to Paris, Mouchez added five from the English-speaking world – Greenwich, Oxford, Cape Town, Sydney, and Melbourne – which he thought could be disciplined through the British navy. The remaining partners in the project would be pulled along by inertia or national pride.

The tale unwound otherwise. Among the French overseas astronomical satrapies, only Algiers completed its task within a generation of starting it. Santiago de Chile, La Plata, and Rio de Janeiro defaulted well before the First World War. Tacubaya persisted, completing its assignment on the eve of the Second World War. Paris collaborated with the British empire to pick up most of the defaulted areas.

The sky-chart project was military science writ large across the heavens. It left a mixed legacy. The burden of the project undermined the health of French astronomy by focusing attention on routine operations related to stellar position instead of on exciting new areas in astrophysics. By 1920, with the French portion of the *carte du ciel* still incomplete, the French astronomical community felt

fatigued and discouraged. The Académie des Sciences commissioned astrophysicist Henri Deslandres (1853–1948),
director of the Meudon Observatory, to undertake a systematic and secret enquiry into the malaise. He concluded that
France would do well to set off on a new path by allocating
half of its astronomical resources to astrophysics, following
the pattern of astronomers in the United States.

We must not imagine that the sky-chart work in itself
impeded creative research. Stellar inventory underlay
analyses of proper motion and stellar parallax – studies
directly related to cosmological issues – and this is precisely
how American and British observers used their catalogues.
Furthermore, as Deslandres pointed out in his secret report,
metropolitan France had ten major research observatories;
nine belonged to the state, and one (Abbadia) to the Académie des Sciences. Of the six observatories not involved in
the star-chart project, Lyons and Nice had been created
at the end of the nineteenth century and endowed with
reasonable facilities; by the early twentieth century, Lyons
concentrated on stellar photometry. Meudon, in greater
Paris, comprised a large complex devoted exclusively to
astrophysics. None of these observatories had enormous
telescopes under ideal observing conditions, but the means
were certainly available for significant, new undertakings.

The routine nature of positional astronomy, which
absorbed the lion's share of funding at French observatories
up to the middle of the twentieth century, was well suited
to the social structure of stargazing in France. The military
connection remained ubiquitous. Observatories affiliated
themselves only loosely with local universities, when,
indeed, they had any university affiliation at all. Astronomers advanced in their career through apprenticeship,
generally working their way up in the observatory hierarchy; to obtain a *doctorat ès sciences* in observational astronomy, one had to be accomplished in the art of observing,
and this meant spending years in a junior and poorly paid
assistantship. Graduates of the Ecole Normale Supérieure

such as Félix Tisserand (1845–1896), Benjamin Baillaud (1848–1934), Ernest Esclangon (1876–1954), and André Danjon (1890–1967) did rise to astronomical heights in France, as did officer graduates of the Ecole Polytechnique like Deslandres, but the recruitment of talented science students remained something of an accident. Many astronomers found their calling in the army or navy, and others rose to positions of authority through dogged perseverance.

What is the cause of the decline of French astronomy? It has little to do with problems of recruitment. The guild system in astronomy is not special to France. It flourished in the British empire and in the Americas, giving way early in the twentieth century to the German and Dutch model of academically trained astronomers. France, ironically, preceded the English-speaking world in requiring the doctorate of senior observatory staff.

Nor can attenuation of the astronomical *discipline* be the cause of the insecurity endemic to French astronomers. It is true that no clear path led to excellence in the field. The summit could be attained by starting in the army, the navy, the provincial universities and *lycées*, or the observatories, but in each kind of institution a prospective astronomer had to navigate around numerous obstacles and spend a great deal of time on things unrelated to the stars. And if one did discover new stars or mathematical equations, there was no respected, disciplinary journal to receive the discovery – the *Bulletin* of the French Astronomical Society being a house organ appealing in large measure to amateurs. But diplomas, societies, and journals need not be tied to a discipline for disciplinary knowledge to flourish. Early nineteenth-century Britain lacked the former while nevertheless fostering a strong community of astronomers.

There is one special feature of French astronomy that may account for what happened. It is the strong military presence, which discouraged collegial cooperation with foreign colleagues, urged against publication in civilian periodicals, and generally entrapped researchers in a quagmire of paperwork. Centralization, an abiding characteristic of science in France, compounded with military regime to create inertia.

Comparison with the Netherlands provides a perspective. Every feature of French astronomy applied to the Dutch astronomical community with the exception of the military connection, and early in the twentieth century, the Dutch community rose to world prominence. In the Netherlands, although university professors collaborated with the navy on a principled basis, they retained their independence and freedom of enquiry.

In a different vein, tedium and sterility in France derived from rigid comportment and inward focus. With the pervasive presence of military discipline and with all eyes turned toward Paris, few astronomers took the trouble to keep abreast of publications in German and English periodicals. Here the contrast with the Netherlands could not have been greater. Most Dutch astronomers would have been literate, if not fluent, in German, English, and French; Dutch astronomers prized individual initiative and outward vision. By the late nineteenth century, French astronomers withdrew into collective work that contributed to the glory of institutions in Paris – not the least of which was the naval command.

Military mappers

The complexities of celestial navigation led the world's navies to take special interest in astronomical questions. The army, too, has been an alembic for refining scientific talent. The army's war schools have traditionally been on the cutting edge of knowledge about the world. The finest example of this interest is found in the Ecole Polytechnique – the military school erected two centuries ago by the Revolutionary Convention in Paris. Officer candidates were taught by the Republic's leading *savants* – giving lustre and prestige to their brevet of command. This kind of sinecure for *érudits* continues in the present-day incarnation of the French school, still under the control of the armed forces and still a fount for abstract wisdom.

The army also has a use for practical knowledge, most notably in the design of weapons and in provisioning and protecting its men under arms. There are well-known instances of impractical spin-off from rather mundane military tasks. The study of ballistics, for example, commissioned by military hopefuls during the Italian Renaissance, led (by a tortuous path) into Galileo's prescient mechanics. The association between scientific research and military needs is found in our own time in high-energy physics. It is instructive to trace the military thread through the nineteenth century – the period that saw the rise of today's great, secular institutions of higher learning. As before, we focus on France.

Just as the navy provided careers for nineteenth-century French scientists, so the army provided special employment for scientifically inclined officers. The navy needed charts, the army needed maps. It assigned surveying to its Dépôt de Guerre. Maps issued in a continuing stream across the first three quarters of the nineteenth century, both for overseas territory and for the metropolis, but the army surveyors enjoyed nothing approaching the cachet of their naval counterparts. François Perrier (1833–1888), a senior military surveyor, set out to rectify the situation. In 1881 he succeeded in assigning geographical work to a section under the army's general staff, which he captained. Six years later the Dépôt was abolished in favour of a geographical service directed by General Perrier. Under Perrier's guidance, the army began to undertake gravimetric surveys; these continued with Perrier's successors, although the service focused increasingly on its North African possessions, linking them by topographical triangulation to French meridians.

Wherever the French army camped, it sent a team of topographers to chart the surrounding terrain. Sometimes the military geographical services received orders from the council of a local regent (Tunisia), a governor general (Indochina and Madagascar), or an occupying army (Syria), but even in cases where civilians came to carry out the surveying (Morocco, Algeria, and Tunisia), military norms were omnipresent. Military geographers attempted to engage projects relating to pure science. The most ambitious of their projects – an early twentieth-century remeasuring of the eighteenth-century arc of longitude in Ecuador – did little more than advance the military career of those, such as Robert Bourgeois (1857–1945) and Georges Perrier (1872–1946), son of François Perrier, who slogged over South American hill and dale. The Ecuadorian experience soured the geographical service on geophysics and astronomy. Bourgeois, who had commanded the expedition to Ecuador, rose to direct the service in 1911. He began the First World War with thirty geodesists; by the war's end, twelve had been killed, three had died through accident or disease, and a large number had retired or been transferred; of the pre-war staff, only Georges Perrier remained. During the interwar years, Perrier laboured in vain to reconstitute a geophysical programme for the military surveyors. Scientific progress eroded his claim to head the army's scientific elite.

Whereas in the nineteenth century the military surveyors had pioneered construction of portable telescopes and analysis of astronomical data, advances in technology lay at the root of the diminished scientific status of the geographical service. Two related developments reduced the geographers to little more than mapmakers: aviation and wireless. Aerial photography, which came to subtend mapping, depended on collaboration with the independent-minded air corps; furthermore, the corps's national meteorological office, which provided weather reports for airports, freely involved itself in geophysical projects. Closely tied to aviation was communications technology. Wireless, which provided the precise time signals that were essential for mapping, was the domain of the army's communications branch, whose resident expert was Gustave-Auguste Ferrié (1868–1932). He became the army's senior scientist, and he set the tenor for much of military involvement with scientific innovation during the twentieth century.

After having graduated from the Ecole Polytechnique and the Ecole d'Application at Fontainebleau, Ferrié served as a lieutenant at Grenoble and Besançon. Soon after he made captain in 1897, he was transferred to the central office of military telegraphy. In 1898 he began experimenting with the newly discovered Hertzian waves, and the following year he assisted Guglielmo Marconi (1874–1937) in sending wireless signals between England and France. The Ministry of War lost no time in having Ferrié organize a military radiotelegraphic service; over the next thirty years Ferrié directed the service's expanding fortunes. He wrote technical manuals, installed radios on airplanes and balloons, established radio communication with French colonies abroad, and lectured tirelessly to popularize his medium. At Ferrié's insistence, radio played a part in France's invasion of Morocco in 1908, and at his command the Eiffel Tower became a giant radio mast for broadcasting time signals. In close collaboration with astronomers at the Paris Observatory, he made use of the time service to determine longitude differences in France and between Paris and Washington. Appointed lieutenant colonel in 1914, he set up a scientific research centre that during the war employed a number of talented physicists, among whom were Henri Abraham (1868–1943), Eugène Bloch (1878–1944), Léon Brillouin (1889–1969), Marius Latour (b. 1875), Lucien Lévy (b. 1853) and Edmond Rothé (1873–1942). Colonel in 1915, brigadier in 1919, General Ferrié received the Osiris prize of the Académie des Sciences (F 100,000) and then took explorer Alfred Grandidier's (1836–1921) chair there in 1922.

By this time Ferrié belonged to every major French committee exercising authority in the exact sciences. As Daniel Berthelot's (1865–1927) successor in presiding over the section of terrestrial magnetism and atmospheric electricity in the Comité Français de Géodésie et de Géophysique, Ferrié exercised discretionary power over funding. He collaborated with Charles Maurain (b. 1871) – France's premier geophysicist and dean of the Sorbonne's Faculty of Sciences – about most geophysical mapping projects in the 1920s. Maurain and Ferrié's schemes also involved key French physicists, such as Berthelot, Louis Dunoyer (1880–1963), and Aimé Cotton (1869–1953). Among French scientists only the death of Marie Curie and her son-in-law Frédéric Joliot-Curie (1900–1958) produced grander funeral services than Ferrié's.

Military weathermen

Maurain and Ferrié worked together for nearly twenty years on questions relating to aviation and radio, and from their mutual interests they formed common cause against outsiders who might threaten their authority. Notable rivals were military geophysicists in the French air corps. Among kinds of warrior there is no solidarity.

The air corps retained the weather as a prize of the First World War. Before then, meteorology (and related global phenomena such as magnetical and seismological disturbances) had passed from the astronomers to a Central Meteorological Bureau in Paris. The central bureau, under Eleuthère Elie Nicolas Mascart (1837–1908) and his successor Alfred Angot (1848–1924), recorded data worldwide by leaning heavily on consular agents and military physicians, the latter of whose meteorological interest outlived the early nineteenth-century miasmatic theory of disease.

With the appearance of trench warfare, poison-gas attacks, and military aviation during the First World War, meteorology came to interest the high command. After the disastrous Champagne offensive in 1915, the army had the director of its geographical service, General Robert Bourgeois, take effective control of the Central Meteorological

Bureau and its network of observing stations. The army also established a meteorological service of its own in the air corps under naval lieutenant Jules Rouch (1884–1974), a graduate of the Ecole Navale who had been charged with oceanographical, meteorological, and magnetical measurements on the second Antarctic expedition of Jean-Baptiste-Etienne-Auguste Charcot (1867–1936). In 1917 Commandant Rouch took charge of the navy's meteorological service, the army's being reorganized under Colonel Emile Delcambre (1871–1951). In 1918 Bourgeois assimilated the army's meteorological service into his civilian network.

The weather was an enemy that would sign no armistice. Jules Rouch wrote during the summer of 1918:

> To use a comparison from current affairs, the synoptic map, showing us the disposition of atmospheric forces, is analogous to a map giving the exact position of enemy forces. The latter document would not perhaps reveal the true intentions of the enemy in a precise way; it would not always allow predicting with certainty the succession of events. But where is the general who, having it in his possession, would neglect to consult it.[6]

To defeat this inhuman foe – to predict its next move – required a continuing, military effort in synoptic meteorology. Alfred Angot's retirement in 1920 allowed the military to keep in peace what it had gained in war, albeit with a few administrative modifications. To the new Office National Météorologique, directed by Delcambre and placed with the undersecretariate of the air in the Ministry of the Interior, went not only Angot's bureau but also the meteorological service of air navigation and the central services of military aviation.

Colonel (subsequently, General) Delcambre watched over French weather everywhere. His rise, and that of his office, stemmed directly from the remarkable growth of military and civilian aviation, especially overseas. In the late 1920s

Delcambre tried to dominate meteorology in North Africa and the Middle East; he succeeded in placing agents at the head of Algerian and Lebanese weather forecasting. Delcambre's successor Philippe Wehrlé (1890–1965) sent an expedition deep into southern Algeria for the International Geophysical Year of 1932/33. And by the 1930s weather forecasting for air transport had come to absorb much of the attention of geophysicists in Indochina. The French national meteorological office was not the only force behind these developments, but it radiated broad influence because its department, the air corps, evolved into an independent ministry in 1928.

Applications and prestige

The French army and navy underwrote a large part of astronomy, geophysics, and meteorology through the fall of the Third Republic in 1940. The military did more than simply authorize expenditures. They had seats of honour at the Académie des Sciences. In the exact sciences division, geography and navigation was the domain of military and naval engineers. The last nonmilitary member of the latter section, the Madagascan explorer Alfred Grandidier, was replaced by General Gustave-Auguste Ferrié after the First World War. Besides constituting one-sixth of the academy's exact sciences division, military men were admitted to the other sections or divisions.[7] As nearly all the latter academicians were observers of the physical world rather than speculators about ideal mathematical realms, military men exercised a strong voice – and sometimes a stentorian one – in the academy's deliberations about physics and astronomy.

Let us now rephrase the question raised at the outset. If military penetration of research universities during the middle of the twentieth century reflects the desire of generals and admirals to regain their position as custodians of nature and interpreters of natural law, whence the original inspiration? Why should warriors have a professional interest in abstract knowledge?

In the modern world, where science is seen as dispassionate and disinterested learning, we often invoke the notion of 'applications'. Science may be applied to solve practical problems, and the military is a customer with big problems to solve. The evolution of scientific institutions facilitated the growth of large institutes for military research.

The 'big physics' of the atom smashers cut off teaching from research – a union that lay at the centre of academic life since the nineteenth century. The rupture brought Germany, Great Britain, and the United States into line with France. Traditionally in France, scientific research had only a small role in the academic certification of scientists. Large laboratories were usually nonteaching, government-run installations devoted to particular problems, such as crystallography, radioactivity, wireless, and so on, and these often came under the control of the army (wireless) or the navy (astronomy). The French pattern of particular, task-oriented research campaigns conducted with military discipline was just the model invoked by the great powers during the two world wars.

To a dispassionate observer in the 1920s, the decision of scientists to jump into bed with the military would have seemed inauspicious. In the nineteenth century, the military of many nations financed grand scientific projects, including tropical and polar explorations. The *scientific* results were unimpressive. None of the military investments led directly to new laws of a general kind. It is true that Charles Darwin's theory of evolution was the eventual outcome of one naval exploration. As his story suggests, however (Darwin was not the *Beagle*'s official naturalist), science comes as a bonus to the grand designs of military professionals.

Academics capitalized on military readiness to underwrite scientific research. The First World War produced a national orchestration of scientific effort still largely centred on universities. Belligerents created national research committees directed by university professors, and they drafted academics to design and manufacture weapons of destruction – from artillery-rangers and submarine detectors to airplanes and poison gas. Then the Second World War, with its extraordinary commitment to technological secrecy, cut science free from university moorings. By 1950, the military had effectively reasserted its authority, established during the seventeenth and eighteenth centuries, as an arbiter of scientific law.

Remilitarization proceeded initially in the name of precision, the reigning scientific ideology, and it centred on bomb sights, radar, and explosives. After the Second World War militarization extended to ballistic missiles and their guidance systems, along with artificial satellites. In one area, however, precision counted for little: the construction of an atomic bomb. The calculations were straightforward and unsophisticated. Required was a tremendous source of electricity for producing uranium 235 or an atomic pile large enough for manufacturing plutonium. The military erected enormous research facilities to manufacture both uranium and plutonium, as well as to test a prototype of the bomb. Precision came later, when the hellburners and their thermonuclear offspring were miniaturized for encapsulation in ballistic missiles.

Clausewitz knew that science may provide technology for military ends. To remain connected to the source of potential gain, then, generals and admirals keep scientists on retainer. The observation is insufficient, however, to explain military acquiescence to supervise extraordinarily expensive projects of little practical import – such as measuring an arc of longitude in Ecuador, sending dozens of tons of scientific equipment into the southern Sahara, or placing a man on the moon. The connection between science and the military goes beyond material improvement.

Certainly military-sponsored research has had a practical impact on the life of ordinary people. During the last half of the twentieth century, military funding of apparently useless and intricate projects such as atomic clocks led into solid-state physics in the form of transistors, lasers, and printed circuits. The new objects changed the nature of personal experience and transformed medical diagnosis and therapy. Indeed, modern medicine and the modern army exchanged metaphors. United States physicians conducted a 'war' on cancer, and military conquerors referred to committing atrocities with 'surgical precision'. The head of American public health is indeed called the surgeon-general, and the position comes with a military uniform.

Notwithstanding the connection between scientists and soldiers, knowledge produced under military tutelage has remained abstract. There is no apparent bloodthirsty quality to the lectures on differential calculus that have been delivered at the Ecole Polytechnique over the past two centuries. America's first modern physicist in the nineteenth century, Joseph Henry (1797–1878), did not tailor his research at the West Point military academy to the survival requirements of calvary officers prosecuting Indian wars. Much of knowledge sits outside a large part of the circumstances behind its rise. After all, not every event in a person's life has common magnitude or import.

A remarkable scientific innovation, wrestled into its final form during the most savage of twentieth-century wars, provides testimony of this otherworldly quality. It is Einstein's general relativity, given its classical expression late in 1915. Einstein worked on it in wartime Berlin. He was against the war, and his neutral Swiss nationality allowed him immunity from the inconveniences visited on pacifist German nationals. Let us suppose for the moment that Einstein saw general relativity as a way of preserving the international spirit of scientific inquiry, which he valued; let us imagine that the criterion of general covariance – eventually a cornerstone of the 1915 field equations – represents an assertion of universal reason over mere tribal animosity; let us believe optimistically that every line in the tensor calculus was set down as a rejection of European militarism. These interpretations are attractive to some historians. But none of them, if any should turn out to be persuasive, accounts for the particular path taken by Einstein in his calculations. The path, as Robert Merton, Derek Price, and Thomas Kuhn would have affirmed, was clearly indicated by disciplinary norms in theoretical physics.

Evidence impels us to seek the science-military interface in a notion no more complex than that of 'potential utility'. It is the prestige of pure learning. In 1970, setting down second thoughts about his book on science in seventeenth-century England, Merton appealed to one of the greatest English architects:

> To appreciate the power of the drive for ethnocentric esteem, we need only paraphrase that epitaph to Christopher Wren: '*Si exemplum requiris, circumspice* [if an example is needed, look around],' as we contemplate the billions of dollars happily expended by the United States to win the race to the moon.

Merton's book detailed the significant military-command component of early modern science. Just as important in recent times, he thought, has been tribal pride and glory, 'Americans contesting with the Russians for scientific pre-eminence.'[8]

Combining the two strands of Merton's discussion, the military patronage of science appears as indulgence money for the military, who would like to persuade credulous onlookers that an army and a navy exist to preserve and strengthen the very institutions they destroy. Chaperoning an eclipse expedition or measuring atmospheric ozone – morally inoffensive projects undertaken with extensive military support – offer an inverted and distorted image of military rule. These kinds of undertaking serve to mask the terror that reigns in times of war. It is just this 'prestige racing' (to borrow historian Walter A. McDougall's expression) by disinterested research that enhanced

American military fortunes during the disastrous Indochinese war of the 1960s.[9] Viewing the Apollo moon landings on television, or hearing a military concert band, we are lulled to feel against all reason that, as the early nineteenth-century moralist Joseph de Maistre contended, war is the most effective cultivator of science, art, and 'the best fruits of human nature'.[10]

III

SENSIBILITIES

Participating:
Beyond Scientific Societies

Whether high society or hoi polloi, Cajun or Creole, social organization in southern Louisiana is dominated by Mardi Gras. For inhabitants of states north of the Mason-Dixon Line, this last day before Lent is unromantically termed 'Shrove Tuesday', a day associated with serving cold pancakes in church basements. For festival-mad Louisianians, however, 'Fat Tuesday' caps a round of balls and parties that began on the day of epiphany. Even this long period of 'carnival', as it is better known in Latin countries to the south, is the fruit of year-long preparations by the teams – known as krewes – who build and ride on their floats.

Louisianians fear that now that talkshow host Jay Leno has brought national attention to Mardi Gras revelry, especially the exchange of baubles for bared flesh, local customs will fall victim to Hollywood commercialization. This would not only threaten a rich cultural tradition, but could also obliterate a remarkable mechanism for promoting co-operation and harmony among a broad spectrum of the population. In the frenzied atmosphere of Mardi Gras festivities, religious affiliations and political allegiances are forgotten. The season's royalty must command remarkable financial resources to underwrite the cost of the beads and 'throws' (sundry items tossed from the floats), breakfasts for their court, and lavish regal headdresses and capes. Whether the king has made his mark in town by reason of his prowess in the oil business, insurance, or the law becomes irrelevant. All that matters is that a good time is had by all: his krewe, his friends, and the revellers shouting 'throw me something, mista' from curbside.

The social harmony promoted in South Louisiana Mardi Gras, as well as the collective amnesia and group indifference towards politics, religion, and wealth, makes the krewe similar in its cultural functions to a new form of scientific society created in the eighteenth century, the literary and philosophical society. Such organizations responded to Bacon's vision that called for a democratization of effort, designed to enrich the amateur's first efforts by association with experienced investigators. Science was an enterprise of shared contributions, with participation assured for all who were willing.

Such an ideal conflicted with the elitism of established scientific organizations. While new ideologies promoted political and economic equality, scientific academies – associated with prestige, privilege, and exclusion – appeared increasingly anachronistic. As historian of science James McClellan puts it, they were 'institutionalized centers of power in science, peopled by the ranking men of science of the age, organizing and controlling science itself'. In other words, they functioned as bastions of 'scientific correctness'.

The pretensions of the academy had become distasteful not only within the context of an increasingly egalitarian society, but also to a number of vocal and visible groups within that society. Dissidents – who had effectively been excluded and marginalized before – rebelled, as McClellan emphasizes, against the 'dictature of the academies'. No longer were they willing to sit on the sidelines. They, too, craved power and representation; as a result, they aimed to establish new institutions that reflected the concerns of a broader constituency. In so doing, they believed themselves to be restoring science to its originally open and accessible nature.

The rise of literary and philosophical societies

Yesterday's rebels often become today's establishment. As continental academies slumbered toward the end of the eighteenth century, a new dawn broke over scientific associations throughout the Anglo-Saxon world. A range of voluntary associations dedicated themselves to the advancement of science. Dissidents (with an active core of Dissenters from High Church dogma among them) appropriated and revitalized academic science by forging open and democratic forums. Generally speaking, in politically decentralized countries, the impetus for new kinds of scientific societies came up from the local citizenry, not down from national governments. These new organizations are more likely to be found in smaller provincial towns than in capital cities. Many of the towns were experiencing the initial stages of industrialization.

As Arnold Thackray describes the situation, until 1781 the Royal Society was the only English institution dedicated to the pursuit of science; a mere sixty years later 'the scene was crowded beyond recognition'. There were sixteen metropolitan disciplinary societies, at least as many provincial societies covering the whole of science, and more than two dozen provincial disciplinary societies. As the oldest of the literary and philosophical societies, Manchester's became a model for those in other cities and towns of the industrial hinterland, including Liverpool, Bristol, and Birmingham.

By subjecting the membership of the Manchester Literary and Philosophical Society to the 'versatile if sometimes barbarous art of prosopography' (that is, collective biography), Thackray draws a number of conclusions about why Mancunians turned to the pursuit of natural knowledge in general and why they joined the local society in particular. He distinguishes seven reasons why members embraced science: for its possibilities as polite knowledge, as rational entertainment, as theological instruction, as professional occupation, as technological agent, as value-transcendent pursuit, and as intellectual ratifier of a new world order. The Society offered a means of social legitimation for 'marginal men', whether religious dissenters, gentleman physicians, or *nouveau riche* manufacturers. In Thackray's view, then, the 'lit and phil socs' responded to social concerns unique to the emerging industrial centres of the Midlands and north of England. Their cultivation and advancement of science was largely incidental to these broader cultural functions.

Associations for the advancement of science

Scientific societies elsewhere actively sought governmental patronage. In 1822, the Gesellschaft Deutscher Natur- forscher und Aertze was founded as an omnibus scientific association to promote the interests of scientists in all the various German states who envied the centralization of scientific talent in London and Paris. Scientists in England were inspired, in turn, by the German society. They believed that a similar organization on their own shores might help them obtain government sponsorship and combat the mori- bund Royal Society.

English scientists took the lead in institutionalizing this new form of scientific organization. It was to have no per- manent headquarters, thus shedding the traditional insti- tutional trappings of rooms, libraries, and museums. Members would meet annually for a week, with a new president elected each year. Part of its official platform called upon the society to spurn the scientific glamour of London for locations throughout the country. Exemplifying these characteristics, the British Association for the Advancement of Science (BA) was founded at York in 1831.

The British Association became associated with the Indus- trial Revolution, since many early meetings were held in new industrial towns. The BA sought out serious scientific practitioners for its annual meetings, in order not to become a gentlemen's social club like the Royal Society. By the mid 1830s, seven sections (which grouped together related disciplines, such as mathematics and physical science) defined the association's scientific concerns. Within two decades of the establishment of the German example, similar national scientific organizations – peripatetic and with broad membership lists – had been set up in France, Italy, and the United States.

Government-subsidized research and publication functioned as an important part of the *raison d'être* of the associations for the advancement of science. At the same time, they became important instruments of science popularization. By incorporating theatre and spectacle into their annual gatherings (thereby supplying their detractors with ammunition to ridicule them), they made science visible to a broad spectrum of the population. The staid national academies of science had introduced pomp and perquisites into their proceedings, but never before had these rituals been brought to the attention of so many spectators.

A BA meeting held at Montreal in 1884 indicates the level of public display and popular interest that the annual meetings generated. Two thousand people attended the inaugural session; a Saturday night lecture on comets (accompanied by limelight illustrations and delivered by the Astronomer Royal of Ireland) was open to anyone who paid the ten cent admission charge. An evening *conversazione*, held in the flower-bedecked natural history museum of McGill University, was packed almost to suffocation. The museum's lecture theatre showed experiments and apparatus to the curious, for example, a polariscope for projecting spectra on a screen and a machine to liquefy gases. At one end of the museum's Great Hall, the Governor General of Canada and his wife received guests to the strains of an orchestra playing waltzes, polkas, and marches. Prominent citizens hosted garden parties; excursions toured nearby natural wonders (rapids, waterfalls, and mountains) and technological marvels (the Victoria Bridge and Grand Trunk Railway workshops).

Festivities reached their peak on the last day of the Montreal meeting. The city fire brigade mounted a virtuosic display: after driving through the McGill grounds they shot streams of water over the roofs of Redpath Museum and Molson Hall. More local colour came from a lacrosse match that pitted the 'Montrealers' against the Indians of nearby Caughnawaga. In the evening, 2000 guests attended a reception at the Victoria Rink. Geological maps decorated the walls of the skating establishment, while electrical instruments were displayed alongside tables laden with refreshments.

According to two historians of the British Association, what the association actually did 'counted for less than what it was believed to stand for'. They contend that its activities were so heavily laden with cultural connotations that its history can be best understood as a set of symbols, images, and evocations, all skillfully manipulated by its directors. The BA transformed science into a 'visible cultural resource' by holding annual meetings at different locations throughout Britain, creating local committees to take charge of the arrangements, and by carefully recruiting members from a wide range of constituencies. With thousands flocking to its meetings, the association became a 'huge unwieldy monster' altogether without precedent in the annals of British learned societies. Its visibility, in turn, imbued BA activities with special significance and permitted its managers to disseminate their own particular ideologies concerning religion, politics, and even what counted as science.

Arnold Thackray and Jack Morrell contend that the British Association fostered social cohesion under the banner of science. Yet, far from being run as a democratic 'parliament' of science as its rhetoric suggested, the BA was controlled by a closed, liberal Anglican and politically conservative oligarchy. Initially, the editor and writer David Brewster (1781–1868) was a central actor in this clique; his replacement was Vernon Harcourt (1834–1919), who subsequently handed the reins of power to the geologist Roderick Murchison. The group carefully steered a course between those who wanted government to take a strong hand in countering an alleged decline of science and those who thought British science was in the pink of health. By sidestepping the 'declinist' crusade to create paid positions for scientists, these 'gentlemen of science' avoided associating themselves with the cause of professionalization. They viewed science as an *avocation*, and as an opportunity to establish an intellectual reputation.

Whereas the BA establishment acted to undercut the traditional authority of London, it still worked to reinforce the interests of other centres of learning and culture, such as Cambridge, Dublin, and Edinburgh. Fundamentally, the ruling clique distrusted the common scientist; likewise, they hesitated to enter his home territory in the unpredictable hinterland, whether a 'coal hole' like Newcastle or a textile town like Manchester. The function of provincial philosophers, according to this view, was to swell membership rosters and thereby fill the organization's coffers with subscription fees. Since these funds would then be channelled in the form of research grants to the directors' pet projects and protégés, the provincials handed over their hard-earned shillings to an indifferent and metropolitan-inclined leadership. Although official rhetoric endorsed applied science and technology, grant allocations and government lobbies in fact enshrined pure 'Humboldtian sciences', especially astronomy, tidology, and terrestrial magnetism. When these concerns began to lose their initial lustre and the BA entered a period of decline coinciding with the rise of the new 'red brick' universities in the 1880s, the organization turned to exploit popularization, education, internationalism, and imperialism.

The common scientist

Whereas any organization, especially one of considerable size, will tend to be dominated and controlled by a smaller group, an examination of the membership and activities of scientific societies still supplies the best glimpse of the interests and concerns of the 'common scientist'. Theodore Hoppen has used the Dublin Philosophical Society to this end, which he sees as belonging to 'the submerged nine-tenths of seventeenth-century science', rather than to the historically visible residue.

Other historians have attempted to describe the common scientist, but detailed descriptions tend to founder on an inability to gather information about the group or even to define it persuasively. For this early period, for example, Derek Price has called attention to scientific instrument makers: surveyors, teachers of navigation, and makers of magnetic compasses. He argues that a close connection existed between scientific instrument-making, for example in the crafting of astrolabes, and copper engraving, such as that required for the production of anatomical plates. By the time of the creation of the Royal Society, hundreds of these artisans were working throughout London. As Price describes, if the instrument-makers were to avoid poverty and starvation, they had to become expert salesmen. Their livelihood depended upon persuading scientists that 'it was a smart and cultivated thing to buy a microscope or a slide rule'.[1]

The object of their sales pitch was another variety of common scientist, the 'science-loving amateur' at the other end of the social spectrum. Such individuals emerged from among the English and Dutch aristocracies during the second half of the seventeenth century. They cultivated science, particularly its experimental varieties, outside the halls of the universities. A good example is Robert Boyle, who completely devoted his time to chemistry although it never became a source of livelihood. (Nor was this an important consideration, as Boyle was one of the richest men in Britain.)

In contrast to the instrument-makers and the gentlemen virtuosi, whose characteristics can be gleaned only in an impressionistic way, much can be learned about the common scientist by exploring the rolls and activities of societies like the Dublin Philosophical Society. The typical member was one who tended to justify his scientific work by means of the Baconian glorification of empiricism, fact-gathering, and cooperative research. Spectacular, if trivial, experiments gained popularity; especially favoured were those related to chemistry and medicine. Consistent with this Baconianism, the Dublin society promoted the importance of utility, whereas more sophisticated scientific practitioners of the day saw such concerns as irrelevant to the advancement of abstract science. Elite scientists tended to cultivate the less experimentally based sciences, like mathematics and mechanics, which were seen as proper objects for solitary contemplation.

Hoppen suggests that the movement towards specialization in science developed from the approach to the natural world adopted by the common scientist, even if it derived from his weaknesses. Unlike a Newton or a Boyle, these individuals were less proficient at moving 'with skill and dexterity' among the various scientific disciplines. They were unable to master the spectrum of natural and physical sciences, and increasingly they were fortunate to be able to keep abreast of the complexity within a single realm. By virtue of this 'enforced specialization', argues Hoppen, the common scientist of the seventeenth century appears to be more 'modern' than his brilliant colleagues. Hoppen uses this development to support his hypothesis that 'methodological change in scientific studies sometimes proceeds from the submerged majority of moderately competent and usually forgotten practitioners'. He argues, moreover, that the Dublin Philosophical Society should be seen as typical of the 'mass of contemporary scientists' both in England and on the Continent.

The common scientist on the verge of our own time is revealed in David Allen's description of the evolution of the field club in Victorian Britain. These were local societies dedicated to the natural history sciences exclusively, thus further accentuating the tendency towards specialization described by Hoppen. The prototype was the Berwickshire Naturalists' Club, whose features would be imitated widely. Meetings lasted all day, not just an evening, and they took place in the field. No permanent premises were required, because meetings were held in various outlying parts of the region. The peripatetic character of these clubs meant that expenses were minimal, limiting the cost of annual subscriptions to around six shillings. Women were admitted as honorary members.

By around 1860, the field club moved into the scientific establishment, with major societies being founded in England's industrial cities. In Liverpool, Manchester, and Bristol, huge clubs emerged with high subscription fees, reflecting their lavish scale of operations. Traditional scientific austerity increasingly deferred to dinners, dances, and receptions. Field clubs became part of local routine, bringing together individuals on a regular basis. Membership was both permanent and undemanding; national societies were scarcely so comforting and available. As Allen summarizes the world of the natural history club: 'It was a gentle world, a world full of tenacious loyalties, that could be dull and uninspiring but that was normally sure of its usefulness and, come what may, was nearly always busy.' The scientific society had come full circle; its routine had become quotidian.

Scientific clubs for everyone

From the middle of the nineteenth century onwards, there seemed no limit to the increase in number and membership of scientific societies. This growth mirrored the burgeoning size and remarkable diversity of the scientific enterprise; both specialist professionals and avid amateurs could find organizations that responded to their particular interests. Some societies, like the Paleontographical Society, were created to answer the publishing needs for a tiny circle of experts. Others, like the Quekett Microscopical Club (organized by the editor of *Science Gossip*), offered a venue for the exchange of ideas to those intimidated by the prestige of the Royal Microscopical Society. On occasion, informal and nearly secret clubs were formed to serve the interests of an elite: such were the student-like 'Red Lion's Club' formed by Edward Forbes (1815–1854) and his friends in London or the Society of Arcueil outside Paris.

In some instances, shared disciplinary concerns were of paramount importance. The Geological Society of London, for example, nurtured and protected offspring whose scientific interests complemented its own. The Paleontographical Society, which grew out of the fossil-collecting London Clay Club in 1847, held its meetings in the rooms of the Geological Society. Many of its local secretaries were fellows of the older society. Some ten years later the Geologists' Association was created by Thomas Wiltshire (1826–1902), James Tennant (1808–1881), and Samuel Joseph Mackie, all members of the Geological Society. In the view of its founders, the new organization would diffuse geological knowledge to those who possessed neither the time nor means to master the subject and become fellows of the parent institution. Unlike the shortlived Junior Geological Club, started in London in 1864 by other Society members, the Geologists' Association achieved lasting success by its popular excursions to rock formations and fossil beds throughout the British Isles and abroad. The president of the Geologists' Association reaffirmed its symbiotic relationship to the Geological Society on the occasion of its thirtieth anniversary celebrations: 'If the society represents the fountain head of geological wisdom, we rather represent the cistern which receives this knowledge from its source and distributes it in a convenient form, rendering it accessible to many who would find it less easy to drink directly from the primal fount.'[2]

In other cases, local interests weighed as strongly as scientific ones in bringing individuals together. The Lunar Society of Birmingham, for example, held its monthly meetings at the time of the full moon so that members could return home on country roads by moonlight. The 'Lunatics' included Erasmus Darwin (1731–1802), James Watt (1736–1819), Matthew Boulton (1728–1809), Josiah Wedgwood (1730–1795), and Joseph Priestley. Some early societies limited membership to those living within a radius of thirty miles of the society's headquarters. The legendary 'conviviality' of the proceedings in Scottish cities reached nearly absurd lengths when members were awarded doctorates 'of mirth and merriment'.

The overseas extension of European models

Just as academies and associations for the advancement of science gave scant regard to national frontiers, the model of local, voluntary scientific societies inspired amateurs outside Britain. Here we will examine in detail the early history of one such society, which is representative of the larger group. In 1827, a group of professional men and entrepreneurs founded the Natural History Society of Montreal. Over the next several decades the fortunes of the Society waxed and waned, according to the degree of enthusiasm of the moment. But when John William Dawson, newly appointed principal of McGill University in 1855, arrived in town, an era of prosperous stability ensued.

Dawson assumed the presidency of the Society within a year, and the organization underwent a metamorphosis as he imposed a new rationality on its structure and activities. A variety of committees and subcommittees were struck to deal with important issues. The membership was supposed to be divided by discipline, with each group taking charge of its respective division of the Society's museum. The size of the council and executive committee was enlarged; a curator was hired to oversee the museum. Whereas previously the proceedings of the society had been published in local newspapers, under Dawson's guidance the Society acquired a journal where its proceedings and reports appeared regularly. As a result of Dawson's efforts, the Society inaugurated an imposing new building in a fashionable part of town in 1859.

The motto of the society, *Tandem fit surculus arbor*, proclaimed the great accomplishments that could be expected as a result of bringing together the modest efforts of individual members. An introductory lecture that Dawson delivered for a popular natural history course reiterated this theme. He called upon all Montrealers to take up the cause of natural history, since 'in truth a large proportion of the new facts added to natural science are collected by local naturalists, whose reputation never becomes very extensive, but who are quoted by larger workers'. Whereas 'good works of art are rare and costly, good works of nature are scattered broadcast around our daily paths'. Dawson proceeded to list 'the most promising local fields of inquiry'; the only interruption to his full-blown Baconian rhetoric came when he considered the geological domain (one of his own specialities), which he believed presented the greatest temptation to 'vagaries'.[3] The important activity of compiling a catalogue of the local environment would bring together gentlemen and scientists, old and young, amateurs and professionals – perhaps even English and French, Protestant and Catholic.

By the end of Dawson's first decade in Montreal, the Society entered a new era of prosperity. A series of popular lectures attracted numerous auditors. A vigorous membership campaign brought in nearly a hundred new members in 1862 alone, increasing the rolls by nearly one-third. The Society inaugurated a well-received annual *conversazione* the following year.

At these gala affairs, displays of microscopes and other scientific instruments regaled spectators, while a military band played from the gallery of the Society's museum. Tables showed specimens of the puzzling fossil fragment called *Eozoön*, a prehistoric human skull, and bones from a mammoth. Among the apparatuses demonstrated were a mechanical pendulum, as well as an electric and fire-alarm telegraph. An exhibition of chemical experiments was cancelled as a result of the concern that 'the gases emitted in the performance of the experiments might not tend to improve the ventilation of the room'. In 1865, amidst a general economic depression, more than four hundred people attended the annual conversazione. Soon the details of these affairs were more finely polished. A ladies' committee decorated the rooms. Flowers were exhibited from the private conservatory of a member; another designed permanent ornaments 'to recall to mind the names of the leaders in different departments of science, emblazoned with mottoes and emblems'. More music came in the form of the Germania glee club and a church choir; soap bubbles were released, filled with gas. Live animals were shown, as well as an aquarium with sea anemones. The eighth conversazione welcomed Prince Albert; on that occasion, geological maps and sections lent by the Geological Survey of Canada were displayed. By the late 1880s, the conversazione ceased to be an annual affair, but the gathering remained 'the greatest means of popularizing science in Montreal'.

In what was intended to be an even more dramatic display of popular involvement, in 1868 the society voted to open its museum to the public on Saturday afternoons. Initial vandalism persuaded the mayor to supply two policemen for these occasions; even so, no more than 130 visitors attended in a weekend (and sometimes as few as thirty). The inanimate collections of the museum failed to fire the popular imagination like the circus conversazione did.

More successful were annual field excursions to outlying areas like St Anne-de-Bellevue, St Helen's Island, and Mont St Hilaire, which awarded prizes to the most zealous collectors for the day. A special train (supplied by railway magnate Cornelius Van Horne, whose daughter was among the excursionists) took the group to Joseph Papineau's magnificent estate and personal museum at Montebello, where they were joined by a contingent from the Ottawa Field Naturalists' Club. On another occasion, the town of St Eustache greeted participants with a main street decorated with flags, bunting, and banners proclaiming 'HONOUR TO SCIENCE and BE THEY WELCOME.' At a fall field trip to horticulturist Charles Gibb's extensive apple orchards in Abbotsford, Dawson lectured to the geological party atop nearby Yamaska Mountain, despite a snowstorm that had intervened during the group's ascent.

Dawson's presidential address to the second conversazione reveals why Montrealers found the activities sponsored by the Society so attractive. The Society aimed to promote social harmony; as Dawson expressed this attitude, 'a true naturalist is never an ill-natured man'. It cultivated the study of 'things that make for peace . . . for the common benefit of all', amidst whatever 'perturbed social and political elements' might prevail. Furthermore, the Society's umbrella protected against the ill winds of modern, industrial civilization. The members kept up 'a testimony in behalf of nature' during 'these artificial days'. In an interesting metaphor (given the Quebec context and the nearly complete disregard of the French element therein), the Society stood for 'the lily of the field' against 'all the glory of modern art'. Nor were the utilitarian functions newly embraced by the Society to be ignored. The Society came to the rescue of anyone 'puzzled' by some 'unaccountable phenomenon in air or earth' or menaced if 'any impertinent insect or fungus ravages your farm, garden, or orchard'; it comforted those threatened by some 'perversion of mining enterprise'.[4] Perhaps to justify their recently obtained government grant, the Society began to take an active role in these socially important issues around this time. They lobbied for the passage of legislation to protect insectivorous birds, set up a committee to examine the reasons for the decline of apple orchards on the island of Montreal, and explored the use of Canadian fibres in paper and fabric manufacture.

Notwithstanding this prosperity, the Society fell upon hard times beginning in the 1870s, entering, in Dawson's terms, into a 'slow and languid condition of our progress'. Longtime enthusiasts had little time and energy left over after their other commitments. The proceedings of the meetings (printed in the *Canadian Naturalist*, the Society's journal, and recorded in the minute books) became perfunctory, registering only a few donations to the museum and the reading of an occasional paper. Obituaries of leading members began to appear with increasing frequency. Meetings were poorly attended (during the late 1880s, attendance declined to about a dozen), and few members came forward with new data or collections. The editors of the *Naturalist* commented upon the 'scantiness' of the material that they had received, and urged the membership to execute more 'scientific work' and record their results in the journal. Around the same time, the Society's curator of thirteen years tendered his resignation, the provincial government reduced and eventually eliminated their grant-in-aid, and the Geological Survey – itself the child of the Natural History Society – moved its staff and museum to Ottawa.

Dawson consoled himself for the Society's lacklustre performance by reflecting that 'scientific societies in a country like this are of slow growth'. None the less he believed that 'after an existence of half a century, and after having held up the torch of science for that long time in a community, this society should have acquired greater strength'. Among the alternatives that he pondered with a view to reinvigorating the Society during 'its second half century' was to invite the American Association for the Advancement of Science (AAAS) to meet in Montreal in 1882.[5] This was intended to regenerate flagging enthusiasm for the Society's pursuits both among the membership and the community at large. It would also celebrate the golden anniversary of the incorporation of the Society, and the silver anniversary of the first Montreal meeting of the AAAS in 1857.

The announcement of the Americans' return to Montreal gave an enormous boost to the Natural History Society's morale, a reinvigoration that only slowly dissipated throughout the 1880s. The *Canadian Naturalist* proclaimed that the 'Society is feeling the influence of the good times upon which our country is now entering'. Their museum, newly opened to the public before the Sommerville lectures, attracted thousands of visitors on those evenings. City schools and colleges were invited to visit the museum on Saturdays; as a result, the number of annual visitors increased from 451 (1888) to 1192 (1889) to 2094 (1890). The subsequent extensive renovation, repair, and rearrangement of the museum seemed to bear out the earlier contention that the Society might benefit from the departure of the Survey museum, as there would be less competition for public attention in the city. The Natural History Society, as a result, approached the 1890s with renewed confidence. This assurance reflected not only popular support but also financial wellbeing, coming about from the abolition of debt, increased rents from rooms, and the reinstatement of a provincial grant.

The Montreal Natural History Society illustrates the high degree of dependence of local scientific societies on individual initiative and enthusiasm. They prospered so long as volunteers actively supported their endeavours; they waned whenever that spirit collapsed. The proliferation of such societies across the landscape of nineteenth-century science testifies to the strength and persistence of the tradition of participatory science. Ultimately this vision – predicated upon the important role of the common scientist – would be subsumed by the forces of professionalization.

Women in science

Despite the broadening base of popular participation in the scientific enterprise, one important social group remained excluded from the deliberations of national, most major disciplinary, and even some local societies. These were women. A new era began with the establishment of natural history societies in Britain, but even then increase in the representation of women occurred with almost glacial slowness. As David Allen points out, since their foundation during the early nineteenth century, the zoological and botanical societies of London had admitted women as full fledged members. But this procedure made these societies highly exceptional in the metropolitan context. The Royal Society organized a special Ladies' Conversazione in 1896, but even the more democratically-inclined British Association refused to allow a woman to sit on a committee the following year. The situation was different among provincial societies, who opened their doors to women in the 1870s and 1880s.

The policy of excluding women from the 'academy' – whether scientific societies or universities – enjoys a long tradition in Western civilization. Historian David Noble has argued that the clerical ascetic culture of the Latin Church hierarchy became the culture of learning by the time of the High Middle Ages. This development had been initiated as early as the fourth century, with the emergence of male monasticism and its emphasis on total clerical celibacy. Western science developed amidst this orthodox culture; women had no place in it.

Noble sees the subsequent sweep of Western history as oscillating between periods of orthodox authority, when this vision remained triumphant, and rebellion, when women and other *exclusés* tried to assert a different ideology. Periods of anticlericalism became associated with women achieving some representation, but such times were relatively infrequent despite the promise of the Reformation and Enlightenment. It was not until the nineteenth century, and particularly in 'frontier' societies far from the established order, that women began to find permanent positions in science. This change was associated with the democratizing effect exerted by the advent of coeducation, the stirrings

of religious revival, and the inexorable progress of industrialization.

Nineteenth-century advances occurred when capitalism
fundamentally transformed social production by abolishing
distinctions between women and men in the labour market.
If both sexes could work together on the assembly line, the
compulsion to exclude women from learned societies and
universities lost its force. Yet the subsequent halting progress of women as they confronted the power of the male
academy and its associated professions reveals, according to
Noble, the misogynist origins of Western scientific culture.

The uneven advancement of women's fortunes in scientific institutions is mirrored by the kinds of science that they
chose to cultivate or, according to radical feminists, were
permitted to pursue given the barriers erected by men. The
landscape of elite science – the 'frontiers of discovery and
controversy', in the words of Marina Benjamin – has been
dominated by male practitioners, but women, like other
scientific 'minorities', are revealed when a broader perspective is taken. By the eighteenth and nineteenth centuries,
women worked in disciplines like astronomy, natural history, and even as science popularizers. Voltaire's mistress,
the *marquise* du Châtelet (1706–1749), expounded Newton's world system to the *philosophes*; the English mathematician Mary Somerville (1780–1872) returned the favour
by explicating the nuances of Laplace's *Système du monde* to
her countrymen.

Unlike the situation a century before (and probably a
century later), around 1700 women began actively to pursue the physical and especially the mathematical sciences.
In England, the *Ladies' Diary* dedicated itself to posing sophisticated 'enigmas and arithmetical questions' to its female
readership. The science of mathematics was eminently
accessible to the noblewoman virtuoso; neither special
equipment nor an extensive library was required for its
cultivation. Mathematics even seemed to possess utility in
an age that attributed the success of Dutch commerce to

the computational skills of the Dutch housewife. Consonant with this orientation, in Germany nearly 15 per cent of astronomers were women. Londa Scheibinger contends, furthermore, that craft traditions on the Continent gave women limited entree to the world of science. To the extent that modern science recognized and embraced these traditions, women found support; to the extent that the alternative university model (successor to Noble's clerical monasticism) was cultivated, women were excluded.

One of the earliest women naturalists, Maria Sibylla Merian, daughter of a Frankfurt artist and engraver, was reared in the culture of the European craft guilds. At the age of thirteen, she began to serve an informal apprenticeship in her stepfather's shop, where she learned the techniques of illustration. She married one of his apprentices, moved to Nuremberg, and established a business selling fabrics decorated with her own flower paintings. Before 1700, she wrote and illustrated two scientific treatises, one on the metamorphoses of caterpillars (predating Marcello Malpighi's [1628–1694] pioneering work by a decade), the other on flowers, both of which portray 'a particular and interconnected process of change'. In the view of a recent biographer, Natalie Zemon Davis, Merian's works are remarkable for emphasizing 'interactions in nature' and 'transformative organic processes', as exhibited in her paintings of larvae, caterpillars, and even the holes that these insects chew in leaves.[6] Around this same time, Merian left her husband in Nuremberg (rejecting, as well, her married name), and joined a Labadist religious community in West Friesland. (The Labadists were radical members of the Dutch Reformed Church who espoused renunciation of worldly pleasures.)

The Friesian terrain of heaths and moors fostered new nature studies by Merian, who introduced frogs into her studies of insects and plants. Still, membership in the austere community meant restrictions on the flow of ideas and publications from the outside world, as well as the contraction of publishing opportunities.[7] After ten years with the Labadists, whose colony had by this time begun to disintegrate, Merian moved to Amsterdam. There she established the same kind of business that she had operated in Frankfurt, this time producing watercolours of natural objects for the wealthy burghers. The cultural resources of the city of more than 200,000 inhabitants overwhelmed her, particularly the natural history specimens brought back from the Indies in its museums and botanical gardens. She met important collectors and naturalists, thereby expanding her entomological studies. Merian yearned to see the exotic objects alive, in their natural habitat. In 1699, at the age of fifty-two, Merian and her younger daughter set out for Surinam, to study exotic insects. She stayed for two years in the stifling tropical climate, formerly home to a Labadist colony, where she laboured without patronage by business or national interests. Malaria forced her to return to Amsterdam in 1701 with abundant specimens and paintings in hand, enabling her to write and publish in Dutch and Latin her major work, the beautifully illustrated *Metamorphosis of the Insects of Surinam*, four years later.

Unlike the static representations of the day, Merian (as J. J. Audubon would do a century later for birds of the New World) portrayed her subjects in violent and even frightening poses. Similar to Audubon's later naturalistic approach, Merian placed insects of different species and animals from several orders in one composition if they fed on the same vegetation. For example, a single illustration might contain various butterflies and moths, wasps and flies, a frog or lizard, and European plants alongside American plants. Merian's enduring success came from her novel emphasis on 'breeding, habitat and metamorphosis', unlike traditional 'expository' renderings. Merian was at once an iconoclast and an anachronism; her accomplishments were less a reflection of the times than a result of unorthodox ambition and unbridled individualism. In the fascinating assessment of Natalie Zemon Davis, Merian was no mere naturalist, but a woman driven to exceptional paths by a communal religious experience that she embraced only later to reject.

A contemporary of Merian's, Maria Winkelmann (1670–1720) was the most remarkable of the German women astronomers. She married Gottfried Kirch (1639–1710), thirty years her senior, who became astronomer to the Berlin Academy of Sciences. She earned some recognition as his unofficial assistant. Especially important to her reputation as an astronomer was her discovery of the comet of 1702; in addition, she produced inventories of celestial events, wrote tracts (as much astrological as astronomical), and prepared calendars. Upon her husband's death, she was denied the position of assistant calendar-maker, turned out of the observatory, and forced to take refuge with her four children in the private observatory of a nobleman. Winkelmann eventually returned to the Berlin Observatory as assistant to her son, Christfried Kirch, although she was driven out by observatory officials once again a year later.

One place where women were allowed to cultivate scientific interests was the French *salon*. In this atmosphere of cultured hospitality, it is hardly surprising that women should play a leading part: they ran them and set their tone. As hostess, the *salonnière* determined the guest list and the topic under discussion. These enclaves of noble talent took the name of the aristocratic patroness at their helm: the salon of Madame Lambert (Anne-Thérèse de Marguenat de Courcelles *marquise* de, 1647–1733), of Madame Tencin (Claudine Alexandrine Guérin *marquise* de, 1681–1740), or of Madame Marie Thérèse Rodet Geoffrin (1699–1777) for example. These ladies were veritable 'uncrowned queens', perhaps esteemed more highly than the actual queen for their charm and grace. Charles de Secondat *baron* Montesquieu (1689–1755) described the salonnières as forming 'a kind of republic . . . a new state within a state; and whosoever observes the action of those in power, if he does not know the women who govern them, is like a man who sees the action of a machine but does not know its secret springs'.[8]

Despite Montesquieu's assertion, we cannot yet affirm whether these women were true 'intellectual power brokers', as Londa Scheibinger has maintained. Did they indeed make or break academic careers, or was the exercise of their 'power behind the throne' more imagined than real? It is difficult to resolve the issue because the salons left no permanent, published record. Scheibinger comments, furthermore, that the salonnières failed to act as patrons to aspiring women, only to promising men. That the French language coined a unique term to recognize the role of these women may be testimony enough to their intellectual significance.

Certainly the salon afforded women who were excluded from the influential institutions of society a means of self-improvement and intellectual stimulation. The salons helped to erode distinctions based on rank and class; similarly, differences based on sex became less important. As historian Carolyn Lougee puts it, in the 'melting-pot' ambience of the salon, social distinctions blurred; a uniform code of behaviour eliminated friction and made the salons 'internally unstratified'.[9] The democratic egalitarianism inside the salons resonated with the *philosophes'* commitment to open Baconian science to all talents.[10] Expressing the significance of the salon, historian Evelyn Bodek calls it *'the* major channel of communication', which 'served as newspaper, journal, literary society, and university' for the leisured and well-to-do. The gathering of people brought together in the salon was so important that they were referred to as *le monde*, literally the world.[11]

The success of the salon owed much to its adaptation to the needs of eighteenth-century French aristocratic women. Most of them had passed through convent schools, where they were poorly trained to deal with modern philosophical issues, whether literary or scientific. The salons, in contrast, encouraged learning about a range of worldly matters, affording many women their first opportunity to engage these concerns. In the view of Dena Goodman, the salonnières were intelligent women committed to educational and Enlightenment ideals.[12] At the same time, the salons incorporated the conventions of courtly life – essentially, education was not divorced from pleasure. According to Bodek, philosophical discourse offered a happy diversion to women of leisure, an escape from the boredom of daily life and its stifling routines which respected the constraints of their limited education and larger expectations. That the salon accommodated itself so well to the reality of eighteenth-century life makes it no less important as a serious and socially useful institution that came to sponsor productive intellectual activity.

The example of Madame du Châtelet

One of the finest examples of the powerful and scientifically adept salonnière is Gabrielle-Emilie Le Tonnelier de Breteuil *marquise* du Châtelet. Du Châtelet is best known for her sixteen-year association with Voltaire (1694–1778), with whom she retired to her estate in the Champagne countryside, at Cirey-sur-Blaise. The liaison with his beloved Emilie allowed Voltaire to escape arrest for the publication of his *Lettres philosophiques* and enabled him to exploit du Châtelet's important connections at court. Du Châtelet, in turn, used Voltaire to make Cirey into a sparkling salon, despite its location far removed from the intellectual orbit of Paris.

Together du Châtelet and Voltaire explored the nuances of the Newtonian system, assisted by a small circle of French Newtonians, including Pierre de Maupertuis (1698–1759) and Alexis-Claude Clairaut (1713–1765). Apparently du Châtelet was particularly inspired by Maupertuis's lessons in mathematics; soon her facility in physics and mathematics overshadowed that of her mentor and lover, Voltaire. She wrote a 'Lettre sur les éléments de la philosophie de Newton' for the *Journal des sçavans*, which reviewed parts of

Voltaire's *Eléments de la philosophie de Newton* (1738). There she discussed the section of the work that dealt with Newtonian attraction. When Voltaire subsequently entered a prize competition sponsored by the Académie des Sciences on the nature of fire and heat, du Châtelet decided to submit her own essay, which she had written surreptitiously. When neither paper won the prize, du Châtelet admitted her subterfuge and Voltaire arranged to have both memoirs published.

When affection between du Châtelet and Voltaire waned, du Châtelet continued her pursuit of Newtonian mechanics. By the late 1730s, she became a convert to Leibnizian notions, particularly the doctrine of *forces vives*. She found new intellectual guidance from Samuel König (1712–1757) and Jean Bernoulli the Younger (1710–1790), which assisted her completion of the *Institutions de physique*, published in 1740. Her espousal of the Leibnizian system embroiled her in controversy with the leading eighteenth-century mechanists, including Euler, Musschenbroek, and Willem Jacob 's Gravesande, a quarrel, according to René Taton, 'that she had helped to sharpen but that surpassed her competence.'[13]

Du Châtelet is best remembered for translating the *Principia Mathematica* into French and supplying a commentary. The job was a natural extension of her interests and talents, combining her love for Newtonian mechanics with her skill as a translator (she had earlier translated the *Aeneid* into French). The work was published in 1759, some years after her death from childbed fever at the age of forty-two. Even the accomplished du Châtelet, then, is remembered as a synthesizer, translator, and popularizer, not as an original thinker in her own right.

A recent study of du Châtelet's metaphysics explains why she was forced to assume something less than a pioneering role in the history of science. For one thing, her relationship with Voltaire constrained her scientific work. She functioned as his amanuensis during his times of illness and seclusion. The task involved handling his extensive correspondence and editing his works to avoid further clashes with censors. These frequent crises took precedence over her own intellectual activity, as did the obligations of her social status. She spent years renovating the dilapidated hereditary estate at Cirey and was forced to make repeated trips to Brussels to represent her family's claims in a complicated lawsuit. Janik concludes that, far from working in academic isolation, du Châtelet was constantly interrupted by demands of the external world. She often had to resort to correspondence to contact contemporary thinkers, rather than consulting or studying with them in person. That she could accomplish so much was made possible only by 'her exceptional energy and stamina'. The projects at which she excelled – translation and commentary – could be carried on intermittently, unlike systematic scientific treatises which would have required her undivided attention over long uninterrupted periods of time.[14]

Janik also maintains that historians have underestimated the originality of du Châtelet's contributions to late eighteenth-century metaphysical debate. She contends that scholars have failed to recognize that du Châtelet thoroughly reinterprets Newtonian mechanics from a Leibnizian perspective. Her principal concern was to consider metaphysical issues, particularly the relationship of God to the universe and the necessity or rationality of scientific truth, although she also undertook to demonstrate the compatibility of Newtonian and Leibnizian mechanics.

Women elsewhere

Whatever influence women may have wielded in the salons did not compensate for exclusion from the corridors of power in national academies and universities. Within these institutions, they could be found only in the lower echelons: as illustrators and preparators, particularly in fields related to the medical and natural history sciences, and as assistants to researcher husbands, especially in astronomy. On occasion, their work appeared in society proceedings and received prestigious awards and prizes, but they were named only to lesser scientific societies. Intellectual women were clearly without honour in their own land. Madame du Châte- let, for example, was elected to several Italian learned soci- eties but effectively barred from the Académie des Sciences in Paris. (Her attack on the Cartesian doctrines of the Academy's perpetual secretary, Jean-Jacques de Mairan [1678–1771], did nothing to promote her academic cause).[15]

If women were denied access to the institutional pinnacle of the scientific enterprise, did they also fall prey to hier- archical distinctions amongst the scientific disciplines? That is, were they more likely to cultivate the biological rather than the physical sciences, directing their attention to the lower levels of the Comtean pyramid where application to the human situation took precedence over abstraction?

During the eighteenth century, women seemed to be as comfortable with mathematical calculations and stargazing as they were with collecting fossils and sketching plants. To adopt modern parlance, 'doing science' in any of its guises became a fashion statement. As literacy levels and leisure increased for all classes of society, the popular appetite for scientific diversions correspondingly grew. No longer was science the exclusive domain of polite society, but partici- pation broadened to include the rising bourgeoisie. If salon women acted as 'purveyors of culture', women generally became equally enthusiastic consumers of that culture. As Joan Landes puts it, 'a lively market in cultural goods' pro- vided enjoyment for a newly enlarged public, in science as well as in literature and art.[16] Because women were largely excluded from the elite enclaves of high scientific culture, they embraced the cultivation of science as a pastime, avidly reading all manner of popular scientific publications.

Eighteenth-century women also became enthusiastic par- ticipants in public scientific experiments and displays. Franz Anton Mesmer (1733–1815) purveyed his electrical and magnetical treatments to female patients in particular, with whom he enjoyed special success. They were so smitten with Mesmer and his cures that they prevailed upon Queen Marie Antoinette to allow Mesmer to stay in Paris, when he was about to be driven out by the joint efforts of the Académie des Sciences, the Royal Society of Medicine, and the Faculty of Medicine. A commission struck by the Academy to investigate Mesmer's practices was particularly concerned with the delicate moral question of the relationship between Mesmer and his women clientele. Critics accused him of putting older women to sleep while subjecting younger ones to *titillations délicieuses*.

It is not surprising that women associated themselves with a popular scientific movement under attack by the government and established science. As Robert Darnton emphasizes, Mesmer offered his Parisian clients 'a sort of fashionable parlour game for the wealthy and the well bred'. Places at his magnetizing 'tubs' had to be booked in advance, like seats at the opera; the tub for 'ladies of breeding' was marked by flowers. Mesmerism fed the scientific curiosity of the public, just as did the Montgolfier brothers' balloon flights. (Any woman who had missed the ascension from the Champ de Mars could recapture the moment with a souvenir snuffbox, fan, or brooch bearing the 'sign of the balloon'.)

For women, perhaps even more so than for men, Parisian fashion dictated embracing the latest scientific fad. Engravings show women as equal partners with men in the enjoyment of all manner of scientific pastime, but they are not prime movers in the scene. They do not control the action, whether directly or at a distance. This movement towards the periphery occurred just as the gap between the professional and amateur scientist widened. As a result, women were forced into the amateur ranks from which they could exit only with the greatest difficulty a century later.[17]

Against this background of limited success and extraordinary perseverance on the part of scientifically inclined women of the eighteenth century, the nineteenth century seems to represent a step backwards. In England, Victorian standards dictated different behaviour and divergent modes of activity for men and women. As one writer on natural history points out, women were not expected to achieve anything, in an age when men set for themselves remarkable tasks of physical and mental endurance, even denial, which bordered on masochism. Women were encouraged to cultivate dilettante interests, but real intellectual assiduousness was discouraged. The most important scientific role for women was in the promotion of museums, zoological and botanical gardens, and local natural history societies. Society approved of women as handmaidens of science, not as scientists in their own right.

There were exceptional peaks across this rather bleak landscape. One was the Scotswoman Mary Somerville, whose *Mechanism of the Heavens* (1831) set out to demystify Laplacean mechanics. Despite her mathematical interests and prodigious talent, Somerville had been systematically discouraged from intellectual pursuits by family and friends. Only with her second marriage – to her cousin, an unremarkable medical doctor – did she find the opportunity to emphasize her previously clandestine study of physical scientific writing above the incessant demands of a busy domestic life and myriad social obligations. Her career blossomed in the fluid, 'companionable' scientific circles of London, where learned societies and the homes of fellow practitioners were within easy walking distance. With the support of colleagues like the astronomer John Herschel, Somerville produced another major treatise, *On the Connection of the Physical Sciences* (1834). This became a 'scientific bestseller', going through nine editions and 15,000 copies in her lifetime. In William Whewell's review of this work for the *Quarterly Review*, the gender-free term 'scientist' was first coined.

Mary Somerville became the darling of the group that wanted to reinvigorate British science; she was persistently courted by the British Association for the Advancement of Science, although she always declined their entreaties to become an active participant. She was made an honorary member of the Royal Astronomical Society (which she visited only by special invitation and with an escort), but she was denied the highest honour accorded to her countrymen who were markedly her scientific inferiors: membership in the Royal Society. They admitted no female member until 1946. (Somerville, for example, sent her husband to use books in the Royal Society library on her behalf.)

The Académie des Sciences, which waited until 1979 to admit a woman, became notorious for its refusal to elect Marie Curie in 1910. At first Curie had hesitated to stand, remembering the distress of her late husband, who had been defeated the first time he ran and only narrowly elected the second time. Her opponent, however, was the sixty-six-year-old Edouard Branly (1844–1940), whose accomplishments in wireless telegraphy had been insufficient to assure his election on two previous occasions. Curie obtained the important support of Academicians Henri Poincaré and the secretary Gaston Darboux (1842–1917); she was placed at the top of the list by the physics section. None the less, in the end, fellow Nobel laureate Branly beat Curie in a second ballot by two votes. This rebuff posed no minor setback to Curie's career and especially to her self-esteem. Never again was she willing to stand for election to the Académie. Moreover, she refused to allow her work to be read to the Académie or published in its *Comptes rendus*, contributing instead to specialist journals in physics and radioactivity. Perhaps the first major national academy to admit a woman member was the Russian Academy, which elected the mathematician Sonya Kovalevsky (1850–1891) in 1889.

The role of women in the development of Western science has been understood from a number of perspectives. One view has emphasized institutional barriers to women's advancement; another has shown that women gained entry to certain disciplines only by dint of unorthodox approaches and only insofar as males tolerated their presence. A third kind of argument has been put forward by feminist theorists to explain why the history of women in science was less remarkable than it might have been. Evelyn Fox Keller maintains that the characteristic language of science promotes a network of gender associations; in particular, the notion of objectivity is equated with masculinity.

Fox Keller's argument may help to explain the persistent bias of the scientific enterprise against the efforts of women. Ever since the early years of the twentieth century, women have flooded into university courses in the sciences. Often talented students married each other, and a predictable pattern ensued: the woman subordinated her scientific interests to those of her husband. The history of science is about to be rewritten as the important contributions of wives of the famous are revealed. No longer can we ignore the work of Tatyana Ehrenfest-Afanassjewa (1876–1964) in statistical mechanics, Mileva Einstein-Marić (1875–1948) in relativity theory, or Grace Chisholm Young (1868–1944) in mathematics, all of whose scientific insights and interactions were important to their spouses' fame.

Appropriating: Science in Nations Beyond Europe

Pick up a basic physics textbook in Yogyakarta or Buenos Aires. It will be a glossy paperback in Bahasa Indonesia or (as they say in Argentina) *castellano*. There are diagrams and illustrations, equations and problem sets. Apparent even after a casual glance is the near isomorphism of the material with textbooks published in the Northern Hemisphere. Physical laws and principles appear in identical form. Tests of the laws manifest close resemblance, too: balls rolling down inclined planes, simple harmonic pendula, mirrors and lenses, all arrive at an identical result. Electrical circuitry follows identical equations; radioactive sources decay according to the same exponential law; X-rays produce one kind of photograph when diffracting through a crystal. The principles and practice of physics seem to be insensitive to culture and language.

We have seen that this convergence in approaches to the natural world was not always so (and it is still not so for fields such as medicine and economics). In the present chapter we examine the path that led the science of Western Europe to flourish in all parts of the world.

Colonial scientific societies

Among the earliest institutions beyond Europe to sponsor scientific research were voluntary associations or societies. For the most part the overseas societies took inspiration from the European movement for promoting broad-based scientific associations. The first of these associations, the American Philosophical Society, exhibits characteristics seen in many other locations.

The American Philosophical Society in Philadelphia emerged from nearly a generation of informal and inconclusive interest in associations for promoting scientific and practical endeavour. The first formulation of the Philadelphia society expired in the late 1740s, to be revived in 1766 by propertied, Anglican burghers. It merged with a related society run by mercantile Quakers to become in 1768 the American Philosophical Society for the Promotion of Useful Knowledge. Benjamin Franklin, the architect of perennial, corporate schemes for self-improvement, presided over the society from 1769 until his death in 1790, and Franklin's extraordinary visibility in Europe (notably during the 1770s and 1780s as American ambassador) projected the society as a place of serious ambitions. Modelled on the Royal Society of London, it operated by membership dues and Pennsylvania state grants. It achieved an early success in coordinating American observations of the transits of the planet Venus in 1769, which it published in the first volume of its *Transactions*.

The Philadelphia society became a clearing house for practical projects. Many of these concerned exploration of the continent. In 1794, for example, the first natural history museum in America, established by Charles Wilson Peale (1741–1827), moved to the society's rooms. The society also became an inspiration for related initiatives in Boston (the American Academy of Arts and Sciences, founded in 1780), and it spawned a host of lesser copies in Kentucky, New Jersey, and Connecticut, as well as the short-lived Académie des Sciences et Beaux-Arts des Etats-Unis de l'Amérique at Richmond, Virginia.

Four years after the inception of the Boston academy, a group of French planters and physicians established the Cercle des Philadelphes at Cap François on Haiti. Like its North American homologues, the Cercle enjoyed the patronage of the local government. It issued a torrent of publications, almost exclusively on matters relating to agricultural exploitation and public health. In 1789 the Cercle received recognition as the Société Royale des Sciences et des Arts du Cap François, and it obtained a confederate status with the Paris Académie des Sciences. The newly named society perished during the Haitian wars of independence of the 1790s.

The pattern followed in Philadelphia and Cap François is manifest in the Spanish colonial world, where there flourished a number of *sociedades económicas de amigos del país*, the economical societies of the friends of the country. They began in Europe with a Basque regional academy of this name, founded by Manuel Ignacio Altuna in Azcoitia and taking inspiration from Altuna's friend, the philosopher Jean-Jacques Rousseau. The economical societies played a leading role in promoting scientific activity. An economical society came to Lima in 1787 under the name of the Academia Filarnómica. One of the Spanish societies, with government assistance, sent Juan José d'Elhuyar (1754–1796) and Fausto d' Elhuyar (1755–1833) to the Freiburg Mining Academy for mineralogical study. The brothers, natives of Logroño whose parents were French Basques, had previously studied science and medicine in Paris, and the Spanish government wanted them to improve Spain's armaments in preparation for war with Great Britain. In 1783 Juan José became director of mines in New Granada; Fausto, an avid researcher, organized metallurgical missions to Peru and Mexico and founded the Royal Mining Seminary in Mexico City in 1792. The brothers, who are known as the discoverers of metallic tungsten, helped spread general interest in societies for practical innovation across the Spanish colonial realm. Interest went beyond the Americas. In 1781 one society was founded in Manila. Before it expired near the end of the century, the Manila economical society undertook practical projects similar to those of the Cap François natural philosophers.

Other Asian scientific societies emerged at the same time

as those in the New World. They shared organizational features with American cousins, but they manifested special interests. Typical is the earliest of the Asian societies, the Batavian Society of Arts and Sciences (Bataviaasch Genootschap van Kunsten en Wetenschappen), founded in 1778 at Batavia (today, Jakarta), the capital of the Dutch East Indies. It provided an occasion for administrators of the Dutch East India Company to assemble and circulate information about the colony. Like the scientific societies in Philadelphia and Cap François, the Batavian Society issued a learned publications series that surveyed local resources and endemic scourges. But unlike the New World academicians, who viewed their surroundings as something of a cultural *tabula rasa*, the Batavian philosophers of nature focused much of their attention on ethnographical matters. They sought to document local civilizations, and they provided descriptions of antiquities.

Within several generations the so-called 'Indological' focus of the Batavian Society dominated other interests, such as natural history. The motivation for this turn is not hard to locate. To consolidate its control over a very much larger, heterogeneous population, a small colonial elite must understand local customs. Ethnographical interest also received stimulation from the nature of civilizations confronting the Dutch rulers: Arabic and Sanskrit traditions, modified from their form in the Mediterranean and in South Asia, were everywhere apparent. There were countless local books and epigraphs to be fathomed.

The babel of languages gave pause to European savants. Arabic, of course, was well known (Edward Pococke [1604–1691] was the first Laudian Professor of Arabic at Oxford's Christ Church college in the seventeenth century), and Sanskrit had entered university philological seminars by the middle of the eighteenth century, but very little of the enormous extent of Oriental wisdom was known to the men who, as foreign invaders, presided over it.

In British India, the situation changed with the arrival in

1783 of William Jones (1746–1794), appointed judge at the Supreme Court of Calcutta. Within four months Jones had set up the Asiatic Society of Bengal on the model of the Royal Society of London. Jones took as his goal the analysis of all features of South Asian civilizations, ranging from literature and religion to knowledge of natural sciences. In the early volumes of the society's *Asiatick Researches*, in fact, about 60 per cent of the articles concern the sciences rather than the humanities. Much of the society's accomplishment derived from the labours of anonymous pandits – local Brahmins who provided translations of texts but who were excluded from society membership.

An analogous situation came to Egypt. In 1798 Napoleon Bonaparte disembarked in Alexandria, claiming to support the Ottoman sultan's attempt to regain authority from the local Mameluke regime. Napoleon's victorious army entered Cairo, and there Napoleon inaugurated a scientific society, the Institut d'Egypte, modelled on the reorganized French academies of sciences and letters, the Institut de France. Following the destruction of his fleet by the British, Napoleon spent more than a year in Egypt, trying to secure his political base and extend his domain. This involuntary sequestration allowed him to promote the scientific work of the Egyptian Institute. Much of the institute's interest went toward archaeology, but description of the physical surrounding is also evident in its proceedings. The discovery of a multilingual epigraph at Rosetta led to a dictionary for hieroglyphics and inaugurated the discipline of Egyptology. The French interregnum ended in 1801 with a British victory, and the British withdrew in favour of local regents in 1803. Napoleon's institute survived into the twentieth century as a forum for archaeological research, and it served as an inspiration for subsequent French colonial academies in North Africa, Madagascar, and Indochina.

Early colonial universities

The earliest overseas universities can be found in Spanish America during the sixteenth and seventeenth centuries. Entirely tied to the mission of conquest and conversion, the universities generally had the three higher faculties – theology, law, and medicine – as well preparatory arts courses. Science instruction, largely in medicine and the arts, was traditional and conservative. A measure of Spanish conservatism in the New World is the defence of Copernicanism by José Celestino Mutis before an academic-inquisitorial tribunal in 1774. The situation is not as bleak as may be imagined. Early in the seventeenth century, Galileo petitioned the Spanish ambassador in Florence to be named master pilot of Spain, in which capacity he proposed to train mariners in navigation and the use of the telescope. Galileo may have imagined teaching at a New World university to escape his clerical oppressors in Italy.

One of the brightest university settings was Lima. The first chair of mathematics at the University of San Marcos went to Francisco Ruiz Lozano (1607–1677), a Creole (or native son), in 1657. Seven years later, Ruiz observed a comet and published a book on his observations. His successor, Ramón Koenig, studied in Paris under astronomer Jean Picard and published an almanac, *El conocimiento de los tiempos*. Koenig's successor, Pedro de Peralta Barnuevo (1663–1743), a virtuoso scholar, observed a comet in 1702 and aided the French-Spanish expedition of 1735–44 to measure the arc of longitude. Peralto's successor, Louis Godin (1704–1760), was a member of that expedition. After the expulsion of the Jesuits in 1767 (following suppression of their order), the university became a college teaching the work of leading seventeenth-century natural philosophers. But not until 1793 did religiously subversive material, above all Copernicus's *De Revolutionibus*, find acceptance.

The Scientific Revolution nevertheless obtained a place in Creole society. Newtonian issues relating to action-at-a-distance and universal gravitation were less controversial than Copernican heresy. Early in the eighteenth century, the Jesuit University of San Pablo in Lima had a full cabinet of scientific instruments, including a Newtonian reflecting telescope and the master's *Principia* to accompany it. There was generally no interdiction against telescope, microscope, barometer, air pump, electrical machine, sextant, and pendulum clock. In the Jesuit universe beyond Europe, as Umberto Eco relays in his novel *The Island of the Day Before* (1994), machines were tools to be used for extracting truth from nature.

Another colonial university, at Córdoba in Argentina, arose from a Jesuit college early in the seventeenth century. In the eighteenth century, Jesuit professors José Sánchez Labrador (1717–ca. 1799) and Tomás Falkner (1707–1784), the first from Spain and the second from England, published surveys of natural history, which included paleontological observations. Falkner, for example, described the *Glyptodon* for the first time. The end of the Jesuits brought the university under Franciscan administration. Until 1829 science lectures at Córdoba were delivered in Latin.

By 1800 the Spanish universities in the New World had granted more than two thousand doctoral degrees, most in the professional faculties. They did not enjoy a monopoly on higher learning. Religious orders also ran colleges or *colegios*, schools to train teenagers for the higher faculties; early in the eighteenth century, in fact, the University of San Marcos functioned for the most part as an examining body and degree-granting institution for various colegios in Lima. In Mexico, the Jesuit colegios were responsible for diffusing Newtonian thought; with the suppression of the order, the Newtonian project found few non-Jesuit takers. These situations suggest that, just as in Europe, some colegios rose to surpass the level of instruction in many universities.

An exemplar is the collection of colleges at Puebla, Mexico. The colleges of San Juan Bautista, San Pedro Apóstol, and San Pablo were founded to implement the resolutions of the Council of Trent, that is, to train the clergy. By the seventeenth century they became known as the Tridentino Seminario Palafoxiano, after their patron Archbishop Juan de Palafox y Mendoza. Palafox designated San Pedro as a religious seminary and San Pablo as a degree-granting institution. In 1646 Palafox endowed a magnificent collegiate library, growing to nearly 100,000 volumes by the eighteenth century, many of these on themes in natural philosophy. Puebla benefited from an unusual patron, but the pattern exhibited there was typical. The College of San Ildefonso in Mexico City, for example, 130 kilometres away from Puebla, provided sustained science instruction and awarded bachelor's degrees. Another distinguished college in Mexico City was the Royal Mining Seminary, in the view of Alexander von Humboldt without peer as a scientific institution in the New World.

Higher education in eighteenth-century North America resembles the picture in Latin America. Among the richer institutions, notably the College of William and Mary and Harvard College, there were instruments and collections for science, along with a master to instruct in their use. Philadelphia had a concentration of scientific luminaries, and a number of these men – for example, Benjamin Smith Barton (1766–1815), Caspar Wistar (1761–1818) and Benjamin Rush(1746–1813) – taught at the College of Philadelphia and its successor, the University of Pennsylvania. Following British tradition, the College of Philadelphia offered a Bachelor of Medicine, and in 1768 ten students received the degree. In Philadelphia and elsewhere, colleges and universities soon awarded a doctorate of medicine, of which one requirement was a dissertation; again following European custom, the dissertation topics ranged across the natural sciences. In 1792, for example, chemist James Woodhouse (1770–1809) wrote a medical dissertation on the chemical properties of the persimmon tree; as late as 1844 paleontologist Joseph Leidy (1823–1891) took a medical doctorate for a thesis on the comparative anatomy of the eye of vertebrates.

Administrators of New World science considered European training essential, despite the examples of illustrious autodidacts like Franklin, Bartram *père* and *fils*, and later Nathaniel Bowditch (1773–1838). Barton, Wistar, and Rush studied in Great Britain before beginning their career in America, and Barton almost desperately sought to obtain a European doctorate *ad eundem* (he succeeded with the Academy of Lisbon). When Thomas Jefferson sought someone to map his new land, he turned to a Swiss savant, Ferdinand Rudolf Hassler (1770–1843). Spanish America continually returned to Iberia for university professors. Yet America's great tropical botanist Thomas Horsfield (1773–1859) obtained a medical doctorate from the University of Pennsylvania late in the eighteenth century, as did the captain of the first great American exploring expedition, Meriwether Lewis (1774–1809). And Antonio Alzate (1738–1799), Mexico's leading botanist, took degrees at Mexico City's university and San Ildefonso College.

Independent universities

New World universities and colleges multiplied over the first two-thirds of the nineteenth century. Science instruction followed a common pattern, whether in antebellum United States or in post colonial Latin America. The primary purpose of teaching – variously at religious, secular, or military schools – was practical. Professors trained men and women in the arts of calculating, analysing, measuring, building, and surveying. The graduates sold their expertise on the open market. Many worked on geological surveys or as engineers, whether in Vermont, Alabama, Mexico, or Chile. Research of an abstract and dispassionate nature was avocational.

A measure of New World practicality, as well as an indication of the complex relation between institutions in

Europe and America, may be found in the background to one of the grand schemes for an academy of learning, the New Harmony experiment. The motive force behind New Harmony was William Maclure (1763–1840), a Scottish merchant who had amassed a fortune in the transatlantic trade late in the eighteenth century. A firm believer in the ability of reason to better the condition of humanity, Maclure settled in the United States in 1796 and undertook extensive natural history expeditions. In 1815 Maclure engaged the naturalist-illustrator Charles-Alexandre Lesueur (1778–1846), veteran of Nicolas Baudin's (1754–1803) expedition to Australia of 1800–4, as a naturalist, and the two spent several years exploring the West Indies and the eastern seaboard. The latter expedition included Thomas Say (1787–1834), a self-taught Quaker naturalist, and Gerard Troost (1776–1850), a Dutch geologist with a medical doctorate from Leiden – both of whom, like Maclure, early members of Philadelphia's Academy of Natural Sciences.

In 1825, after having spent five years in Spain, Maclure joined British industrialist Robert Owen in purchasing the town of New Harmony, Indiana. They planned a utopian community, and they proclaimed that science would be its basis. The keelboat *Philanthropist*, a 'boatload of knowledge', left Pittsburgh and navigated down the Ohio and up the Wabash rivers. On board were Maclure, Lesueur, Say, Troost, Owen's son George, and a collection of books and instruments. The utopian schemes of Owen senior and Maclure foundered, and Maclure resettled in Mexico. But George Owen, Say, and for a time Lesueur stayed on as productive naturalists; Troost became professor of geology at the University of Nashville.

As the New Harmony story suggests, throughout the first half of the nineteenth century, scientists in the United States drifted in and out of academic settings. Formal credentials were not indispensable for a university appointment. Salaries even at the most ambitious institutions – the College of New Jersey at Princeton, Transylvania University in Lexington, Kentucky, and the Military Academy at West Point – were low, and the tasks were demanding. Natural history surveys, financed by federal or state authorities or by private patrons like Maclure, offered significant attractions to scientifically inclined minds trapped behind a lectern. When all else failed, one could anticipate earning a livelihood as an independent naturalist-entrepreneur – the path chosen by Owen, Say, and the epoch-making J. J. Audubon.

The situation changed dramatically during the last third of the nineteenth century. The practical streak of science in academia, encouraged by the vogue of positivism in Latin America and by American inventiveness generally, found vindication in the Second Industrial Revolution and especially in the way that technological progress determined the course of local wars. Useful benchmarks include the Morrill Land Grant Act of 1862 (providing federal land grants to finance state-controlled higher education in the United States), the reform in 1869 of the University of Córdoba in Argentina, and the founding in 1874 of the Ouro Prêto Mining School in Brazil.

The dramatic upgrading of university facilities led directly to the establishment of a particular kind of science. As in the course of the nineteenth-century higher learning progressively abandoned its traditional affiliation with church doctrine, the abstract principles of practical knowledge emerged as a new catechism for educated men and women. Familiarity with chemistry and physics became a sign of higher culture. The German doctorate of philosophy and its attendant ideology of *Wissenschaft* provided inspiration for New World educators – in sciences and in humanities. Doctoral programmes came to new universities such as Cornell, the University of California, and the University of La Plata in Argentina, as well as to older institutions like Harvard, Yale, and the University of Córdoba. With the doctoral programmes came an emphasis on faculty research.

The research university in the United States

The rise of the great American research university was as paradoxical and uneven as it was spectacular. It was paradoxical because research and research diplomas came to institutions whose primary aim was professional training (law, medicine, theology, and later pedagogy and engineering). It was uneven because the development of costly 'graduate' education came at the expense of relatively cheap undergraduate instruction, and hence tended to impoverish the institution. This has continued to be a significant question in higher education. Nicholas Murray Butler (1862–1947), who supervised the expansion of graduate work at Columbia University during the twentieth century, worried that research would compromise undergraduate instruction. Indeed, universities originally emphasizing graduate studies in the nineteenth century – Johns Hopkins, Clark, and Chicago – all eventually made peace with a large contingent of undergraduates.

Yale University was the pacesetter in graduate science instruction. It cobbled together a 'scientific school' in 1854, renamed 'Sheffield' after a donor in 1861, and in that year began awarding doctorates; the first American doctorate of philosphy went to a Yale physicist, and physicist Josiah Willard Gibbs obtained another one in 1863. Four influential American universities founded at the end of the nineteenth century – Cornell, the University of California, Johns Hopkins, and Chicago – were headed by Yale graduates.

The great engine of university science in America was the private university; the engine ran on endowed money. Universities with substantial endowment, for example Chicago's Rockefeller largesse and the New England fortunes behind Harvard and Yale, rose dramatically in the twentieth century. Those with a less secure financial base, such as Johns Hopkins, tended to mark time. Several bright nineteenth-century stars, like Clark and the Rensselaer Polytechnic, nearly faded from sight. An infusion of money transformed Princeton and the California Institute of Technology into major scientific players. Money made late entries, like Brandeis, into contenders from the outset. Careful marshalling of resources allowed small colleges like Brown to emerge with strengths in several scientific fields; other colleges, like Amherst and Swarthmore, deliberately held back from entering graduate work.

The elite stratification of research universities, coincident with the consolidation of scientific medical training in the wake of an influential report of 1910 by Abraham Flexner (1866–1959), stood in apparent opposition to democratic ideas, notably that of secular, public education. Before the First World War, scientific research did find a home at public institutions, notably in the Midwest and in California. Biologist Jacques Loeb and astronomer William Wallace Campbell (1862–1938) at Berkeley, astronomer William Joseph Hussey (1862–1926) at Ann Arbor, and physicist Jakob Kunz (1874–1938) at Urbana all illustrate ambitious beginnings. Yet it was only during the interwar period that major research enterprises arose at universities under state control. In physics, Berkeley led the pack under J. Robert Oppenheimer (1904–1967) and Ernest Orlando Lawrence (1901–1958), exporting 'big' science to state universities at Urbana, Bloomington, Ann Arbor, Columbus, and West Lafayette. In fact, a survey of cyclotrons in 1940 reveals that public universities owned eight of twenty-four cyclotrons in the United States; private universities (including MIT) accounted for fourteen, and private foundations had only two. Selected public universities successfully entered the elite ranks of higher learning. Doctoral statistics are revealing in this regard. In 1920 there were 31 doctorates in physics; in 1930 there were 106. The decade of the 1920s counted 729 physics doctorates, almost all awarded at 15 institutions, both public and private.

Scientific migration

The expansion of university science in America – indeed in the New World generally – owes much to foreigners. They came in part through the romance of an American tour, harking back to Humboldt and stretching forward to British physicist John Tyndall (1820–1893). The romance continued at century's end with international meetings, like the British Association for the Advancement of Science convening in Montreal in 1884 and Winnipeg in 1909, or the various World's Fairs and Exhibitions (Chicago in 1893 and St Louis in 1904). French, German, Italian, and British science masters set out to make their fortune in the New World. The change was not always happy. Mathematician James Joseph Sylvester, excluded from the academic world of England by his Jewish origins, briefly took the measure of the University of Virginia in the early 1840s and then returned home; Italian physician Pedro Carta Molino (ca. 1797–1849) and physicist Octavio Mossotti (1791–1863) lectured at the University of Buenos Aires between 1826 and 1834. But after mid century the trickle became a torrent. American Ethel Fountain Hussey (1865–1915), who effectively shared direction with her husband of the astronomical observatory at La Plata shortly after 1910, noted about the German academic émigrés: 'There isn't any crop more overdone ... than the supply of learned "doctors" who starve for something to do.'[1]

A discipline enjoying exponential growth, like physics in the years 1890–1910, experienced a European diaspora. To the University of Illinois, as we have seen, came Max Abraham and then Jakob Kunz. European physicists found positions at new universities in China. Many of the latter, for example St John's University and the Aurora University in Shanghai, were often under foreign religious authority, although secular institutions, such as the German-Chinese Institute of Technology in Tsingtao and the Tung-Chi University in Woosung, also had foreign (in this case, German) patronage. Well into the twentieth century, the metropolitan seat of the British Empire sent out scientist missionaries to staff universities in New Zealand, Australia, and South Africa. Physicists William Henry Bragg (1862–1942) at Adelaide and Ernest Rutherford (a recycled New Zealander) at McGill University in Montreal – both of whom won Nobel prizes for work begun beyond Europe – are exemplary in this regard. Around 1900 the United States government also sent out scientists to its colonies, notably chemist Gilbert Newton Lewis (1875–1946) to the Philippines.

Early twentieth-century scientific exchanges with the New World exhibit a special character. Europe provided synthetic vision, while America provided luxurious working conditions. In the 1920s, European theoretical physics flowed toward the United States: Paul Epstein (1883–1966), George Uhlenbeck (1900–1988), Samuel Goudsmit (1902–1978), Otto La Porte (1902–1971), Gerhard Dieke (1901–1965), Fritz Zwicky (1898–1974), John von Neumann (1903–1957) and Eugene Wigner (b. 1902) all found permanent posts in American universities, and many senior men, from Einstein down, occupied temporary posts. Younger American physicists had absorbed much of the new quantum mechanics by 1933, when National Socialism in Germany transformed the stream of scientific émigrés into a torrent.

South America lagged about a decade behind the United States in recruiting Europeans. Argentina especially, by virtue of its old university tradition and its neutral stance during the Second World War, took in hundreds of scientists through the 1940s – men and women of all political colours. Argentina, then ruled by Juan Domingo Perón, gained a certain amount of prestige through its scientific acquisitions. Nations of the Northern Hemisphere gave some credence to Austrian émigré Ronald Richter's (b. 1909) claim to have initiated controlled thermonuclear fusion in the early 1950s.

Richter received unusual privileges from Perón, among them viceregal powers over Huemul Island, located in a lake in the southern Argentine Andes near San Carlos de Bariloche. Money and resources were generally the motor of the exchange between the Old World and the New World. Overseas salaries were many times grander than those in Europe, and despite differences in cost of living, overseas life styles were often opulent.[2] Overseas facilities matched salaries. The largest Dutch observatory in the interwar period was at Lembang, on Java, and Dutch astronomers vied to spend time there. California observatories and American summer schools attracted young European talent. Facilitating this flow was the initiative of American philanthropy, notably the Rockefeller Foundation's General Education Board, which provided support for scientists to travel in both directions across the Atlantic and which underwrote significant expansion of key European scientific facilities. The Institut Henri Poincaré in Paris, Niels Bohr's Institute for Theoretical Physics in Copenhagen, the Mathematical Institute in Göttingen, and the Leiden Observatory all benefited from Rockefeller funding.

Emigrés Rutherford and Bragg received prestigious posts in England, a tradition begun in the nineteenth century. In 1875 Jean Abraham Chrétien Oudemans retired after seventeen years of surveying the East Indies to a chair of astronomy at Utrecht. Ten years later mathematician Horace Lamb (1849–1934) left a chair at Adelaide for one at Manchester. In 1927 Nobel laureate John James Rickard Macleod (1876–1935) left Toronto for Aberdeen. Gustav Angenheister, director of a colonial geophysical observatory on Samoa during the First World War, became director of the Göttingen geophysical institute in 1928. Joanny Philippe Lagrula (1870–ca. 1942), briefly director of the Quito Observatory early in the century, received direction of the Algiers Observatory (in greater France) in 1931. Richard Gans (1880–1954), director of the physical institute at La Plata, received the chair of physics at Königsberg in 1925. Jean Coulomb (b. 1904), director of the Algiers geophysical institute, took charge of the Paris geophysical institute in 1942, and Charles Jacob (1878–1962) rose from directing the Indochinese geological service to occupy a chair in Paris. Not all returnees had been missionary émigrés. Georg Balthazar Neumayer (1826–1909), for example, became director of the Hamburg Naval Observatory after having directed the Flagstaff Observatory at Melbourne from 1857 to 1864; nearly a century later, in 1950, Leopold Infeld (1898–1968) left a chair in physics at Toronto for one at Warsaw. In the twilight of life, physicist Emilio Segrè (b. 1905) returned home to Italy from Berkeley, as did mathematician André Weil (b. 1906) to France from Princeton.

Australasia

By 1914 there were a handful of universities in Australia. The oldest of these, in Sydney and Melbourne, had taught science since their foundation in the 1850s, and science figured prominently in newer universities (Adelaide in 1874, Hobart in 1891, Brisbane in 1910, and Perth in 1912). But teaching reigned supreme. Research was an avocation discussed in scientific societies, of which there were many. The instance of Horace Lamb, one of the first professors at Adelaide, is typical. His research in hydrodynamics derived from work undertaken at Cambridge.

The 1890s mark a new departure for academic science in Australia. First there were in 1891 the '1851 Exhibition' scholarships for allowing students from the British Empire to carry out postgraduate work in England; Cambridge obliged the vogue in 1896 by creating a BA for Research (the honours BA, instituted early in the century, was based only on examinations). Academics proudly exhibited their colonial status as the seat of empire sucked the life out of research in the antipodes. The pattern continued through the doldrums of the 1920s and the proliferation of war-related work beginning in the late 1930s. Doctorates in physics came to Australia only after the Second World War.

New Zealand blossomed early in the twentieth century

with universities in four locations. Science instruction was prominent in all of them; research was not, despite the best efforts of Ernest Marsden (1889–1970), an alumnus of Rutherford's Manchester who ran the national Department of Scientific and Industrial Research. Yet the sharpest intellects in the island nation were nevertheless apprised of the latest word. An indication of this spirit is found in the first volume of a textbook on physical optics published in 1908 by Richard Cogburn Maclaurin (1870–1920), Professor of Mathematics and Mathematical Physics at Victoria University, Wellington. Maclaurin knew that a revolution was brewing in light and matter, but he chose to ignore mechanics entirely in favour of a dynamical approach.

Explanation, Maclaurin avowed at the beginning of his book, was illusory. The fall of an apple was not 'explained' by Newton's law of gravitation. The popular explanation of the wave theory of light depended on 'the ether, which, for aught anyone can say, may be a mere figment of the imagination'. Arriving at laws by Baconian induction was entirely uncertain, even though the assumption of lawful behaviour in nature produced results that were not absurd. This did not mean that scientists sought to reveal naked reality. Scientists who claimed as much – declaring, for example, that only matter and energy were real – acted 'worse than the conduct of the artist who, after devoting his life to portraiture, declared that the only real thing was paint'. Science did not seek to eliminate appearances; rather, 'these are what it specially strives to know and to master'. This is the phenomenological account of science, advocated in Austria by Ernst Mach (1838–1916).

Away with all philosophical difficulties, urged the pragmatic New Zealander in the liberating spirit of Ernst Mach's iconoclastic positivism. 'Science is interested only in experience; its end is to know and to communicate.' Its aim was to 'introduce definiteness and order' into the 'confused blur' of 'a mass of facts that no mind could grasp and no memory retain'. Science was 'a premeditated art, not only aesthetic in its intensity, but consciously aesthetic in its aim'. It did not matter, consequently, whether or not hypotheses corresponded to reality. Maclaurin quoted Descartes: 'I believe that it is as useful for living to know the causes imagined in this way as if one had knowledge of the true causes.' He incorrectly cited Andreas Osiander's (1498–1552) preface to 'Kepler's books *De Revolutionibus Coelestibus*', that hypotheses needed only to reconcile observation and calculation, not to be true or probable. Maclaurin's key hypothesis was in fact the electromagnetic ether.[3] His work found him a following in the United States, a bastion of opposition to Einstein's relativity. Maclaurin rose to direct America's finest engineering school, the Massachusetts Institute of Technology.

Scientist missionaries in South America

For generations after South American nations achieved political independence, they looked abroad for models of scientific organization. The models chosen nearly all concerned applying science for technological development. Just as in the United States, knowledge was to be useful. The philosophical inspirations for this practical ideology were Saint-Simon, Jeremy Bentham, Auguste Comte, and Herbert Spencer. The innovations contributed to foreign investment in engineering marvels (notably railways and mines) and improvement in literacy (notably through the pedagogical ideas of Joseph Lancaster [1778–1838]). The rhetorical interest in practicality coupled with a persistent appeal to European talent.

ECUADOR

The earliest Europeans to chronicle the New World were religious missionaries, for whom natural knowledge was an adjunct of religious fervour. A late nineteenth-century manifestation of the sixteenth-century impulse took place in Ecuador. The prime mover was Gabriel García Moreno

– devoutly Catholic, decidedly technocratic, and ruthlessly efficient. In 1870 he became president for a second time and set about turning Ecuador into a benevolent theocracy.

Inspired by the Jesuit empire in colonial Paraguay, García Moreno charged a dozen German Jesuits with transforming the University of Quito into an institute of technology. The result was the finest science school in the Americas. Juan Bautista Menten (b. 1838), a former assistant with Friedrich Wilhelm August Argelander at Bonn, came as dean and director of Quito's astronomical observatory. Teodoro Wolf (1841–1924), another Bonn alumnus, came as geologist and zoologist. Luis Sodiro (1836–1909) came from Innsbruck to teach botany. Joseph Epping, one of the Jesuit mathematicians, began his pioneering examination of ancient Babylonian ephemerides at Quito. The Quito experiment expired, just as the New Harmony, Indiana, experiment had expired a generation earlier. Menten and Wolf, trained to refute Darwinian theory, went over to the side of the enemy; the university's patron García Moreno fell to an assassin in 1875.

The University of Quito never recovered, despite the efforts of several French missionaries. In 1899 the Paris Académie des Sciences in collaboration with the Geographical Service of the French army, decided to remeasure the eighteenth-century arc of longitude surveyed by Pierre Bouguer (1698–1758) and Charles-Marie de La Condamine (1701–1774). An expeditionary force arrived in 1900, accompanied by an astronomer, François Gonnessiat, (1856–1934) who took charge of Quito's observatory. When the French army departed, Gonnessiat soon followed them. He persuaded the Ecuadoran government to hire four French science teachers to renovate the university; a political revolution persuaded three of the four men to return to France. Astonomy went to Luis G. Tufiño, an Ecuadoran who had been trained in Paris. In 1913 Tufiño was helped by Milan Ratislav Štefánik (1880–1919), a Slovak astronomer naturalized in France who tried to set up a French coaling station in the Galapagos Islands under the cover of establishing a programme of scientific cooperation in wireless technology.

Foreign scientific missionaries entirely dominated Chilean science from independence through the early decades of the twentieth century. French geographers, the most renowned of whom was Claude Gay (1800–1873), mapped the country and taught at its university in Santiago. In 1849 the United States navy sent James Melville Gilliss (1811–1865) to Santiago for measurements of the solar parallax. The Chilean government purchased his equipment and hired Karl Wilhelm Moesta (1825–1884), a former student of Marburg astronomer Christian Ludwig Gerling (1788–1864). The importation in 1883 of several German astronomers produced a fiasco. Chileans looked south of the Rhine, and in 1887 three French astronomers arrived to take charge and provide data for the great photographical sky chart project. Civil war and precarious funding prevented the work from proceeding. Then, in 1893, the 'D. O. Mills Expedition' arrived in Santiago from Lick Observatory in California. Their aim was to measure radial stellar velocities, and they manned an independent observatory until 1927.

Fin-de-siécle Chile, however, was nevertheless enamoured of things German, not least because a German officer had advised the victorious side in Chile's civil war of the early 1890s. The Chilean educational ministry called in dozens of German science teachers to staff the nation's schools, including its national pedagogical institute and university. At the observatory, Friedrich Wilhelm Ristenpart (1868–1913), an astronomer from Berlin, arrived in 1904 to replace French astronomer Jean Albert Obrecht, (1858–1924) who instead of observing had organized a French-language scientific society. Ristenpart appointed German associates and began research. But, as in Quito, the programme foundered with the death of its patron President Pedro Montt in 1910. Ristenpart committed suicide in 1913. Obrecht, who had been appointed to the local university returned once more to guide the attenuated observatory. Chilean astronomy rose to world prominence again during the last third of the twentieth century, when universities in the Northern Hemisphere erected extraordinarily expensive telescopes high in the Chilean Andes.

ARGENTINA

The pattern of foundation, decline, and instauration under the eyes of foreign talent is typical of scientific communities in Latin America during the period before the First World War, even in its most sophisticated academic environment, that of Argentina.

In 1869 President Domíngo Sarmiento, a onetime resident of New York, renovated education in Argentina from bottom to top. He instituted a network of national schools and imported schoolteachers for them. He also brought a number of German scientists to Córdoba, where they eventually taught at the university. And he appointed United States astronomer Benjamin Apthorp Gould as director of the Córdoba Observatory; Americans ran the observatory through the late 1920s. Just as in Ecuador and Chile, the observatory was loosely affiliated with the local university.

Foreign naturalists ran late nineteenth-century museums in greater Buenos Aires, and French physicians dominated the medical faculty there, but the most interesting site for scientific missionaries was the University of La Plata, situated a short distance from the national capital. The architect of La Plata's scientific prominence was Joaquín V. González, who as a federal minister of education had created a national normal school staffed with German teachers. As president of the university, González called in a troop of foreign savants beginning in 1905. La Plata's observatory had already begun life under a French naval astronomer, Victor François César Beuf (1834–1899). In 1906 González named Italian astronomer Francesco Porro di Somenzi (1861–1937) as observatory director and professor. Porro's successor, in 1910, was William Joseph Hussey, director of the observatory of the University of Michigan. Hussey imported staff from the United States and directed La Plata along with his Michigan charge – commuting half way around the globe. Hussey's eventual successor, between 1922 and 1932, was the director of the University of Göttingen observatory, Johannes Hartmann (1865–1936).

Complementing Hartmann at the University of La Plata were accomplished researchers in other fields, for example, the anthropologist Robert Lehmann-Nitsche (1872–1938). There was also a strong group of German physicists. These included the former editor of the *Physikalische Zeitschrift* Emil Bose (1874–1911) and his wife Margrete Heiberg (1865–1941), Einstein's first scientific collaborator Jakob Laub, Konrad Simons (1873–1918, a student of both Emil Warburg [1846–1931] at Berlin and Thomas Edison [1847–1931] in Schenectady, New York) and the expert on ferromagnetism from the University of Tübingen, Richard Gans. The German colony was large enough to sustain a lecture tour by Nobel laureate Walther Nernst (1864–1941) in the spring of 1914.

The high density of foreign academic talent in Argentina generated an indigenous research community. The most notable early achiever was the Nobel-laureate in medicine biochemist Bernardo Houssay (1887–1971) – the creator of a tradition that produced two more Nobel laureates; Enrique Gaviola (1900–1989) attained eminence as a physicist, astronomer, and administrator. Foreign stimulation also evoked a strong school of history and philosophy of science. Spanish mathematician Julio Rey Pastor (1888–1962) and Italian historian of science Aldo Mieli prepared the ground for José Babini (1897–1984) and Mario Bunge (b. 1919) to achieve scholarly eminence. Momentum from foreign erudition produced Latin America's earliest scientific community of international distinction. Remarkably, the community emerged during the years 1930–70, a generation of intense political and economic turmoil.

Science at American universities

Any focus on learned immigrants tends to obscure the fact that several university settings beyond Europe rose to world prominence early in the twentieth century, and they did so with native talent. Generalizations across cultures are risky, but over the years between 1870 and 1920 it appears that talent imported from Europe did not shine as bright as young men and women educated in the new universities. Despite an invasion of German, French, and Italian professors, arguably the finest sometime university scientist in Latin America, during this period, was Houssay, a native of Argentina. The finest university scientist in New Zealand was the future president of the Massachusetts Institute of Technology, Maclaurin.

The maturity of new institutions in the United States, especially, is striking. One may consider, as an indicator, the list of residents of the USA elected to the Royal Society of London and the Paris Academy of Sciences in the fifty years before 1920. Twenty-one men were recognized by London, and sixteen by Paris (eight received recognition from both societies). Nearly all were university professors. Taking both sets together, we may count the scientist immigrants on one hand: Simon Newcomb (1935–1909), William Osler, Jacques Loeb, and Leonard Eugene Dickson (1874–1954) – all the others were native Americans, excepting Albert Abraham Michelson who arrived before adulthood.

The situation changed dramatically in the period beginning in 1920. Through 1988 no fewer than thirty-three Nobel laureates in science were naturalized United States citizens when they received the prize; the vast majority held university posts. By way of contrast, only a handful of laureates *became* Europeans – Ben Mottelson (b. 1926, a

374 · *Servants of Nature*

Dane) and a few Australasians who technically held British citizenship from the beginning (Rutherford, William Lawrence Bragg [1890–1971]). The immense destruction of the First World War, the twilight of the Romanovs in Russia and the material deprivations of communism, and the rise of fascism all projected the United States as a land where elite institutions preserved and extended nineteenth-century values, notably the ideology of pure learning.

Science at Japanese universities

Like the young American republic, Tokugawa Japan also received foreign scientists, largely through the Dutch port at Decima near Nagasaki; in the final fifteen years of the Tokugawa regime, more than two hundred foreign technicians and scholars were active, among them eight French, sixty Dutch, and thirty Americans. The system grew under the Meiji, beginning in 1868. Between 1872 and 1898 the government hired more than six thousand foreigners in all; perhaps two times as many foreigners worked in the private sector. But because foreigners were extremely expensive, the government progressively dispensed with their services, reducing its commitment from 527 in 1875 to fewer than a hundred in 1894. Throughout the Meiji regime, natural science teachers constituted the largest class of foreign experts engaged by the government – some 46 per cent.

With the consolidation of various schools for foreign science into the University of Tokyo, it was natural to hire foreign savants as professors. In the Faculty of Science in 1877, for example, there were five American professors, two English, two French, and one German, in addition to nine Japanese. Through the 1870s instruction proceeded in French, German, and English, finally moving into the vernacular in the 1880s. By the eve of the First World War, Japan had projected itself to the forefront of science with two new universities, Kyoto and Tohôku Imperial. At this time older university researchers like physicists Aikitsu

Tanakadate (1856–1952) and Hantaro Nagaoka (1865–1950) were world-renowned. Younger researchers – physicist Jun Ishiwara (1881–1947), bacteriologist Shibasaburo Kitasato (1852–1931), and microbiologist Hideyo Noguchi (1876–1928) – had climbed to the heights of their field by travelling and publishing in the West.

Special circumstances lie beyond the rapid rise of academic science in Japan. Part of the explanation, historian Kenkichiro Koizumi has emphasized, is demographical. The first generation of researchers in physics, he notes, belong overwhelmingly to samurai families who had been displaced by the end of feudalism. These scientists transferred their cultural allegiance from Chinese imaginative literature (the measure of educated samurai) to English, French, and German scientific literature. They were sent to the finest Western universities for advanced training at government expense. Part of the explanation, also suggested by Koizumi, is institutional. Like their American counterparts, Japanese universities were able to introduce portions of curricula and doctrines from a number of foreign settings, avoiding what appeared to be superannuated programmes. They were notably unencumbered by the somewhat rigid distinction in Europe between practical technology and pure science.

During the First World War, Japanese university scientists created a number of new science chairs related to military needs like aeronautics, and they arranged for a new, para-academic Institute of Physical and Chemical Research, along the general lines of the Imperial Institute for Physics and Chemistry in Berlin; young Yoshio Nishina (1890–1951) worked at the Tokyo institute, spreading the new doctrine of quantum mechanics that he had absorbed from a stay of nearly seven years in Göttingen and Copenhagen. Nishina's example stimulated outstanding Japanese colleagues Hideki Yukawa (1907–1981), Shin'ichi Tomonaga (1906–1979), and Shoichi Sakata (1911–1970). Despite obvious differences in scale, developments in Japanese academic science paralleled European and North American trends. New universities came in line (Nobel laureate Yukawa initiated physics instruction at Osaka University in 1931), and university professors increasingly depended on grants from national organizations. Furthermore, following the pattern in European colonies, Japan established universities in its overseas territories. Three technical and medical colleges opened in Japanese Manchuria in 1922, and branches of Japan's Imperial University System came to Seoul in 1924 and to Taipei in 1928. Furthermore, during the Pacific War, Japan managed a number of foreign universities on mainland Asia and in the Dutch East Indies – precisely the practice of Germany in Europe. The global prominence of science and technology in Japan in the last thirty years has roots in the striking modernity of Japanese academia at mid century.

British India and Dutch Indonesia

India and Indonesia, strong scientific communities emerging from the two largest European colonies in Asia, both enjoy a heritage of assimilating foreign doctrines. South Asia absorbed Hellenistic learning, elaborating celestial mechanics with trigonometric functions; then it received the Islamic sciences. Early on, Indonesia received Buddhist and Hindu learning, and then (at about the same time India did) Islam. Both civilizations fell under European rule beginning in the sixteenth century. At the very end of the eighteenth century, Western institutions of higher learning came to each country.

Satpal Sangwan has contended that the British rulers of India, up to the time of the Revolt of 1857, created educational institutions to discipline the local population and to train people who could exploit local resources for the benefit of Great Britain. Exploitation implied training in techniques of colonial administration, notably mastery of English, as well as in technologies of imperial control, such as topographical surveying. No fewer than twenty-seven colleges emerged in Calcutta, Benares, Delhi, Agra, Bombay and elsewhere in the generation before 1857.

S. N. Sen shows in a masterful survey of science education in nineteenth-century India that the colleges sustained a respectable level of instruction. Calcutta, which had seen the creation of the Asiatic Society, was the initial educational focus. It hosted a medical college, founded in 1835. Thomas Oldham (1816–1878) came to teach geology at the Calcutta Civil Engineering College in 1855. In 1857 there were chairs of natural philosophy and astronomy and of natural history and geology at Presidency College, also in Calcutta, and in 1845 a chair of natural and experimental philosophy came to the Hindu College there. By the end of Company rule in 1857, medical colleges were producing Indian physicians at Madras, Bombay, Hyderabad, and Chudderghaut. And in 1857, in a final act of colonial patronage, the Company inaugurated universities at Calcutta, Bombay, and Madras.

The Indian universities took their first breaths as examining bodies, following the mandate of the University of London. Initially they focused on the classical syllabus of *literae humaniores* at Oxford. Systematic science certification arrived in the 1870s. The BSc came to Bombay in 1879; up to 1895, 43 students took the degree. The BSc and DSc came to the University of Calcutta in 1899. In 1901–2, 1476 students registered at Calcutta for the BA degree, but only 14 for the BSc and 16 for the Bachelor of Engineering. Research remained foreign to university settings. Typically, Indian-born Sir Ronald Ross carried out his Nobel-prize work on malaria in the Indian Medical Service during the 1890s. The British professors were for the most part uninterested in undertaking original investigations into natural phenomena.

However the expatriate professors presented themselves in colleges and universities, the general message was not lost on wealthy Indians, who believed that national renewal and eventual independence would derive from Western science. Mahendra Lal Sircar (1833–1904) was the foremost nineteenth-century advocate for science. Sircar received one of the first medical doctorates from Calcutta in 1863. A convinced homeopathic physician, he became a moving force behind the Indian Association for the Cultivation of Science; the association led the Indian National Congress, a nationalist lobby, to place science on its agenda. It would be a mistake to imagine that Sircar's optimistic and universalist vision of Western science sustained no opposition. Foremost among the traditionalists was Mohandas Gandhi, who beginning in 1917 promoted traditional technologies and sciences as a way to stimulate nationalist fervour. Historian Jagdish Sinha has emphasized: 'Gandhi maintained a hostile indifference to modern science, attacked modern civilization and looked upon machines as an evil.'[4] With Jawaharlal Nehru's ascension around 1936, however, a recognizable technocratic impulse emerged to direct nationalist fortunes. This impulse has largely guided independent India.

Gandhi's distaste for Western science and technology had a limited effect because by the eve of the First World War a pattern of private endowment and public acquiescence had already taken root. Industrialist Jamsetji Nasarwanji Tata endowed the Indian Institute of Technology, which opened in Bangalore in 1911 on the model of Johns Hopkins University. In 1916 Madan Mohan Malaviya, a key player in the Indian National Congress and a follower of Sircar's, was instrumental in providing a strong science programme for the new Banaras Hindu University. From 1912 to 1921 Calcutta received large endowments for science chairs and fellowships from Taraknath Palit and Rashbehari Ghosh. The money went to a number of scientists, including chairs for Acharya Profulla Chandra Ray (1861–1944), Chandrasekhara Venkata Raman (1888–1970), Meghnad Saha (1894–1956), and Satyendra Nath Bose (1894–1974). Yet the university career was often insufficient for ambitious professors. Distinguished researchers retired from universities to occupy a sinecure in government or industry. Typical of dissatisfaction with academic indifference to research is the career of Jagadis Chunder Bose (1858–1937), who taught physics at Presidency College in Calcutta from 1884 to 1915, when the Bengali poet and Nobel laureate Rabindranath Tagore raised funds to maintain Bose in a private laboratory.

Scientific institutions in Indonesia, India's mirror-image twin, share certain features of the South Asian experience. The nineteenth century saw establishment of a number of scientific institutions on Java, notably the Indigenous Medical School at Batavia (known by its acronym in Dutch, STOVIA). Beginning around 1900, the Dutch inaugurated a policy of promoting education generally. In the 1920s, STOVIA became a fully fledged Dutch medical school, and an institute of technology (complete with the prerogative of awarding doctorates) emerged at Bandung. These educational institutions and a handful of others offered the very highest level of instruction. Among the professors were Nobel laureate physiologist Christiaan Eijkman (1858–1930) and physicist Jacob Clay. Among the graduates was Sukarno, an architect of independence.

Scientific education in colonial Indonesia was for a restricted elite, and very few indigenous Indonesians published research under their own name. Yet fifty years after independence Indonesia enjoys a sophisticated system of public and private universities. The nation of hundreds of languages and dozens of creeds is self-sustaining in agriculture. It manufactures commercial airplanes and supports a space programme. Indonesia is one answer to those who are inclined to doubt that Western science may flourish in harmony with extra-European traditions.

The variety of scientific traditions beyond Europe matches the profusion of scientific institutions within Europe. Some nations, like Austria and Italy, have opened the university gate to almost any determined adolescent; others like

Greece and Germany invoke a *numerus clausus* to restrict enrolment severely. Research remains firmly attached to English universities, whereas new ideas emerge only occasionally in French universities. National research institutes dominate science in Russia but not in Denmark. The protocols of the European Union have left much latitude for individual variation. Cross-cultural integration and multi-lingual scientific cooperation may well be a dominant theme in the new millennium.

Believing:
Science and Religion

What is the *aim* of knowledge? What is the *purpose* of seeking to understand the natural world? *Why* do we strive continuously to deepen our knowledge of nature? Questions like these probe into the realm of motivation, a topic more likely to raise questions than to provide answers. Someone's motivation for understanding the natural world may derive from social or economic concerns, or it may seem to defy materialist accounting altogether.

Among the most common reasons ascribed for undertaking scientific investigations, apart from raw intellectual interest, is the fulfilment of a religious quest or the articulation of a philosophical point of view. This attribution occurs despite the fact that we have tended to associate the advent of a scientific world view with the decline of religion. Science appears to be the natural ally of secularism because it is the staunch foe of anything that smacks of mysticism. Moreover, mysticism can shelter intolerance and repression, as symbolized by Galileo's trial before the Inquisition. The seventeenth century, the century of the Scientific Revolution, has even been characterized as a battleground, where 'moderns' fought 'ancients' as part of a general struggle between progressive science and retrogressive religion.

Increasingly, however, this interpretation has been seen as providing only a partial, if not misleading, analysis. Religion motivates much human activity, including science. One must not confuse anticlericalism with a universal antipathy to religion, nor scepticism with a lack of faith.[1]

During the eighteenth century, as many natural philosophers were motivated by deism as by agnosticism; during the nineteenth century, natural theology was as important to some scientists as materialism was to others.

Richard Greaves has suggested that a clearer picture of the relation between religious experience and science derives from separating the origin of modern science from its early elaboration. Modern scientific thought emerged in Catholic Europe during the late sixteenth century and early seventeenth century; one thinks of Andreas Vesalius exploring human anatomy and Galileo inventing mechanical and mathematical principles in northern Italy. Pierre de Fermat and Blaise Pascal, who helped to forge the philosophical and mathematical tools of modern science in France, also come to mind. But by the end of the seventeenth century, the Protestant countries of northern Europe provided a secure home for the nascent scientific movement against the fury of the Counter-Reformation. French renegade René Descartes wrote and published from exile in Holland, as did the Portuguese exile Benedict de Spinoza; there is not a single English or German *savant* of note who sought a safe haven in Paris or Rome. In this chapter we examine how religion permeated modern science, both during its foundation and over later centuries.

Science in the Counter-Reformation

The challenge delivered to religious authorities in both Canterbury and Rome has been transformed into an anthem of modern science – knowledge based on observation, experiment and measurement – which has been opposed to the authority of hierarchical churches deriving from revealed truth. But unthinking obedience to higher authority is a red herring in early modern science, for much of Christendom was highly intolerant. Following a party line is common in Protestant sects as well as in the Catholic Church. We remember John Calvin's systematic interrogation of the citizens of Geneva – ten at a time – to verify their allegiance to his view of the world; we wilt before John Knox's fiery denunciation of Presbyterian schismatics and deviants. Open-mindedness was hardly the coin of the Protestant realm: Martin Luther expressed hostility toward Copernicus and his heliocentric theory. Thus, although northern Europe (and its Protestant tradition) was more tolerant of new ideas than the Catholic south and west, it does not follow that Protestants, who counted a fair share of dogmatics among their number, were completely open-minded.

Jesuits, in particular, were interested in new knowledge about nature. Galileo's tribulations at the hands of Roman religious authorities have passed into the realm of fiction, Bertoldt Brecht's play being a case in point. Nevertheless, Galileo's principal accusers, the Jesuits of the Collegio Romano, were anything but opposed to science. Beginning in the last half of the sixteenth century, the Jesuit order took training in understanding nature as one of its goals. Over the next two centuries, Jesuit *collèges*, whose curriculum and intent corresponded to today's undergraduate education, spread throughout Catholic Europe; in German-speaking lands, especially, the colleges merged into or absorbed universities. Jesuits were often engaged with new ideas.

Numbers tell the story. In Austria, for example, the single Jesuit college that existed in 1551 had multiplied to fifteen in 1663 and to thirty-eight in 1767; Germany and Austria together had 107 Jesuit colleges in 1750, and France had 113 in 1762. From the beginning, the colleges emphasized mathematics as a way of sidestepping diabolical menaces to doctrine – such as the Copernican heresy – but by the end of the seventeenth century, Jesuit colleges in Rome and Paris openly discussed the mathematics of heliocentrism. There was no shortage of Jesuit science professors. By 1700, France saw fifty physics teachers and twenty-one mathematics teachers in its Jesuit colleges.

Jesuit interest in science transcended elaboration of speculative hypotheses. Jesuits were among the most constant investigators of electrical and magnetical phenomena. Observation of nature extended to the heavens. We have seen that on the eve of the suppression of the order, in 1773, Jesuits ran a large number of the world's 120-odd observatories. Both Bacon and Leibniz took instruction from the organization as well as the publications of Jesuit men of science. Jesuit textbooks, progressively freed from Aristotelian chains, found ready readers in Descartes and Newton.

Jesuit natural philosophers had one of their most challenging constituencies beyond Europe. In 1577 Matteo Ricci (1552–1610), a Jesuit student at the Collegio Romano, left for Goa and Macao. In 1583 he established a mission in Guangdong Province in China, and in 1601 he settled in Beijing. With the help of Chinese collaborators, Ricci translated works of his teacher, Christoph Clavius (1537–1612), into Chinese, notably Clavius's version of the first six books of Euclid's *Elements*; he also introduced Chinese readers to the European notion of the world's geography. The principal aim of Ricci and his successors, however, was to transform the Chinese view of the heavens. Ricci admired the equatorial armillary spheres and instruments of Kuo Shou-Ching (fl. 1270), but when he became one of the emperor's astronomers, he replaced equatorial Chinese equipment with the ecliptical instruments common to Europeans at the time.

The papal condemnations of Galileo (in 1616 and 1632) cautioned inventive minds over the following generations, but few scientific works found a place on the infamous index over the seventeenth and eighteenth centuries. Jesuit polymaths *in partes infidelium* quickly embraced new technologies. The telescope was introduced to Chinese readers by a Jesuit hand – that of Emanuel Diaz (1574–1659) – in 1615. In 1618 Johann Schreck (1576–1630) brought a telescope to China; Adam Schall von Bell published a picture of a Galilean telescope in a Chinese treatise of 1626; Michael

Boym (1612–1659), bound for China, received a set of the Rudolphine Tables from Johannes Kepler in 1627. In 1640 Schall finally introduced Galileo, Tycho, and Copernicus by name, and in the same year Wang Hsi-Shan (1628–1682) published a work elaborating the Tychonic system (where the sun moves around the earth and the planets circle the sun). By this time, however, Ptolemaïc views, taught since 1611, had become a matter of Papal dogma.

Ferdinand Verbiest personally instructed the Manchu emperor in traditional, Mediterranean astronomy during the 1660s, even though Beijing had enjoyed a bureau of Islamic-based astronomers since medieval times; this good will allowed Verbiest to reconstruct the instruments of the Beijing Observatory in a traditional, European mould. Ptolemy did not obstruct Verbiest's instrumental revolution, for heliocentrism is no practical help to earthbound astronomers. But Copernicus entered China again only with Protestant missionaries during the nineteenth century – a century after the Dutch had brought him to Japan. The Chinese Jesuits were more circumspect, focusing on laboratory phenomena. In the 1750s the great historian Antoine Gaubil (1701–1777) was able to report significant experiments that led Franz Ulrich Theodosius Aepinus (1724–1802) to demonstrate the electrical polarization of glass. Michel Benoist (1715–1774) rejected Aristotelian prejudice against experimentation by presenting an air pump to the Chinese emperor in 1773.

The activity of the Chinese Jesuits manifests a separation of scientific enquiry from religious doctrine. This extends a European tradition. Copernicus himself, a devout Catholic, asked a Protestant colleague, Andreas Osiander, to see his heliocentric *opus* through press. The Minim monk Marin Mersenne animated a European-wide network of scientific correspondents during the first part of the seventeenth century; many of his contacts were heretics.

A subtle change in the relationship between science and religion nevertheless occurred over the course of the seventeenth century. Science, freed from the chains of tradition, responded to a new theological imperative, which had been reformulated as a result of the Protestant reformation. Historian Charles Webster argues for the 'deep interpenetration' of scientific and religious ideas. As he expresses the case, to see the development of science as 'unrelated to the Reformation and the Puritan Revolution or to the socioeconomic framework of which Puritanism was a constituent element' is 'to eliminate vital factors in the explanatory mosaic'. After all, Puritanism and modern science achieved rapid development simultaneously, and by their geographical proximity almost symbiotically.[2]

The Merton thesis

To Protestants, Roman Catholic ritual, natural magic, and Aristotelean philosophy all belong in one category: they seek final or ultimate causes beyond the realm of human experience. Such supporters of 'remote science', of unverifiable explanation, were arch enemies to the followers of Francis Bacon and the empiricists, who predicated the validity of knowledge on its derivation from sensate experience. When Protestantism emerged as an alternative to Catholicism, it struck an easy partnership with the new attitude towards the natural world epitomized in Bacon's philosophy. As John Hedley Brooke explains, the two reformations (in science and in religion) had much in common: 'Each prized the original copies of God's two books, nature and the Bible, bypassing the corrupting influence of scholars and priests.' Indeed, this is precisely the argument made by Robert K. Merton, who posited a connection between ascetic Protestantism and the emergence of modern science. Giving this formula its tersest expression, Merton argued that 'Puritanism, and ascetic Protestantism generally, emerges as an emotionally consistent system of beliefs, sentiments and action which played no small part in arousing a sustained interest in science.'[3]

Merton's 'thesis', as it has become known by historians, is

a natural extension of the German sociologist Max Weber's argument about the rise of capitalism in European centres where Calvinism held sway. Weber emphasized the importance of three attitudes associated with Calvinist theology: diligence in religious practice, renunciation of material satisfaction, and constructive use of time. A society guided by these ethical norms would be predisposed to nurture an economic system based on capitalism. In such a society, making money functioned less as an end in itself than as the symbol of possessing important spiritual virtues.

Merton built on Weber's interpretation by suggesting that in societies like that of seventeenth-century England (where a form of Calvinism emerged as a dominant force under the reign of the parliamentarian Oliver Cromwell and his Puritan followers), scientific activity could be seen as a demonstration of one's faith. In the words of Bacon, the 'rule of religion that a man should justify his faith by works applies also in natural philosophy; knowledge should be proved by its works'.[4] Merton even proposed that his results extend beyond the case of seventeenth-century England. The association of Protestantism with scientific and technological interests and achievements, he writes,

> is largely understandable in terms of the norms embodied in both systems. The positive estimation by Protestants of a hardly disguised utilitarianism, of intramundane interests, of a thorough-going empiricism, of the right and even the duty of *libre examen*, and of the explicitly individual questioning of authority were congenial to the same values found in modern science. And perhaps above all is the significance of the active ascetic drive which necessitated the study of Nature that it might be controlled. Hence, these two fields were well integrated and, in essentials, mutually supporting not only in seventeenth-century England but in other times and places.[5]

Elsewhere Merton comments that if the 'Puritan ethos' did not directly influence the development of science, in its 'psychological compulsion towards certain modes of thought and conduct', it favoured empirical science. The result of this tendency was to direct talented young men to choose scientific careers over other, more traditional pursuits. Merton's overriding concern was not to emphasize the letter of Puritan doctrine, but rather to raise the issue of how its 'system of beliefs, sentiments and action' promoted scientific inquiry.[6] Like Weber, Merton set out to evoke the Puritan 'ethic' and to show how it helped to give rise to the spirit of modern science.

Merton found evidence for his view in statistical data. According to his count, 62 per cent of original members of the Royal Society were 'Puritan', an especially remarkable figure considering their minority position in the English population as a whole. Even in France, the rolls of corresponding members of the Académie des Sciences up until the end of the nineteenth century yield only eighteen Catholics, in contrast to eighty Protestants. Scientific publications taken from the mid seventeenth century placed particular emphasis on applications of science that could lead to the betterment of humankind, as exhibited in a flowering of literature on agriculture, mathematics, and medicine. Merton interpreted this orientation as a reflection of the Puritan faith in human progress assisted by science and technology. The scientometric calculations of other historians (among them Dorothy Stimson and Raymond Stearns) have corroborated Merton's position.

Upon closer scrutiny, however, Merton's thesis is less successful. The most distinguished natural philosophers associated with the Scientific Revolution – taking Newton, Descartes, and Boyle as cases in point – were not ascetic Protestants. Even considering those who were, a compelling case has not been made for religious beliefs motivating their scientific work. As a sceptical A. Rupert Hall expresses the case, 'Iteration and exemplification cannot ... make the proposition "many scientists are Protestants" equivalent to the statement "men are scientists because they are Protestants".' The demonstration of 'parallel value-complexes' for Puritans and scientists also poses a significant methodological hurdle: it is far from clear, for example, that concepts like utilitarianism, experimentalism, and commitment to the operation of *libre examen* carry identical meanings in scientific and religious contexts.[7] Certainly there were many Calvinists among the general population who remained uninfected by science.

A careful analysis of the early Royal Society suggests, moreover, that Anglican members were in the majority, and that only a very small proportion of fellows could be identified as products of the Puritan middle classes. Deciding who was and who was not Puritan is not a trivial exercise. Consider John Wilkins (1614–1672). His Puritan background was sufficient to win him the hand of Cromwell's sister; during the Interregnum, he was appointed Master of Wadham College, Oxford. With the restoration of the monarchy, however, he was ordained as an Anglican and subsequently became Bishop of Exeter.

In the opinion of Charles Webster, none of these objections is insurmountable. He views the cultivation of science as 'an important social phenomenon' – not necessarily an intellectual achievement. Puritanism, as a social movement, valued 'the active study of nature' for a variety of reasons. Webster agrees with Merton on this point, but he emphasizes a different aspect of Puritan doctrine. Merton (like Weber before him) explored 'the social implications of Calvinist soteriology', that is, how the doctrine of 'good works' supported activities that brought about the accumulation of wealth and the advancement of applied science. Webster, in contrast, is interested in how eschatology – the omnipresent belief in an impending Judgement Day – furthered scientific pursuits and spawned other kinds of intellectual movements 'to exploit the natural environment for the health and wealth of mankind'. As he puts it, 'the Puritans genuinely believed that each step in the conquest of nature represented a move towards the millennial condition'.

Webster maintains that the English Civil War exerted only

'a minor disturbing influence on the groundswell of scientific activity'. The scientific flowering following the restoration of the monarchy in 1660 came at the hands of people recruited to science during the Puritan years. From the standpoint of science, the dividing line between the Interregnum and Restoration appears to be an artificial separation. Disciplinary prejudice is behind low estimates of Puritan science. Many Puritans cultivated natural history and chemistry, sciences traditionally less esteemed by historians than mathematics and mechanics.[8]

The Webster thesis: millenarianism and science

Webster's extensive study of science in seventeenth-century England takes Merton's work a step further. Webster suggests that historians have been misled by Royalists who minimized the extent of intellectual change under the Puritan Revolution. Webster argues that English science benefited from 'the catalytic influence of the revolutionary intellectual and political situation'. In his view, 'committed Puritans and parliamentarians' dominated the scientific community of mid seventeenth-century England. They were the active members of the Royal Society and the authors of the majority of scientific texts of the day.

Royalists were responsible for creating a myth that minimized the Puritan contribution to the Scientific Revolution (Sprat's early history of the Royal Society is silent about Puritan virtues), and, according to Webster, modern historians of science are to blame for perpetuating it. Among them are those who advocate an important role for Baconianism, as well as those who give pride of place to Continental mechanical philosophy. In Webster's opinion, modern thinkers downplay the role of theology in general and Puritanism in particular. During the seventeenth century, however, 'no dimension of human speculation or action', including the formulation of scientific concepts, was untouched by religion.

In Webster's view, religious affiliations mattered greatly for the seventeenth-century natural philosopher, whether Catholic, Anglican, or Puritan. Within this mixed community, Puritans assumed a critical position due to their 'numerical strength, dominant social position, aggressive confidence and scientific awareness'. As Webster puts it, 'the entire Puritan movement was conspicuous in its cultivation of the sciences'; moreover, each division within the movement developed a view of science 'consistent with its doctrinal position'.

Webster correlates religious orientation with scientific activity, as defined by type of phenomenon investigated and method of approach. The scientific study of agriculture, for example, was cultivated by a group of Puritan intellectuals almost exclusively, whereas experimental physiology, particularly the study of blood circulation, was the preserve of Anglican Royalists. The hermetic philosophy and certain varieties of experimental chemistry attracted Puritans, unlike abstract philosophy, which appealed to Anglicans. On the other hand, Webster contends that Puritan science was no monolith; there was indeed a radical wing of occultists and social reformers, but a conservative flank also entered into dialogue with Continental mechanical philosophers.

According to Webster, then, members of a number of religious groups behaved like Puritans during this time of 'intellectual crisis', constantly 'readjusting' their view of nature by invoking 'general ideological principles'. But the Puritans dominated the landscape, by their 'numerical strength, social importance, their pronounced attention to scientific matters and the explicit manner in which they articulated their ideas about religion, society, and the natural world'. Because of their dominance, Puritans controlled the complexion of modern science and shaped the intellectual outlook of many of their non Puritan peers.

Webster contends, moreover, that Puritanism promoted other values associated with modern science, namely 'a critical attitude towards inherited wisdom [and] an emancipation from scholastic values'. Due to the enthusiasm it engendered for science, Puritanism accelerated and broadened recruitment into science, and it encouraged cooperation among various segments of society. This, posits Webster, provides a different way of understanding the 'marriage' of the craftsman and the scholar, a development that has been treated by other historians. Although Puritanism collapsed as a political and ideological force with the Restoration, its pervasive influence lived on in the rise and development of modern science. As Webster summarizes this influence, Puritanism and the social conditions it generated explain 'the distinctiveness, diversity, and creativity of English science on the eve of the foundation of the Royal Society'.[9]

The Enlightenment

The Puritan belief in the millennium, whereby the vigorous efforts of the elect aimed to install the kingdom of God on earth, spawned a world view that incorporated a commitment to social and intellectual progress. When the millenarian zeal of the Puritan movement dissipated, belief in progress remained. The staunchly religious Puritan fathers of the seventeenth century became unwitting progenitors of the one indisputable article of Enlightenment faith: adherence to progress in all its forms. Modern science served as a tool to effect this progress; the commitment to progress, moreover, furnished an incentive for prosecuting science. The association of science and progress is one of the fundamental tenets of Western culture.

The Enlightenment took the religious ideals of the seventeenth century to new extremes, even appearing, at times, to stand religion on its head. As one commentator notes, the Christian creed of 'no salvation outside the Church' was replaced with the ideology of no salvation except by means of the use of reason. According to eighteenth-century thinkers, superstition and ignorance were tantamount to original sin; furthermore, 'withholding knowledge became ... a cardinal vice'. Freemasonry functioned as 'a counter-church with its counter-symbols and counter-sacraments'. Historian Amos Funkenstein sees all these characteristics as demonstrating the dialectical relationship of the Enlightenment to Christianity.

As in all dialectical relationships, there is always imbalance between opposing forces. The forces of secularization threatened to overpower religion. Certainly the *Encyclopédie* of Denis Diderot (1713–1784) and Jean le Ronde d'Alembert was dedicated to this end. It enshrined scientific achievement and technological progress and used this god to subvert religion as well as all other manifestations of authority. Elsewhere, however, theists gained the upper hand. They reigned victorious whenever the philosophers went to atheist and materialist extremes. Maintaining the symbiotic relationship between science and religion over the course of centuries was a considerable difference of opinion that split both camps: neither preachers nor *philosophes* could agree about essential matters, such as what constituted religious doctrine or scientific fact.

Generally speaking, inductive reasoning is associated with science; deductive, with theism. For this reason, Francis Bacon became the patron saint of the *philosophes,* followed in popularity by his foremost apostle, Isaac Newton, whose famous *hypothesis non fingo* rang in their ears. René Descartes's endorsement of deduction delayed the exile's triumph in his homeland, given that his reasoning from first principles and his positing the existence of God allied his disciples with the forces of reaction and authority.

Over the course of the Enlightenment, materialists – who appeared in different guises – seemed to gain the upper hand. To some people, materialism meant that the universe was nothing but matter in motion, and that its organization was the accidental result of this mechanical activity. To others, it meant that the operations of the mind might be understood without reference to any spiritual or nonmaterial agency. John Hedley Brooke gives several common examples cited by eighteenth-century materialists. Claims for the ability of tiny organisms to generate themselves spontaneously, deriving from the work of the English cleric John Needham (1713–1781), appeared to buttress the view that life could be introduced on earth without the necessity of divine intervention. The Swiss naturalist Albrecht von Haller's (1708–1777) discovery of muscle irritability, whereby muscle tissue contracted in response to external stimulation, provided evidence that matter possessed the ability to move itself. Radical materialists like Julien Offray de La Méttrie, whose *L'homme machine* (1747) left no place for the human soul, found Haller's work especially supportive. Most notable among this group was the Swiss naturalist Abraham Trembley (1710–1784), who claimed that after he chopped a hydra or 'polyp' into pieces, each part could regenerate itself into a complete organism. This phenomenon appeared to support the view that matter could organize itself, and to further discredit proponents of the inviolability of the soul. By the 1740s, radical materialism seemed to have won the day in France, receiving its ultimate expression, in the work of Diderot and his co-encyclopaedists. Diderot's *De l'interprétation de la nature* of 1753, which explained the progression of earth's living forms in terms of continued rearrangements of organic matter, extended the power of sensibility to inanimate objects. Diderot's system of the world, moreover, left no place for a divine force. Other eighteenth-century philosophers distanced themselves from this radical position, remaining unpersuaded about the completeness of materialist proofs. Voltaire, for example, accused the Catholic priest Needham of fraudulent procedures in his work on spontaneous generation. Haller viewed his own research as supporting his deism: irritability ranked with Newtonian gravitation as a secondary manifestation of God's power as prime mover. And moderate thinkers dismissed Trembley's attention to polyps as irrelevant for higher organisms.

Brooke muses about why – given that these spectacular 'events' might be interpreted for either a materialist or non-materialist point of view – materialism won the day among Enlightenment thinkers. He argues that, in contrast to traditional religious arguments predicated on the doctrine of original sin, materialism seemed to allow for the perfectability of humankind. Materialism was deterministic like Christianity, but it allowed that if circumstances changed, especially through education, then outcomes could vary. According to this view, humankind could escape the ravages of war or the excesses of tyranny. Like Funkenstein, Brooke sees the French philosophers as constructing a virtual 'mirror image' to the pillars of Christian faith.

Systems of 'matter in motion', elaborated by late eighteenth-century mechanists, were inconclusive, in the end, to materialists, but comforting to vitalists and theists. Laplace's nebular hypothesis, for example, appeared to demonstrate the self-sufficiency of Newtonian mechanics for explaining the universe, without appealing to God or teleology. To natural theologians, however, the perfect balance of the clockwork universe proclaimed the existence of a supreme clockmaker. Brooke emphasizes that 'a self-stabilising system' showed 'even greater ingenuity on the part of the Creator than one that required a service contract'.

Further undercutting the solidarity of both the materialist and theistic camps was the emergence of national differences, which began to crystallize towards the end of the eighteenth century. Materialism tended to flourish on the Continent, especially in France, where the Enlightenment had magnified an incipient polarization between religious and philosophical interests. French *philosophes* embraced materialism. In England, religious toleration had produced a wide range of positions that accommodated religion to science in various ways. Deism and natural theology – both of which allowed the free operation of the laws of science on earth, provided that the great Designer was installed behind it all – offered a happy solution to English minds of various persuasions.[10]

Deism

Deism may be considered 'the belief in an omnipotent and benevolent but distant and impersonal creator, who does not interfere with the laws of nature'. According to this faith, 'the fabric of the universe implies the existence of a creator just as a watch implies a watchmaker'. In contrast to the fundamental anticlericalism and irreligion of the materialists, the deists sought to understand God by comprehending his attributes in the universe. Whereas this response to the threat of agnosticism and atheism at first seemed to serve the needs of Christian apologists well, in the end the doctrine was a costly compromise. By eliminating God's supernatural attributes and reducing him to the reign of rationality, his omniscience seemed diminished.[11] With the advent of deist 'natural religion' the old-style God of revelation, miracles, and mystery transmuted into a God of science, law, and order. Now he was accessible to everyone by the common exercise of human faculties.

The steady gains of deism in Britain took place despite a powerful critique constructed around the arguments of the Scottish philosopher David Hume, whose posthumously published *Dialogues Concerning Natural Religion* of 1779 contended that the existence of God could not be rationally inferred from the natural world. In Hume's view, all religions had originated from fear and superstition; in the course of their elaboration, they tended to become fanatical and intolerant. On the one hand, Hume disagreed with the *philosophes'* optimistic enshrinement of reason as the basis of everything positive in human experience, arguing that values like justice and morality arose from habit and custom, not from reason. On the other hand, he disputed that natural theology or any other religious system could be given a rational defence. Hume maintained that ultimate causes remained unknowable.

Proponents of design inferred a divine intelligence behind the workings of the universe by making an analogy to the operation of reason on earth. In Hume's view, this approach, instead of working to glorify God, minimized the deity's attributes by making them more human and less omniscient. Furthermore, since no one had witnessed the creation of the universe, there was no way of testing the hypothesis of God as motive force. There was no compelling argument for choosing a mechanical analogy for God's activity over any other analogy, say, an organic one.

As part of his demolition of design arguments, Hume covered the terrain that would later become the battle-ground for natural theologians of William Paley's (1743–1805) persuasion. This was the realm of contrivance or the adaptation of structure to function, exemplified by the eye, the woodpecker's beak, and the giraffe's neck. Whereas Paley eventually identified each example as providing inde-pendent and cumulative evidence for design, Hume main-tained that these instances were misleading. He turned design on its head by suggesting that particular behaviours had resulted from the existence of peculiar structures, reversing the traditional argument that the structures them-selves presupposed an omniscient creator. The *coup de grâce* to the assumption of a beneficent creator, in Hume's view, was the presence of pain and suffering.[12]

Natural theology

Hume's friends prevailed on him to hold back his *Dialogues*. They believed that his critical attack on deism and other religions would be dismissed by the theistic majority as a scurrilous diatribe. But the power of Hume's critique escaped most of his contemporaries, allowing natural theol-ogy to achieve enormous popularity, especially among conservative thinkers. The doctrine received canonical expression in William Paley's *Natural Theology* of 1802. There the Anglican priest Paley argued that the existence of God could be proved by the benign functioning of the universe. Paley emphasized the adaptation of structure to function. One favourite example was the human eye.

Brooke identifies the literature of natural theology as 'the most sustained tradition in which science and religious belief were integrated'. The concept of seeing God's handiwork in the natural world accommodated a range of religious doctrines. To Christian apologists, it supported scriptural accounts by providing evidence of God's existence. To deists, the possibility of inferring the existence of the prime mover from his handiwork diminished the necessity of relying on revelation. In other words, 'natural theology could be used both to attack and defend Christianity'.[13]

Natural theology attracted European clerics and scientists alike, but it achieved greatest success in Britain. Newtonian mechanics had earlier provided an unassailable foundation for the natural theological tradition. Paley extended the approach with his interpretation – epitomized by the timepiece metaphor – by appealing to the eighteenth-century affection for machinery. As Frank Turner notes, 'Paley transformed his deity into a skilled and ingenious English engineer'. Subsequently, natural theology achieved prominence by working to unify diverse religious and political persuasions.

Not every subtle mind accepted natural theological explanations. Like Hume before him, the German philosopher Immanuel Kant challenged the doctrine, taking issue with the way it employed analogical reasoning. In his *Critique of Pure Reason* (1781), Kant maintained that the existence of God could not be proven rationally. Kant spoke in favour of separating scientific explanations from theological purposes.

British thinkers stepped around the Kantian exposition by assuming that Christian theology provided a solid basis for the apparent adaptation of form to function so central in design arguments. As late as the 1830s, natural theology reigned triumphant. It is a leitmotif in the works of the authors of the Bridgewater treatises of the 1830s, among them philosopher William Whewell, chemist William Prout (1785–1850), and geologist William Buckland (1784–1856). In the realm of the life sciences, however, natural theology carried both negative and positive implications. On the negative side, it tended to excuse creationism instead of promoting a dispassionate discussion of evolutionary theory. On the positive side, the belief in design arguments compelled naturalists to delve into other questions related to teleology, such as biological adaptation and the fossil record.

Just as religion forced accommodation in scientific quarters, so the teleologists had to readjust design arguments in the light of new scientific discoveries. The fossil record provided evidence for widespread extinction of organisms, which required natural theology to modify the principle of plenitude (according to which God created and nurtured every conceivable creature) upon which it was based. As a result, three kinds of natural theology emerged during the second quarter of the nineteenth century. One advanced a more sophisticated elaboration of Paley's original doctrine (that the parts of an organism were perfectly adapted to their function) in response to Etienne Geoffroy Saint-Hilaire's concept of unity of structure, which left no place for teleology. The comparative anatomist Richard Owen produced a second variation on design in which all vertebrates were created according to an ideal model, called an *archetype*. To natural theologians, the adaptation of archetypes to the needs of a particular species provided instances of God's benign purpose. A third tradition – articulated by Oxford's Savilian professor of geometry and Anglican clergyman Baden Powell (1796–1860) and exemplified in the work of Charles Darwin – supported the view that the laws of nature (rather than the domain of the inexplicable) provided the most persuasive evidence for design. William Whewell inaugurated this tradition by his contention, reminiscent of seventeenth-century Protestant writings, that the regular operation of natural law provided better evidence of the omniscience of the Creator than isolated acts of divine intervention.[14]

The argument against Darwinian evolution

In Darwin's early formulation of his theory, evolution supplied compelling evidence for design by explaining with detailed examples how organisms had adapted to their natural environments by producing species of increasing complexity. As Darwin elaborated his theory in subsequent editions of the *Origin of Species* (1859) and as the full impact of his arguments were grasped by naturalists, this proposition came into question. The free, mechanistic operation of natural selection left no real place for a God whose best designs might be rendered obsolete by interspecial competition or by a hostile environment.

Some theistic naturalists like Harvard University's Asa Gray (1810–1888) tried to minimize evolution's fundamental contradictions with design arguments by emphasizing that natural selection offered a solution to the vexing problem of pain and suffering. These unhappy occurrences resulted from the operation of natural selection, which drove forward God's broader creative plan. Gray believed that Darwinian evolution resolved this issue better than Paley's creationism.[15]

Other naturalists, like McGill University's John William Dawson (who inherited the mantle of the creationist comparative zoologist Louis Agassiz), were more distrustful of evolutionary doctrines and eventually challenged Gray's teleological rendering of Darwin's theory. Dawson objected to the logic of Darwin's argument. The problem arose, he explained to geologist Charles Lyell, not from 'any defect in the facts of the case', but rather 'in coupling together two classes of phenomena which have no logical connection, and assigning as common causes terms which mean one thing when applied to one class of facts and another when applied to the other'.[16] In other words, Dawson – like many other critics of Darwin – objected to a facile association between artificial and natural conditions.

Dawson contended that Darwin had constructed a misleading analogy between artificial and natural realms. Moreover, in his reading, one could not provide hard empirical evidence of transmutation; at best, one might 'build up an imaginary series of stages, on the principle of natural selection, whereby these results might be effected; but the hypothesis would be destitute of any support from fact'. Acceptance of Darwinism meant that 'our old Baconian mode of viewing nature will be quite reversed': as a result, 'instead of studying facts in order to arrive at general principles, we shall return to the mediaeval plan of setting up dogmas based on authority only, or on metaphysical considerations of the most flimsy character, and forcibly twisting nature into conformity with their requirements'.

In Dawson's view, the faults of accepting Darwin's argument by capitulating to 'showy analogies' were compounded by ingenuousness in giving credence to the imperfect geological record. Paleontology provided no reason to favour evolution by natural selection over any other explanation of why the natural world appeared as it did. Rather than variation and change, Dawson saw constancy over time in both fossil and modern forms.

Given the faults in Darwin's methodology and evidence, Dawson inclined to a benign, teleological explanation. A more 'agreeable' thesis, along the lines of Darwin's, suggested 'that each species finding its means of subsistence and happiness constantly extending, exerted itself for their occupancy, and so developed new powers'. This explanation accounted for 'elevation', Dawson insisted, 'as if nature, like a skilful breeder, were giving constantly better food or pasture, instead of imitating the luckless experimenter who strove to reduce the daily food of the horse to a single straw'. (Here Dawson sided with Jean-Baptiste Pierre Antoine de

Monet de Lamarck, as did many 'theistic evolutionists'.[17])
Working naturalists had even more to gain, in Dawson's
view, if they remained 'content to take species as direct
products of a creative power, without troubling ourselves
with supposed secondary causes'. In essence, *On the Origin
of Species* had provided neither accurate reasoning nor compelling evidence to persuade a sceptical Dawson.

During the early 1870s, Dawson exchanged opinions by
letter and in print with Asa Gray. Dawson reiterated his
view that the hypothesis of evolution ran counter to 'true
science', especially in its 'hypocritical and unfair attempts
. . . to confuse embryology with geological sequence and to
frame classifications looking to evolution and not to actual
affinity'. He maintained that 'the teaching of facts' proved
neither 'derivation' nor, for that matter, even 'the immuta-
bility of species'. Happily, this 'terrible doctrine' had not
been proven true, or Dawson would be bound to believe it
as a scientific man 'even though it should rob me of all I
value most in this life and that which is to come'.

What made Gray's endorsement of Darwin's views so
dangerous, in Dawson's opinion, was an easy extension
from the natural world to the social environment. This mis-
application produced 'such monstrous creations of fancy' as
'the materialistic evolutionism' of Herbert Spencer and John
Tyndall, as well as 'the wretched superstition that man is a
descendant of apes', which 'lay the axe at the root of all
that is valuable even in the social organization of society'.
Huxley and his allies posited that 'man is a mere automaton;
all his notions of virtue, vice, responsibility, and immortality
are mere delusions'. Not only did Darwinism sweep away
both Christianity and natural religion, but a populace
imbued with 'the doctrine of the struggle for existence'
could only 'cease to be human in any ethical sense, and
must become brutes or devils or something between the
two'.

Gray, for his part, challenged the views expressed in
Dawson's *Story of Earth and Man* (1873). Dawson maintained

his belief in fixity of species with 'earnestness, much variety of argument and illustration, and no small ability', Gray wrote. Yet, as Gray correctly observed, Dawson believed that evolutonary doctrines were not only false, but were also 'thoroughly bad and irreligious'. In Gray's view, Darwin could not be accused of eliminating design from nature. His studies of climbing plants and the fertilization of blossoms by insects had 'brought back teleology to natural science, wedded to morphology and already fruitful of discoveries'.

Even in the hands of a balanced and reasonable proponent like Gray, Darwinism remained, in Dawson's opinion, a dangerous doctrine. By the late 1870s – nearly two decades after publication of the *Origin of Species* – Dawson outspokenly opposed the theory. To him, evolution by natural selection remained a simplistic approach to nature. By explaining too much, it explained nothing at all. He viewed evolution as a 'philosopher's stone' (which could 'transmute the viler into the more exalted species') or as a panacea developed for 'specialists and enthusiasts, who ever tend, like quacks in medicine, to refer all effects to the same cause, and to cure all evils by one specific'. Dawson concluded that the history of science was 'strewn with the wrecks of such hypotheses, devised in every age by ingenious men, to serve as a substitute for actual knowledge, and to spare themselves the labour of arduous investigation; satisfying one generation with a comfortable form of words, only to be cast off by the next'. He recognized that he could not accommodate his religious beliefs to Darwinian evolution and that the 'Christian Darwinism' of Gray and others remained chimerical.

Well before the end of the nineteenth century people recognized that natural theology – whether in Gray's muted arguments or Dawson's more dogmatic ones – fettered scientific investigation. As the century wore on, a new group of professionals built their reputations on banning theological considerations from scientific matters. As Frank Turner puts it, the scientific community had become 'self-defining'.

It proclaimed independence from ecclesiastical concerns.

Once again, national differences complicated and undercut the political and religious response to Darwinian evolution. In France, the scientific and political establishment exemplified by Georges Cuvier saw an alliance in catastrophism and creationism, both of which they supported. The transformism advocated by figures like Geoffroy Saint-Hilaire was seen as revolutionary and perhaps even seditious. The association of evolution with materialism and secularism propelled mainstream Catholic scientists to view Darwin's theory with disfavour. In England, the radical materialism of Tyndall and Huxley gave way before more respectable advocates like Charles Lyell, Richard Owen, and even Darwin himself. In Germany, the theories of Ludwig Büchner (1824–1899), Carl Vogt (1817–1895), and Ernst Haeckel (1834–1919) cast biological materialism in its most extreme expression.[18]

Twentieth-century developments

One theme central to twentieth-century intellectual history has been a crisis of faith in the doctrines associated with traditional religions – whether Christianity, Islam, Confucianism, or Hinduism. One sees this crisis in the mid century social upheavals, when many nations asserted political autonomy. In communist China, Indonesia, India, Iran, and Yugoslavia, social progress exploded traditional religious values. The various cultural revolutions of the 1960s and 1970s – in China, Indonesia, and Iran (and the internecine war in fragmented Yugoslavia during the 1990s) – may be read in part as reactions to imported, soulless technology.

Over the past generation, religious thinkers have been eager for a rapprochement with the world of science. Intellectuals at Gadja Mahda University in Yogyakarta on Java have produced texts integrating modern astrophysics with traditional Islamic thought. Spiritual heirs of religious reformers John Calvin and John Knox have claimed once more that biblical scripture provides an acceptable alternative to Darwinian-inspired evolutionary theories.

Creationism is the view that the Bible provides an acceptable explanation for the profusion of life on earth. It draws strength from current notions that science is nothing more than a collection of socially conditioned prejudices and attitudes. Equal time is now sought in state school science classes for biblical accounts of creation. The resulting controversy may be less disruptive than the popular press has suggested, however, for when they work in laboratory or observatory, religiously devout scientists are no less scrupulous in their attention to detail and reproducibility than their irreligious counterparts. Oil deposits can be found and DNA can be sequenced without swearing allegiance to Darwin's theory of evolution.

Less conventional, holistic gurus have also advocated reintroducing spirit into nature. One neospiritual proposal emerged about 1970 under the name of 'Gaia', the mythological earth goddess of classical antiquity. James Lovelock, a chemist studying the possibility of life on Mars for the National Aeronautics and Space Administration in the United States, came to believe that life on Earth evolved from more than a random and directionless, or non teleological, sequence of events. In his view, many complex processes, both organic and inorganic, *favoured* the persistence and increasing complexity of life forms. A parliament of nature voted for changes in geophysical parameters, such as the gaseous components of the atmosphere and various ocean currents, to compensate for phenomena that might extinguish living things, for example the increasing intensity of solar radiation; Gaia was the parliáment's prime minister, its personification. Lovelock's views have been taken up by biologist Lynn Margulis (b. 1938), and they are part of an intellectual movement emphasizing non linear and self-organizing properties of nature.

It is hard to say how many claims of the scientific holists will stand up under scrutiny by sceptical and dispassionate scientific colleagues. In some measure, what the holists assert lies beyond the fabric of science. The claims bring to mind the imaginary properties of animal magnetism proposed by Franz Anton Mesmer in the eighteenth century. They echo the fragmentary messages of many non linear prophets, from the dialectical materialism of political economist Karl Marx, to the hylozoist monism of the spiritualist philosopher Rudolf Eucken, to the occultism of educational reformer Rudolf Steiner.

One need only observe the political success enjoyed by 'Green' movements over the past generation, many of which infuse the Earth with a consciousness, to appreciate the enormous gap separating the existential angst of the 1950s from the goal-directed spiritualism of the 1990s. Whereas organized religion suffered stunning reversals at mid century, knowledge based on divine revelation is today resurgent everywhere. Anthropologist Gregory Bateson, one of the prophets of the new deism, did not avoid seeking allies among the religiously devout.

We have not heard the last word about the place of religion in science. In his recent book, *The Western Canon* (1994), literary critic Harold Bloom imagines that the twentieth century, the 'Age of Chaos', shall give way to a new theological age, where religious values assume paramount importance. Enough is known about science and religion to suggest that in the future the two realms of knowing may coexist. Margaret Mead (1901–1978) expressed the hope a generation ago:

> We must take upon ourselves the task of providing a twentieth-century faith we can put together with empirical knowledge. Theologians must work with scientists to build this new faith, because without faith and love we may destroy the world. With faith and love and no knowledge the world may also be destroyed.[19]

Knowing: Progressing and Proclaiming

James Burke, in his television series *The Day the Universe Changed*, emphasizes the remarkable power that modern science and technology has bestowed upon twentieth-century society. Each of the ten episodes is cast in a mode that instructs by showing how today's invention, discovery, or institution is rooted in past experience. As Burke asserts, it is ever 'onward and upward'. This unbounded enthusiasm for the fruits of scientific discovery and technological innovation is a hallmark of modern civilization, although it has been challenged in the past by luddites and today by postmodernists. Great minds since the seventeenth century have associated scientific and technological achievements with a firm belief in progress.

Burke also raises profound questions of epistemology, taking his cue from Thomas Kuhn's *Structure of Scientific Revolutions* (1962). He argues that knowledge is culturally determined, that it changes as we change. He is, then, a relativist, but a relativist who believes in progress. We savvy creatures of the twentieth century have, as Burke describes, discarded religious and superstitious explanations for natural phenomena. Our insatiable curiosity has driven us to new heights, and we do stand on the shoulders of giants. But, as we enter the last years of the century, we still suffer the stockpile of weapons of mass destruction, the perpetuation of horrific wars whose victims are innocent civilian populations, the scourge of epidemic disease with the potential to decimate huge populations, as well as the glaringly inequitable distribution of wealth and resources.

The Scientific Revolution of the seventeenth century did indeed produce a new attitude toward knowledge. The emphasis on writing in the vernacular, rather than in the dead tongue of scholars, demonstrated a commitment to broader accessibility. The Enlightenment added a social imperative to ameliorate the human condition, an ideology that called for the diffusion of the fruits of scientific and technological advance. Humanity was invested, in addition, with an inalienable, innate ability to reason, judge, and act in its own best interests. Defining those interests – with an appreciation for the formative role of science and technology – has been the story of the past 300 years.

Magic and science

The Scientific Revolution, revising humankind's ability and capacity to know, reflected singular new beliefs and priorities. Historians have pondered, moreover, whether the Scientific Revolution itself was not the product of a more fundamental shift in worldview. We have already considered the interrelation of Puritanism and early modern science, whereby religion motivated scientific work.

Attitudes toward the natural world seem to have changed long before the Scientific Revolution. The belief in natural magic, especially, lay at the heart of Renaissance philosophies. As Keith Thomas describes the situation, the desire for power associated with natural magic produced a fertile environment for empiricism and experiment: 'It marked a break with the characteristic medieval attitude of contemplative resignation.' This magical or 'hermetic' tradition (named after a collection of treatises written around the second or third century AD and attributed to the Egyptian god Thoth, who was named by the Greeks 'Hermes Trismegistus' – 'Hermes Thrice-Blessed') enriched astrology and alchemy with the notion that hidden or occult powers could improve the human condition.

Traditional historians dismissed mysticism, maintaining that it could only be antagonistic to the emergence of modern science, that the scientific imagination could wax only if the infatuation with magic waned. Under the tutelage of Dame Frances Amelia Yates (1899–1981), however, scholars reconsidered the role of the hermetic corpus. In her *Giordano Bruno and the Hermetic Tradition* (1964), Yates identified elements of natural magic, such as its exaltation of the practical or applied arts and its impetus for interrogating the world of nature, as being closely allied with the spirit of modern science.

The works of Paracelsus (1493–1541) and his followers (one of Yates's favourite examples) illustrate a rapprochement between scientific interests and religio-magical concerns on the eve of the Scientific Revolution. For the Paracelsans, alchemy was supposed to bring about redemption and purification in the physical world, just as Christ had redeemed humanity. Hence, the laboratory could produce cures for disease. Alchemical manipulations, moreover, cohered with the Calvinist worldview. Historian Christopher Hill explains that in both these realms, salvation came 'from without', whether by the grace of God or the use of the magical philosopher's stone (which conferred immortality). Yates includes Francis Bacon in the hermetic camp (she has him belong to the secret hermetic society, the Rosicrucian Order, whose earliest extant work appeared after his death). In Yates' view, Bacon's utopian sketch of 'New Atlantis', a land ruled by 'mysterious sages who keep the citizens in tune with the cosmos', is nothing other than the Rosicrucian belief in *prisca sapientia*, lost ancient knowledge that could be recovered by scientific enquiry. Even Isaac Newton – generally seen as the embodiment of the rational spirit of the Scientific Revolution – was motivated by the hermetic, *prisca* ideals. Newton plunged into ancient texts to disentangle their intentionally obscure formulations and reveal their practical wisdom. Newton viewed his theory of universal gravitation as a restatement of what had already been known in ancient times. He was also keenly interested in 'an *occultist* tradition of alchemical, astrological, and magical beliefs'.[1]

Occult beliefs certainly persisted into the seventeenth century (they are alive and well today), but Newtonian mechanics and other triumphs of the Scientific Revolution provided reasons to dispel hermeticism. 'The notion that the universe was subject to immutable natural laws,' Thomas notes, destroyed the intellectual basis of the magical tradition. Partnership between magic and science could only dissolve; by the late seventeenth century, it had collapsed. At that time, we are hard-pressed to find a *savant* who dabbled in magic like the aging Newton. In Thomas's words, the 'epistemological demand for certain knowledge' entailed the steady erosion of natural magic.

Baconianism

The decline of magic left no philosophical vacuum. The writings of Francis Bacon provided an ideology for a new order based on experience, experiment, activism, and the rejection of past dogma. The sterility of traditional, Aristotelian philosophy had spelled its own doom, whereas Bacon's system proved its worth by seeking empirical results. Bacon extracted from the magical tradition the view that the people generally, not just privileged individuals, should be able to help themselves.[2] Knowledge, when placed in the hands of those willing to direct it to useful purposes, could ameliorate the condition of humankind. By the eighteenth century, Enlightenment philosophers hailed Bacon's philosophical system, enshrined virtually every one of its elements, and made him their patron saint.

Historians associating Puritanism with the rise of modern science have often seen the scientific philosophy of Francis Bacon as supplying a connection between the two realms of experience. Christopher Hill relates the Puritan emphasis on individual religious testimony to the Baconian imperative for personal experience and observation. Charles Webster describes Baconianism as a philosophy 'commensurate with the Puritan ethic'. He views the Puritan reform of religion and the Baconian advancement of science as two parallel movements, each leaving 'a permanent cultural legacy'.

Such characterizations may be misleading. Many hardline Puritans viewed Lord Chancellor Bacon's writings with scepticism because of his high court and high church connections. Many people outside the Puritan fold embraced the Baconian philosophy.[3] Furthermore, the attempt to demonstrate 'parallel value-complexes' in the realms of science and religion is fraught with difficulties, as critiques of the work of Robert Merton have shown. A concept like 'utility' or 'progress' may mean quite different things to scientists or to preachers.[4]

Dismissing such objections, Charles Webster has done much to clarify why many Puritans found Baconianism so congenial. Bacon's philosophy appeared to be 'providentially designed for the needs of the Puritan Revolution', adopting a similar biblical and millenarian frame of reference. It gave 'precise and systematic philosophical expression to the anti-authoritarianism, inductivism and utilitarianism' that figured so prominently in the Puritan worldview. Webster argues that the most suitable form of intellectual enquiry for Puritans involved the 'patient and accurate methods of experimental science', which dealt with the secondary manifestations on earth of an inscrutable Calvinist God.

What were the chief elements of Bacon's philosophy? The *New Organon* (1620) presented four major features of his system. It introduced, first of all, the concept that true knowledge gave power to humanity: in Bacon's words, 'knowledge *is* power'. Unlike scholasticism, whose epistemology depended upon fitting new phenomena into a rigid intellectual structure, with Bacon, knowledge derived from increasing humankind's power over the environment. Improved understanding of the natural world, moreover, translated into the ability to command it. Baconianism was utilitarian, unlike traditional scholastic philosophies. Second, Bacon separated science from theology, positing that each system obtained truth about a separate order of phenomena. In Bacon's mind, the ultimate nature of the world remained unknowable from the exercise of the senses. Ultimate truth belonged to the theological realm, which did not traffic with the natural order knowable through science. Third, Bacon's system of reasoning in the *New Organon*, the method of induction, derived general statements from collections of individual facts alone. Such accretions of fact gave rise to general propositions; the predictive power of these generalizations helped to consolidate control over the natural world. Baconianism, then, was a profoundly sceptical philosophy, both in its distrust of unassisted reason and its antipathy to systems.[5] Fourth, in contrast to the static Aristotelian corpus, Baconian science emerged as a cooperative and collective endeavour, by its very nature dynamic and progressive.

Bacon's works contain concrete examples of his methodology. He proposed the idea of compiling a universal natural history, a comprehensive collection of facts drawn from observations and experiments. This was no mere exercise in abstract data collection; a central part was to be a 'history of trades', which would detail improvements in the manual arts. Such a project turned out to be enormously important to early members of the Royal Society, with whose utilitarian and commercial aspirations the Baconian programme resonated resoundingly.[6]

In the view of Charles Webster, Bacon's concept of a collaborative natural history formed his most important intellectual legacy for the Puritans. It provided a cooperative research programme for craftsmen and scholars, directing their talents towards practical ends in economic geography, agriculture, and medicine. It vindicated their commitment to inductivism and their freedom from authority. And it bypassed the slough of atheism and subversion that could arise with the cultivation of other areas of natural philosophy.

Bacon applied his epistemology to celebrate experiment. In his view, experiment captained the attack on authority, for it represented the side of works versus words, nature versus books, industry versus idleness, and doing versus thinking. The physical exertion required by experiment was virtuous, in distinct opposition to the inactivity of reading, a pursuit he saw as slothful.

Bacon's *Great Instauration* proposed a total reorganization of science and society. His cooperative programme brought together a range of people, including artisans, merchants, and philosophers, and embued them all with a new zeal for the undertaking. It democratized science by appearing to lower the intellectual qualifications required for scientific work. This resulted from the Baconian subordination of reason and learning to sensate experience, as well as from the enormous manpower requirements for compiling a universal natural history. Bacon believed that cultivation of the mechanical arts brought immediate benefit to humankind, and he appreciated the extent to which artisans had improved shipbuilding, ballistics, and printing. During the Puritan Revolution, Bacon's philosophy fit well with the values of literate English society. It also entered the popular consciousness, as Bacon's works were reissued in new editions and his name began to appear in inexpensive sources with wide distribution, such as almanacs.

The sociologist Joseph Ben-David associated the rise of science in England with a generalized allegiance to Bacon's experimental philosophy, which gave savants a unity of plan and purpose. This coherence was based largely upon an activist approach towards the natural environment. Bacon recognized not only that knowledge progresses, but also that the growth of knowledge contributes to the betterment of humankind. With Bacon, knowledge became subservient to human needs, making happiness the end of science.[7] This legacy has brought enormous power to the scientific enterprise, but it also has called scientific legitimacy into question whenever it has seemed to forsake this aim.

Encyclopaedism

The Baconian appeal to progress and utility, its rejection of authority, its scepticism, and its methodology based on experiment, observation, and induction supplied French Enlightenment thinkers with the perfect philosophical platform. Baconianism placed people at the centre of the world, and for this reason it was attractive to eighteenth-century social reformers. Unlike Bacon, who had tempered his enthusiasm with the spectre of 'errors' in human understanding, the *philosophes* foresaw no obstacles to human progress. The tradition of English science, 'marked by a freedom from *a priori* assumptions and system-building ambitions' inaugurated by Bacon, almost assumed the status of a cult in late eighteenth-century France.[8]

Ben-David has described how scientific 'hegemony' shifted to France from England during the last half of the eighteenth century. France imported the successful explanations of British empirical science, typified by Newtonian natural philosophy, and wholeheartedly embraced Baconianism, the tradition's most important underpinning. Just as some French Enlightenment thinkers like Voltaire became more Newtonian than the English followers of Newton, others like Denis Diderot became more Baconian than the English Baconians. As historian of science Margaret Jacob explains, Bacon's call for the classification of knowledge 'appealed to the organizers of knowledge, those directly connected with the printed word, who by the late seventeenth century faced the monumental task of simply keeping track of all that was now being published'.

One of the best ways of keeping track was to present what was known, revealing its great expanse and its interconnections. Indeed, the production of encyclopaedias underwent enormous expansion during the last quarter of the seventeenth century and the first half of the eighteenth. More than thirty different works were published during this period, greater than the number produced in the preceding 200 years. There is no mystery about the reason for this proliferation. The rapid expansion of knowledge placed a premium on accessing specialized information across all domains.[9]

The most important undertaking of the French Enlightenment was the *Encyclopédie*, or *Dictionnaire raisonné des sciences, arts et métiers* (1751–72) of Jean le Rond d'Alembert and Denis Diderot. It was a monumental project drawing upon 135 contributors and consisting of twenty-eight volumes (seventeen letterpress volumes and eleven with engraved plates). From a commercial standpoint, it earned more than two million francs in profit for its publishers.

By its inspiration and direction, the *Encyclopédie* derived from Bacon. It was a collaborative project of a 'society of men of letters' – among them the leading *philosophes* – dedicated to the diffusion of the practical and useful arts. In a sense, the *Encyclopédie* functioned as a universal natural history, taking as its starting point the necessity of cataloguing what was already known. But the *Encyclopédie* was no mere compilation of human knowledge; it made moral judgments. In the view of philosopher Rom Harré, the subversive character of the *Encyclopédie* stemmed from its desire to present science as a comprehensive intellectual system, thereby offering an alternative to orthodox religion. Additionally, its promotion of Bacon's beliefs about the scientific amelioration of crafts, particularly the corollary that the skilled trades formed the basis of civilized life, called in special scrutiny from the censors. Unlike any other encyclopaedia of the day, the *Encyclopédie* contains nearly three thousand plates dealing with technology, also the subject of many of its articles. It was the greatest 'how to' manual the world had ever seen.

D'Alembert's 'Discours préliminaire' presents the philosophes' hopes for the enterprise. By showing relationships among ideas, the work aimed to provide an 'encyclo-paedic' picture of the unity, coherence, and order of knowledge. This organization drew inspiration from Bacon's 'Tree of Knowledge'. The *Encyclopédie* was to function, as well, as a 'reasoned dictionary', allowing the reader to consult it easily about any particular branch of the sciences or arts by means of a system of cross-references.[10]

The *Encyclopédie* is self-reflexive about its own history; in addition, it discusses the social utility of dictionaries and encyclopaedias in general. D'Alembert's preliminary discourse refers to Diderot's original prospectus for the work, which attributed the diffusion of scientific ideas and the spread of enlightenment to the publication of dictionaries. D'Alembert argues that dictionaries were designed not for continuous reading, but for occasional reference. The editors' desire to present scientific and technological discoveries in a condensed format aimed to make it easier to learn about progress in these areas. Originally, the editors had set out to translate Ephraim Chambers's (ca. 1680–1740) *Cyclopaedia* (1728) from English into French, but they soon realized that too much essential information had been omitted from his two-volume work. Moreover, insisted d'Alembert, in such a project the omission was worse than a mere imperfection, because 'it breaks the entire series or chain of articles and spoils the form and the substance of the work'.

A second difficulty with Chambers's undertaking was that a single individual could no longer hope to be conversant with the full range of advances in the arts and sciences. Alternatively, in the view of Diderot and d'Alembert, an author experienced great difficulty in treating 'thoroughly and profoundly' a subject to which he had dedicated himself entirely. By making the *Encyclopédie* a collaborative project, then, the editors both avoided the inevitably subjective orientation of a single individual and whole-heartedly embraced one of the tenets of Baconianism; namely, cooperation among all manner of philosophers and artisans.[11]

Both its content and its publishing history soon transformed the *Encyclopédie* into the 'war machine' of the French Enlightenment. Some of its articles were straightforward: natural philosophers like Bacon, Descartes, Galileo, and Newton were presented as inaugurating a new era in thought. Other points were made by omission. Articles on Catholic doctrines of transubstantiation and the eucharist, for example, were absent. An elaborate system of 'philosophical' cross-referencing was brought into play, whereby articles on orthodox theses were ridiculed. In other cases, critical discussions were presented in a persuasive way. Some of these techniques stalled the censor's pen for a short time, but eventually they resulted in the systematic suppression of the *Encyclopédie* in France. Every act of censorship made the volumes more highly sought after and turned it into a *cause célèbre*. It appears that 25,000 copies of the entire work were in circulation on the eve of the French Revolution.[12]

Although they fell far short of attaining the cultural and philosophical significance of the *Encyclopédie*, other eighteenth-century encyclopaedias merit some discussion. As historian Frank Kafker puts the case, without comparison we have no clear understanding of the ways in which the *Encyclopédie* was 'unique, bold, and innovative, and in what ways it was conventional, timid, and even retrograde'. Unlike the *Encyclopédie*, the *Encyclopaedia Britannica* (published in 1771 in three volumes) did not aspire to become a 'book of universal knowledge'. Instead it examined a wide range of individual subjects by means of long disquisitions. It made no attempt to be comprehensive; rather, it was selective, and within any given subject, exhaustive.[13]

In contrast, encyclopaedias of the nineteenth and twentieth centuries, such as the re-edition of Abraham Rees's (1743–1825) *Cyclopaedia* in the early 1820s (originally a revision of Chambers' *Cyclopaedia* published in the 1780s) and the *Cabinet Cyclopaedia* of Dionysius Lardner (1793–1859), aimed to be works of ready reference. Rather than trying to systematize knowledge (a process that made them highly resistant to revision and easily rendered them obsolete), they presented miscellaneous subjects in alphabetical order. Thus convenience may well have abetted a tendency toward specialization and fragmentation, in contrast to the unitary vision of the *encyclopédistes*.

Materialism

If the *Encyclopédie* represented French philosophers' comfortable adoption of Baconian philosophy and the Newtonian world system, it failed to afford them an avenue for the free expression of their metaphysical beliefs. These beliefs had been irrevocably altered by the triumph of Newtonian physics, which championed mechanistic explanations. Human behaviour was seen as no exception to the mechanical worldview, but it remained a dangerous activity to write about humankind too reductively. People were distinct from the rest of the animal kingdom by their spirituality; moreover, all living things, whether plant or animal, were entities possessing something more than simple matter in motion.

Squabbling between the two editors of the *Encyclopédie*, d'Alembert and Diderot, reflected the widening gulf between the *philosophes* loyal to the deistic vision of Voltaire and those attracted to the atheism and materialism of a *baron* d'Holbach (1723–1789).[14] Julien Offray de La Méttrie produced the first systematic materialist treatise, *L'homme machine*. His studies under the tutelage of the great Dutch physician Herman Boerhaave at Leiden led him to attempt to reconcile Newtonian physics with Cartesian physiology. La Méttrie maintained that people stood out from the rest of the animal kingdom only by their complexity, not by any inherent spirituality. People, in La Méttrie's view, were simply machines capable of thought. La Méttrie's uncompromising materialism allowed a distinction between animate and inanimate realms, which was not posited in the work of Diderot. In *Le rêve d'Alembert* (1769), Diderot rejected any categorical distinction between living and dead, arguing for the unity of all matter. Other French materialists shared an outlook in common with their encyclopédiste counterparts. They believed that it was a source of happiness to learn about the material world, unfettered by philosophical systems of the past. This knowledge had to be based on sensate experience, which would ultimately lead to the control of nature.

The Enlightenment, fuelled by Baconian humanism, bestowed an unbounded faith in progress upon the nineteenth century. Technological advances associated with the Industrial Revolution and the ever-expanding explanatory power of the natural sciences lent further support to an unbridled optimism and activism. An overriding secularization formed a concomitant part of this intellectual legacy. It is not surprising, therefore, that the nineteenth century witnessed the articulation of grandiose philosophical systems. The general faith in progress was to be subjected to scientific scrutiny, leading to the discovery of a general law that would explain and direct social development. These philosophies melded a commitment to action with an overriding belief in the power of science and technology to effect unlimited progress.

Positivism

The word *positivism* has been used to denote philosophical systems that assign a central role to scientific and technological agents. More specifically, positivism posits the view that the branches of human knowledge can be classified and ordered according to a hierarchy, thereby parting company with the Encyclopédistes who assign no greater or lesser value to any of its constituents. The *philosophe* contributors to the *Encyclopédie* had sought to present the products of human invention especially, finding inspiration enough in their quantity and utility.

Positivists codified their belief in progress. They proposed that human societies inevitably pass through particular stages, related largely to their cultivation of science and technology. The philosophy of positivism, accordingly, by its inspiration and in its proposed social application is *engagée* – it possesses a constant referent to contemporary society. Unlike the logical positivism of the twentieth century, or even the late nineteenth-century epistemological positivism of Ernst Mach, early positivists eschewed methodological questions for cultural prescriptions. As Auguste Comte stated the case in his six-volume *Cours de philosophie positive* (1830–42), 'knowledge generates foresight, foresight generates action'.

Positivism, then, was defiantly rational and secular, dismissive of both metaphysics and theology. The philosophical movement was launched by the reforming fervour of the Parisian technocrat Claude Henri *comte* de Saint-Simon, who became so enamoured with the recent accomplishments of the physical sciences that he sought to bring social dynamics under their spell. The term positivism, coined more properly by Saint-Simon's student Comte, can usefully be employed to encompass and unite the historical writings of Henry T. Buckle, the socialist theories of Karl Marx and Friedrich Engels, the philosophies of William Whewell and John Stuart Mill, and the sociology of Herbert Spencer. English positivists Mill and Spencer, as well as their French predecessor Auguste Comte, received special attention in John Theodore Merz's masterful *History of European Thought in the Nineteenth Century* (1904–1914).

In his accessible work, Merz describes a number of changes that transformed European philosophy during the nineteenth century. The positivists were part of the process by which philosophy moved from issues of logic and metaphysics and became a weak partner to sociology. Positivists constructed elaborate systems, with a proclivity for the grandiose and the all-encompassing – witness the multivolume philosophical 'course' of Comte, or Spencer's tomes that unite mind and society in a grand evolutionary synthesis. Even major national distinctions, whether the exaltation of inductive empiricism in England or the French penchant for deductive mathematics, were swept aside by the tide of positivism. It triumphed because it attached itself to the prestige associated with science as well as to Enlightenment convictions about the progressive development of humanity and society.[15]

What unites the procession of nineteenth-century positivists is the conviction that there is no separation between the physical and natural sciences, on the one hand, and the elaboration of a science of society, on the other. Positivists believed that the phenomena of human thought and social life were continuous with the phenomena of the organic and inorganic realms. The behaviour both of the individual and of aggregated individuals, therefore, was susceptible to investigation by the methods of science, which would yield completely reliable results. The ultimate function of knowledge was the improvement of society, a view that gives science a central role in effecting social progress. Moreover, positivists sought to discover the law underlying this social progress, whose validity was as sound as the law of gravitation.[16] Merz viewed these concerns as dominating philosophy in France and England, although the social gospel evoked little interest in Germany, almost an anachronistic bastion of metaphysics.

The polemical positivism of Auguste Comte

If positivism is the central ideology of the nineteenth century, its *Zeitgeist*, it is Comte's doctrine as expressed in his *Cours de philosophie positive* that must be fathomed in the first instance. A central tenet of this doctrine is Comte's law of three states or three stages, which he first articulated in 1822. According to this law, human thought has passed through three distinct historical phases: the theological or fictional state, the metaphysical or abstract, and the scientific or positive. The two earlier stages, where explanations invoke gods or metaphysical forces as causative agents, were necessary precursors to the positive state. In this later scientific era, the appeal to causes is repudiated in favour of the precise expression of verifiable correlations between observed phenomena in the form of laws. Knowledge in general, as well as each branch of knowledge, passes through these three stages.

The task of assessing the degree of 'positiveness' of each branch of knowledge called forth the second major characteristic of Comte's doctrine, his theory of the hierarchy of sciences. Comte's taxonomy ranked mathematics (followed by astronomy, physics, chemistry, and biology) at the top and sociology at the bottom of the order. In Comte's view, only mathematics and astronomy had achieved full maturity as positive sciences.

Besides the degree of positiveness, the hierarchy of sciences displays other characteristics in Comte's system. Complexity, concreteness, and dependency increase as one moves from astronomy to sociology. As one proceeds down the hierarchy, appropriate methodologies change and diversify. For example, astronomers can only observe, while biologists can also compare, experiment, and construct analogies. The aim of Comte's intellectual exercise was to establish a science of social physics on a firm foundation. By so doing, society might ameliorate its political inequities by referring to the example of the more advanced state of the physical sciences. A new social order based on an alliance between industry and the 'positive sciences' replaced outmoded theological and metaphysical structures. By extension, the scientific study of politics, from which historical laws could be deduced, enabled explanation and prediction of social affairs.[17]

Comte, like his mentor Saint-Simon, believed that a law of progress could be discovered through historical study and scientific analysis. According to Saint-Simonian doctrine, it was a physiological law of the human species that so-called organic epochs alternated with critical ones. Comte expanded upon this view of correspondence between the social phenomena and intellectual state of a given society. The end result – the final state of humanity – entailed government by *savants*, guided by the principles of positive philosophy. As the early twentieth-century historian John Bagnell Bury observed, Comte foresaw a period of *continuous* progress in order to achieve this state of affairs, but he expressly repudiated the idea of *indefinite* progress.

The eclipse of positivism

Comte's positivism found followers among a range of nineteenth-century thinkers, attracting philosophers like Mill by its antimetaphysical stance and social theorists like Friedrich Engels for its deterministic reading of history and its collectivist orientation. But positive philosophy began to lose its centrality by the end of the century. In its intense commitment to teasing out the operations of universal law in societies, it displayed no interest in individuals apart from their social relations. This left positivism on the sidelines when it came to the elaboration of social sciences focusing on the dynamics of the individual psyche. Positivism held little relevance for nascent behavioral sciences like psychology and psychiatry. John Theodore Merz also writes of the decline of synoptic, philosophical systems by century's end. Philosophy began to lose its role as intermediary between religion and science, involving itself more squarely with issues related to the unification of knowledge. As such, it moved to realms of greater abstraction and lost its human reference point. Finally, positivism's most important underpinning – the almost naïve unquestioning faith in science – was increasingly challenged. Because knowledge brokers moved away from social issues, leaving these to ethicists and moralists, science itself began to lose its compelling grip on Western consciousness.

Knowing: Relativizing

When the authors were students at an Ivy-League university in the United States during the 1970s, a professor of European history noted that the nineteenth century merited special attention because it was unusually long – beginning in 1789 and ending in 1914. From the European perspective, it is indeed tempting to locate the origin of all major features of life today in these 125 years. Before the storming of the Bastille, Europe was feudal and agrarian. There were no national systems of education or finance – there was only one *nation state*, France, and most of its subjects did not speak Parisian French. Europeans had no effective cures for disease and nothing more than vague prejudices about the nature of illness. People dressed themselves in handspun animal hair and plant products, adorned with heavy metals. Armies and navies were composed of mercenaries, who depended on muzzle-loader, sabre, horse, and sail. Candles provided illumination during long winter nights. Communication occurred at the speed of a fast trot or brisk wind.

Then came progress. By 1914, a spectrum of nation states had sophisticated systems of education and finance. Illness was known to be caused by microbial organisms, and chemical serums provided cures for deadly diseases. Complex polymer chains and organic compounds were conscripted to produce artificial plastics and pigments. Power radiated from dreadnoughts, submarines, machine guns, and mechanized armour. An intaglio of steel rails disfigured the industrialized world (and much of the agrarian hinterland); goods and people circulated by coal-burning steam engines, on land and on sea. Petroleum fueled motorcars. Cities blazed with electrical lighting. Heavier-than-air craft surveyed continent and ocean. There was communication by means of electromagnetic waves across the void.

The years since 1914 have seen innovations: new sources of energy (notably from nuclear fission), new vehicles of transport (powered by jet engines and liquid-fuelled rockets) and new forms of communication (through solid-state microelectronics and laser optics). Except for a dramatic decline in family size, however, Europe has undergone few major transformations in how people live. The modern world was defined by 1914, and one of its canons is what the American novelist and diplomatic historian Henry Adams called the acceleration of social and intellectual change. But as Adams sensed, when change becomes an end in itself, eclecticism reigns. People retreat into Plato's cave of shadowy impressions, there to elaborate the flickering images of tribe, marketplace, and theatre that Francis Bacon held to impede true learning. In our own time, critics have begun to wonder whether science has a future. They doubt the existence of ethical standards. Ours has become the era of relativity.

The century of relativity

How does one define an era? Traditional definitions refer to the ascension of a monarch, conqueror, or saviour. We know unambiguously about Thang China, Abbasid Egypt, Paleologoi Byzantium, Tokugawa Japan, Elizabethan England, Jacksonian America, Wilhelmian Germany, and Romanov Russia. And much of the world reckons time by reference to the mythical birth of one-third of the Christian deity (there seems to be little ambiguity in the notion of the 'twentieth century'). Since 1914 personalist designations, apart from those attributed to idiosyncratic tyrants or populists like Stalin, Franco, and de Gaulle, have faded from view. We have learned to live with managers rather than with leaders. Historical identification designates places – the Weimar Republic or Vichy France – and movements, such as fascist Italy and communist China. Relativity captures the essence of the times within living memory, just as the Baroque absorbs all the diverse potentates who dissolutely ruled over late seventeenth-century Europe.

To call the period since 1914 the 'era of relativity' is no more unreasonable than to place Linnaeus and Newton in the Baroque. Epochs cannot be all-consuming and monolithic; our world has many dimensions, and we experience it on a number of levels. Millions of people over the past three generations have felt themselves in supreme possession of absolute certainty – certainty that could be verified paradoxically by faith, by unreason. It nevertheless remains that much of our century's intellectual and political thought derives from the view that there can be no fixed point of human reference and no universally true set of values.

High culture in Europe and North America provides a starting point for twentieth-century relativity. Art, moving from impressionism through cubist, futurist, dadaist, constructivist, and abstract expressionist manifestos, passing into the veneration of kitchen garbage and coming to reside in postmodernism, is clearest in this regard. Music, too – beginning with Richard Strauss's elaboration of Wagner, then on to atonality and automatism, ending in John Cage's celebration of chaos – also suggests the reign of relativity. Poetry quickly divested itself of rhyme; prose lost nearly everything (punctuation, spelling, and grammar). The only rule was that there is no rule. The shock of relativity extended even to reason itself. At the beginning of the last decade of the century, fashionable sociologists (taking their cue from the writings of Paul K. Feyerabend and Michel Foucault) contended that reason had no standards – that what we take to be knowledge is nothing more than prejudice, context, and accident.

We see two results of the relativist era, one of them bright, the other dark. The bright outcome is the destruction of European certainty through the persistent efforts of subject and client civilizations – precisely the reverse of what Europeans had hoped to bring about. Europeans believed that their vision of rational exploitation of nature had made them masters of the world. Colonies and satrapies deserved subordinate status until they could assimilate the norms of European science. Colonialist planners miscalculated by confusing technology with science. Technology – unlike the abstruse and impractical procedures of one or another scientific discipline – often crosses cultural boundaries with blazing speed, and this happened everywhere at just the time that Europeans overran the world. Diverse civilizations from Japan to Mindanao to Algeria to the American Southwest embraced military innovations, especially, with extraordinary success. But even the most obtuse colonial administrator recognized, by 1920, that the world would never be Europeanized completely. Distinct peoples followed their karma, and their way of life had merit of its own.

The dark outcome of the relativist era is the apparent triumph of unbridled avarice and cupidity. During the decades after 1914 capitalism – a mode of life centred on exploiting human talent – was held in check by socialism. The Bolshevik Revolution in Russia provided inspiration – and the Bolshevik state provided resources – for an astonishing variety of social reforms in Asia, Africa, and the New World. As Europe moved from one conflagration to another, only the Soviet Union (notwithstanding its flagrant internal abuses) extended the hope of a better life. There can be no doubt that socialism ended material deprivation in large parts of Eurasia, although the strain of maintaining an enormous military sector – something of no conceivable domestic use – eventually brought the entire socialist economy to an inglorious end. The socialist economies acted to prevent tribal cannibalization of the sort recently seen in the Balkans and parts of Africa and Southeast Asia, but they also stifled innovation and criticism. With their passing from the scene, balance has disappeared from the marketplace. Anything goes in the race to accumulate capital. It is the heyday of economic relativism.

Relativism emerged rather suddenly at the end of the nineteenth century, when European thinkers had been seeking the grail of absolute truth. Once it became apparent that truth had countless facets, speculative thinkers changed to focus on critical epistemology: if they could not say what was forever true, they could establish how current notions of nature had developed. In their view, the process of discovering truths about the natural world marked the superiority of Europe. When other peoples obtained the appellation of civilization, however, the unitary, Comtean scale of human evolution – from savagery to barbarism to literate society – promptly evaporated. Each of the world's peoples excelled in special ways; no one was better than another.

It is evident that relativism did not govern all creative minds over the past one hundred years. Lagerlöf, Hauptmann, Tagore, Romain Rolland, Hamsun, Yeats, Bergson, Lewis, Galsworthy, Martin du Gard, Buck, Eliot, Mauriac, Churchill, Hemingway, Steinbeck, Sartre, Sholokhov, and Neruda (to mention only some Nobel laureates who are read today) all advocated or sought an immovable fulcrum for ethical and aesthetic judgments. Cultural relativism nevertheless finds a place in great works of imaginative literature in our century. Among Nobel laureates again, France, Shaw, Mann, Pirandello, Hesse, Gide, Faulkner, Russell, Beckett, Canetti, García Márquez, and even Camus defended a kaleidoscopic vision of humanity. It is hard to find a master of fiction alive today who has not been affected by the relativist wave.

Mach and Einstein

Late in the nineteenth century, when literature by Ibsen and Nietzsche and Baudelaire and Wilde had begun to question the search for absolute ethical and aesthetic standards, one writer rose to prominence as a champion of relativism. He was Ernst Mach, professor of physics at the German University of Prague. Mach's immediate reaction was to question anything of an absolutist cast. He worked in close contact with Czech colleagues, some of whom aspired to greater political independence under the imperial, Austrian regime. By the 1880s Mach had become a vocal supporter of radical reforms to the secondary-school system, where Greek and Latin were still the indispensable preparation for future scientists. Mach's persistent criticism of orthodox classicism was gratified around 1900 by large-scale changes in school curricula.

Mach was one of the earliest critical epistemologists. Given the ambiguity and uncertainty of even well-established scientific principles, he believed that it was best to view the body of knowledge with a sceptical eye. One had continually to purge knowledge of extraneous and superfluous notions lacking a foundation in human experience. This materialist bias against arbitrary theoretical constructions elicited stern criticism from all sides: both the transcendentalist physicist Max Planck and the materialist social philosopher Vladimir Ilyich Lenin strongly rebuked Mach in the years before the First World War. But Mach's iconoclasm, apparent to scientists and laymen alike, appealed especially to the young Albert Einstein, and in this way Mach's critique of absolute motion entered directly into the philosophical foundations of twentieth-century relativity. Comte's positivism was given new life and a new meaning by an Austrian physicist.

In the years before 1905, when he prowled outside the assembly hall of theoretical physicists, Einstein was attracted above all to Mach's scepticism. Mach's role as an outsider would have been congenial to Einstein (who lacked advanced instruction in Latin and Greek). And Einstein's fundamental questioning of received truths, his persistent attempt to root out contradictions between various branches of the exact sciences, his formulation of new and fruitful hypotheses, all these traits may be found in Mach's writings.

Then Einstein published what became the emblem of our century. With his Principle of Relativity, Einstein brought about a unification of classical mechanics (rigid-body motion following the laws and formulations set down by Newton and eventually codified by Laplace) and electrodynamics (the revision of James Clerk Maxwell's theory of electromagnetism carried out by Hendrik Antoon Lorentz). The principle, simply announced, stated that bodies could only be held to experience motion relative to each other. There was no universally preferred framework against which velocities could be measured. What did remain invariant and absolute was the speed of light in a vacuum. By reconsidering nonaccelerated motion from the point of view of exchanges of light signals, Einstein derived the recently discovered kinematical anomalies of charged particles – that they appeared to change mass and shape with velocity. As a consequence of his approach, he obtained the new result that time, as measured by mechanical clocks, also varied with the velocity of the clock's frame of reference. Einstein did not address *why* the speed of light remained invariant (that question, leading into quantum mechanics, depended on the nature of electromagnetic radiation, and it remained under scrutiny during the first third of the century). He did urge, however, in a Machian vein that the traditional notion of a medium for propagating light waves – the electromagnetic ether – be abandoned as an unreasonable fiction.

Relativity is not the same as *relativism*. In German the words are distinct, although both are borrowed from a French and ultimately a Latin root. Einstein was not the first of his time to use the word 'relativity', and in his publication he referred to the classical relativity principle of Galileo (he did so, he recalled later, because he thought that Galileo was the first person to formulate the kinematics of relative motion). Henri Poincaré also referred to the relativity principle. Einstein was slow in agreeing to dignify his notion with the designation of a 'theory'. For at least five years he referred to this part of his research as work dealing with the 'principle' of relativity. Others – notably one of his earliest supporters Max Planck – called it the 'relative theory'. Only when Einstein began to publish thoughts about integrating gravitation into his ongoing synthesis did he follow a number of his colleagues in writing about the 'theory of relativity'.

Ironically, Einstein's epoch-making work went toward seeking inviolate and rock-solid principles upon which to base the laws of nature. There is little evidence to suggest that he believed that science was merely a question of one's point of view. He was no supporter of Poincaré's 'conventionalism' (the notion that fundamental laws are social conventions). Neither was he a supporter of Mach's more radical relativism, which ended by excising from science all nonverifiable entities and propositions. Apart from matters of *couture* (he wore his shabby garments as the badge of honour of a *distrait*), he was a man of conservative appetites. Women, for example, were in his view objects of stimulation and gratification – much as coffee, tobacco, and sugar were. At least until he separated from his first wife Mileva Marić, he followed the traditional path of German professors in abstaining from political matters – even though his material circumstances (a Jew and Swiss citizen married to a Serb with Hungarian nationality) readily lent themselves to political expression. Einstein and Marić also made a traditional decision to have a child before they married – illegitimacy was an accepted way of life in Central Europe around 1900.

Einstein's relativity triumphed in physics as no theory had ever triumphed. Intellectuals waited generations before agreeing in print with Copernicus's heliocentrism; opposition to Newton's physics lasted well into the eighteenth century – French partisans of Descartes and German adherents of Leibniz seeking to withhold the Englishman's laurels; Darwin's evolution gave rise to fundamental, specialist disagreement well into the twentieth century. But shortly after 1909, when Einstein first aired his views on a connection between gravitation and electromagnetism, relativity was embraced by the most talented of the world's physicists.

There were disagreements about one or another of Einstein's propositions. Some, like Emil Wiechert (1861–1928), sought to recover absolute space; others, like Walter Ritz (1878–1909), offered an emission-theory of light; Max Abraham introduced his own variants on gravitational theory – just as he had proposed his own electron theory in the years before 1905. Felix Klein proposed that Einstein's work be known as 'invariance theory', after Einstein's search for laws that did not change with frame of reference (and also the constancy of the speed of light). None denied the legitimacy of Einstein's search to unify physical laws. With his application of Tullio Levi-Cività (1873–1941) and Gregorio Ricci-Curbastro's (1853–1925) tensor calculus of differential geometry in 1912, Einstein transformed the mathematical level on which physical theory had to be discussed. Einstein worked feverishly toward a satisfying gravitational theory, and in 1915 he narrowly preceded mathematician David Hilbert (1862–1943) in publishing the final form of the covariant field equations of general relativity.

The world's popular press made this unlikely genius into an icon for all the new forces that were transforming the world. According to a woman protagonist in Robert Musil's great novel about Vienna on the eve of the First World War, *The Man without Qualities* (1930–42), modern civilization, created by science, was 'a frustrating state of affairs, full of soap, wireless waves, the arrogant symbolic language of mathematical and chemical formulae, economics, experimental research, and mankind's inability to live in simple but sublime harmony'.[1] To the delight of ordinary people everywhere, the irreverent Einstein appeared as the very antithesis of this arrogance and pretension. Treated as a Swiss citizen in Germany during the First World War, Einstein spoke out against the war with impunity; more than that, he championed socialism and social justice for the ordinary person; and, spurred on by his second wife Elsa Einstein, he became a leading spokesman for the Zionist cause.

The result of this attention was predictable. The merits of Einstein's physics were submerged by pseudo-philosophical debates and vituperative, political harangues. Benito Mussolini enlisted relativity for fascism, just as some of the early Bolsheviks did for communism. Einstein's theories and his persona upset nationalist Germans, who beginning around 1920 did all in their power to ridicule him. Well before the horrific events of the National Socialist revolution, Germany hosted an 'anti-Einstein' movement that caused Einstein to worry about attempts against his life. Einstein's relativity was proclaimed by the popular press as a revolutionary theory at just the time that cosmopolitan socialist revolutions rocked Europe. As a result, it generally became associated, in Europe and North America, with relativist subversion of established moral values. In the Stalinist Soviet Union, relativity was anathema because of its connection with Mach's positivism and Einstein's Zionism. In England and the United States, physicist traditionalists strove mightily, well into the 1950s, to disprove Einstein's relativistic predictions. Despite much publicity, none succeeded in his self-appointed task.

The reception of Einstein's thought

Few among the world's most talented physicists and astronomers denied relativity's place at the centre of natural law, but for more than forty years, from 1920 until around 1960, the theory underwent an eclipse. In the first place, the tensor calculus was a complex formalism that discouraged all except the most motivated researchers, and even they had to spend months to solve equations for the most simply stated gravitational problems. And the tensor calculus was something of an anomaly in physics, for it did not lend itself to other topics. Furthermore, observational tests for general relativity remained limited, in this period, to the three identified by Einstein himself. As it became disconnected with the world of experiment, general relativity was sometimes discussed disparagingly in theological terms. The results of special relativity were easily taken over into atomic physics without appealing to Einstein's theoretical point of view – after all, a number of the equations of special relativity had been known widely before 1905. Indeed, the Soviet school of general relativity, under the inspiration of Vladimir A. Fok (1898–1974), referred to it as 'geometro-dynamics'. Certainly, electrical engineers did not hesitate to continue to invoke the electromagnetic ether – as indeed they sometimes still do – and cosmologists continued searching for a universal, preferred frame of reference. Until the 1960s in Europe and North America, students learned that relativity was a bizarre and useless *cul-de-sac* of physics.

Civilizations freed from the ideological dead weights dragging down Western Europe had little difficulty assimilating Einstein's theories of relativity. Japanese and Arab intellectuals greeted Einstein as a sage – distinguishing his scientific achievements from his persona, from his nationality, and from his political views. South Asians also responded favourably to the complexities of tensor analysis. This transfer across cultures follows directly from the conditions that discouraged the wide diffusion of general relativity in Europe. It required little investment beyond pencil and paper, and its cyphers spoke to all people of learning directly, without depending on the cultural nuance of word or phrase. These characteristics it shared with Ptolemaïc astronomy, the other great cosmological system that had previously spanned civilization and epoch.

The renascence of general relativity over the past generation derives from a number of related circumstances. The routine cataloguing of nuclear physics – the source of a number of Nobel prizes – led to a focus on so-called elementary particles, whose behaviour related to mysterious, quasi-theological visions. The apotheosis of quantum mechanics in quantum electrodynamics created hundreds of young theoreticians who were accustomed to dealing with recondite mathematics. The 'new maths' of the French Bourbaki school – a persistent search for order and organization – inspired a substantial simplification of the tensor calculus in newly rediscovered parts of nineteenth-century differential geometry. American and Soviet military chieftains required 'scientific' justifications for rocket technology (sending a man to the moon was a costly publicity stunt of no scientific merit), and exploring the cosmological frontiers of the universe offered ready-made research programmes. New gravitational phenomena were discovered or predicted – from the Mössbauer effect to gravitational lenses to black holes to the elusive gravitational waves claimed to have been observed by Joseph Weber (b. 1917). A number of Nobel laureates in physics over the past thirty years have devoted much time and effort to problems stemming from Einstein's relativity.

The least tangible contribution to relativity in the 1960s may be the most important one. It was a deep questioning of Western values and imperatives. The questioning derived from a persistent dissonance between the rhetoric and the action of the world's leaders. Why did nations proclaiming liberty and justice engage in brutal wars abroad and condone repressive laws at home? The defence of traditional Western democratic notions – freedom of conscience and enterprise – appeared to be a sham, just as Stalinist repression made a sham of socialism. Great interest emerged in alternatives to traditional faiths and families. Holistic thinking enjoyed a renascence, especially Oriental medicine and religious philosophy, as well as dialectical reasoning in general (from Hegel and Marx to Herbert Marcuse and Mao Zhedong). Mechanical notions of cause and effect were increasingly held in disrepute. General relativity rode the crest of this interest.

The crescendo of interest in cosmological questions led to the construction of enormously expensive instruments for observing the sky – in Chile, on Hawaii, and in the American Southwest – resulting in the crowning achievement of the billion-dollar Hubble space telescope, launched in 1990. It is unclear whether the euphoria for seeing into the eye of God (the transcendental metaphor used by cosmologists to justify construction of their expensive mechanisms) will survive the initial failure of the orbiting telescope and a series of major and minor disasters in space travel generally.

Eclecticism and hope

The end of the twentieth century is sometimes called the domain of postmodernism, but it would be more fitting to call it the reign of relativity. Architecture is an eclectic appeal to diverse schools and traditions; painting and music reflect primitive patterns and traditional rhythms as much as they do abstraction and dissonance; personal experience is limited only by personal taste – the most relative of sensations. Radical relativism even evolved a short-lived philosophy – the so-called strong programme in sociology of science, according to which all ideas derive from social relations. Extraordinary mixes of culture seem to mirror the solipsism of imperial Rome rather than the vitality of Thang China, Abbasid Islam or Renaissance Europe.

A century ago there was general optimism about humanity's ability to understand nature and improve the human condition. Scientific medicine promised cures for plagues and scourges; radiation had just opened new vistas in diagnosis and therapy. Electricity and running water were transforming urban life. The Paris Exhibition of 1900 confidently displayed grand engines and technological marvels. European and American academics, from Göttingen mathematician David Hilbert to Princeton historian Woodrow Wilson, welcomed a new century, confident that knowledge would continue to grow and life would improve. The confidence and hope is expressed in the labour of Paul Tannery, who used the Exhibition as the occasion for infusing life into history of science as an international, learned discipline.

Today, nearly a century later, people of learning question whether science is relevant or even necessary. The world has too many scientists, it is claimed; we must not train so many of them. Scientists frequently disagree among themselves, it is said; they do not have firm bases for their beliefs. Science has produced great engines of destruction, it is observed; let us not fund this kind of highly technical research. And the most damaging criticism: science is simply a cultural product like other cultural products; it is no more true in any of its judgments than, say, Mozart's music is true.

One observer of science in the United States sees signs of a cascading deflation of postmodern relativism.[2] It could hardly be otherwise. After a generation of hoopla, postmodernists have produced no substantial body of analysis and no major work of scholarship. (It is hard to see why a postmodern relativist would be inspired to write in any except an anarchist or ironic vein.) Postmodernists even misunderstand the *oeuvres* of their heroes. The specialists on whose work they draw, for example Sinologist Joseph Needham and anthropologist Clifford Geertz, affirmed the continual deepening of knowledge.

Our century concludes with the majority of the world's people believing in material progress. We see the multiplication of institutions of higher learning around the world nurturing the search for new understanding of such natural phenomena as disease, prime numbers, and war. We read scientific publications in Spanish and Japanese, and we use the findings in our own search without asking about the authors' creed, marital status, or income. The disciplines of

438 · *Servants of Nature*

science do seem to screen out background noise – a screening at the base of the original Royal Society of London. It is possible after all for people to understand each other across language and culture.

We nevertheless threaten to lose a vital attribute of the scientific enterprise. It is the notion of the secular value of pure science. The notion emerged in the eighteenth century. (Even today, when people ask about the good of research, the response echoes Benjamin Franklin's reply: 'What is the good of a new-born baby?') It became a touchstone of universities in the nineteenth century. But it is no longer venerated.

Pure science has fallen victim, in part, to success. It has become highly dependent on costly technology, so much so that the distinction between science and technology is blurred. Because pure science has resulted in spectacular inventions, we speak today about Research-and-Development in one breath. We imagine that research is indissociable from development, whether for new medicines or new machines.

Relativism has driven forward-thinking people away from science and toward technology. Science depends on words, and words have been shown to equivocate. Technology, however, has tangible standards for success. Technology celebrates objects, whose effectiveness is apparent for all to see: a rocket will go up, an X-ray machine can reveal bone fractures, a computer does perform large sums quickly. Practical people whose dreams depend on material progress – that is to say, most of humanity today – have little time for scholastic controversy about whether the laws of electromagnetism are inextricably bound up with patriarchal, Anglo-Saxon society. They are quite insensitive to an identification of technology with the social regime that animates it.

The practical uses of science predominated in many of the cultures that received European scientific institutions during the nineteenth century. The recipients, seeing science as a vehicle for material progress and understanding the spirit (if not the letter) of Comtean social progress, rearranged the bounds of European knowledge. Engineering, in particular, was placed alongside science in America and Asia – quite unlike the traditional distinction between the two enterprises in European institutions. The closing of the gap between science and technology is evident today in the title of institutions of higher learning like Texas Tech University and Jawaharlal Nehru Agricultural University. Even Europe is changing. To the Technische Universität Clausthal in Germany one may add the Kharkov State Technical University of Radioelectronics in Ukraine and the acronymic British university UMIST – University of Manchester Institute of Science and Technology.

Will the evanescence of relativism lead to a renascence in classical form and content? Earlier in our century, the virtuoso historian Arnold Toynbee contended that renascences are a persistent phenomenon. In addition to cycles of repression and liberty, there were cycles of style and taste. In Toynbee's opinion, the proponents of calls for rebirth, whether religious or secular, dabble in the black art of necromancy. They seek to give new meaning to dead notions.[3] If ideas like fashion are cometary and there will be a classical revival in the new century, precisely whose classicism shall prevail? Which golden age – which monument in stone – will provide the reference point?

Above the intellectuals hovers the spirit of Einstein. Some forty years after his death, he has entered popular consciousness as the omnipotent critic of common prejudice. Although he is the eponymous icon for twentieth-century relativism and for the destructive force of science, he laboured mightily in favour of secure foundations for both science and world peace. The monuments dedicated to his memory are located in Israel (the state governed by his chosen people) and the United States (the land where he ended his life). The town of his last residence, Princeton, is finally contemplating a statue in his memory. Einstein's works are now slowly being edited and published, after much delay and indecision. His private life is held up to blistering scrutiny. The public pleas of his mature years for tolerance and understanding, for justice and charity, mask the primitive passion and unthinking impulses of his youth. But the architect of relativity is the last person we should blame for failing to remember on occasion that actions entail consequences.

At the beginning of the twentieth century there was widespread faith in progress and human perfectability based on Western science. Einstein, the foremost 'relativist', wrote to his friend the Queen Mother of Belgium in 1952:

> It is now almost twenty years since I was last able to talk and play music with you. How many harsh and difficult times we have lived through in the years between! Saddest of all is the disappointment one feels over the conduct of mankind in general. Younger people may feel little surprise; but, then, they have never known times of calm and reason. Things appeared rather differently to us when we were young. We believed the brutality of former times had been eliminated forever and had yielded to an age of reason and stability.[4]

By the middle of the century, following two world wars and the development of enormous nuclear arsenals, the hope had eroded.

War and suffering continued over the second half of the century, but material standards of living generally did improve. Today a large portion of the world's population enjoy direct communication with each other; they have access to life-saving drugs and surgery. If they are inclined, they may read these lines. The sciences of electronic information processing and molecular biology underlie new opportunities. Whatever the future holds, it will derive from truths about nature that humankind has carefully, although not infallibly, won.

NOTES

We have endeavoured to keep notes to a minimum, generally limiting them to direct quotations. Bibliographical notes in the section on Further Reading indicate additional sources. Readers seeking more information about a particular point are urged to consult the *Critical Bibliography* published annually by the History of Science Society as the fifth number of its periodical, *Isis*; cumulative bibliographies have appeared for the years 1913–65, 1966–75, and 1976–85. Society members with access to the World Wide Web may obtain the *Critical Bibliography* through the Society's Home Page. In an age when the world's great libraries may be browsed electronically from distant locations, little purpose is served by long lists of titles.

INTRODUCTION

1 E. D. Hirsch, Jr., Joseph F. Kett, and James Trefil, *The Dictionary of Cultural Literacy: What Every American Needs to Know* (Boston: Houghton Mifflin, 1988). Both incidents find prominence in the book.

2 The eighteenth-century proceedings of the Academy of Sciences in Paris, the Royal Society's rival, appeared in an irregularly published 'history'. Natural history, a translation from the French that might better have been rendered as narratives about nature, has become an independent and richly endowed discipline.

3 Over the recent past the word paradigm has inspired a cartoon in the *New Yorker* and even a television advertisement (for the company OfficeMax).

4 The late historian of science Stanley Goldberg related that when he reviewed Price's book *Little Science, Big Science* (New York: Columbia University Press, 1963) for the American periodical *Science*, he first proposed a title, 'Laws, Lord, Laws'.

5 Price, *Little Science, Big Science*, p. 66, in a discussion of invisible colleges.

6 For example, molecular biologist Gunther Stent in *The Coming of the Golden Age: A View of the End of Progress* (Garden City, NY: American Museum of Natural History, 1969), literary critic George Steiner in *In Bluebeard's Castle:*

Some Notes Towards the Redefinition of Culture (New Haven: Yale University Press, 1970), physicist John Ziman in *Public Knowledge: The Social Dimension of Knowledge* (Cambridge: Cambridge University Press, 1968) and novelist Thomas Pynchon in *Gravity's Rainbow* (New York: Viking, 1971).

7 Michel Foucault, cited in David A. Hollinger, 'Science as a Weapon in *Kulturkämpfe* in the United States during and after World War II,' *Isis, 86* (1995), pp. 440–54, on p. 453. This article, the text of a Distinguished Lecture given to the History of Science Society, finds the term 'postmodernism' entirely unproblematic.

8 Leo Marx, 'The Idea of "Technology" and Postmodern Pessimism', in *Technology, Pessimism, and Postmodernism,* ed. Yaron Ezrahi, Everett Mendelsohn, and Howard P. Segal (1994; Amherst: University of Massachusetts Press, 1995), pp. 11–28, on p. 24. Marx provides an illuminating account of postmodernism and science, but his contention of widespread disenchantment with technology is less persuasive.

9 William H. Honan, 'Footnotes Offering Fewer Insights', *New York Times,* 14 August 1996, p. B9.

10 Bruno Latour, 'A Relativistic Account of Einstein's Relativity', *Social Studies of Science, 18* (1988), pp. 3–44; *The Pasteurization of France,* trans. Alan Sheridan and John Law (Cambridge, MA: Harvard University Press, 1988). Peter Monaghan, 'For Scholar Steven Shapin, the "Scientific Revolution" was a Non-Event', *Chronicle of Higher Education,* 6 December 1996, p. A 17, citing Shapin. Historian Margaret Jacob identifies Latour as a postmodernist in 'Reflections on the Ideological Meanings of Western Science from Boyle and Newton to the Postmodernists', *History of Science, 33* (1995), pp. 333–57, a work that provides evidence to challenge the notions of both Latour and Shapin.

11 Chris McClellan, 'The Economic Consequences of Bruno Latour', *Social Epistemology, 10* (1996), pp. 193–208, on p. 205.

12 Mordechai Feingold, 'When Facts Matter', *Isis, 87* (1996), pp. 131–9, quotation on p. 138, reviewing Steven Shapin, *A Social History of Truth: Civility and Science in Seventeenth-Century England* (1994); letters to the editor by Peter Dear, *Isis, 87* (1996), pp. 504–5; Mordechai Feingold, *ibid.,* pp. 505–6; Steven Shapin, *ibid.,* pp. 681–4; Mordechai Feingold, *ibid.,* pp. 684–7.

13 Robert Palter, '*Black Athena,* Afro-Centrism, and the History of Science', *History of Science, 31* (1993), pp. 227–87. Martin Bernal's book is *Black Athena: The Afroasiatic Roots of Classical Civilization, 1: The Fabrication of Ancient Greece, 1785–1985; 2: The Archaeological and Documentary Evidence* (1987–

91). Quotation from Bernal, 'Response to Robert Palter', *History of Science*, 32 (1994), pp. 445–64, on p. 458; Palter comments, *ibid.*, pp. 464–8.

14 M. Norton Wise, 'The Enemy Without and the Enemy Within', *Isis*, 87 (1996), pp. 323–7, reviewing Paul R. Gross and Norman Levitt, *Higher Superstition: The Academic Left and Its Quarrels with Science* (1994).

15 Clifford Geertz, *After the Fact: Two Countries, Four Decades, One Anthropologist* (Cambridge, MA: Harvard University Press, 1995), p. 98.

16 The defenders of traditional scientific norms and values tend to tar all social historians of science with the postmodernist brush. Indicative are physicist Steven Weinberg's comments in 'Sokal's Hoax', *New York Review of Books*, 8 August 1996, pp. 11–15, where the meticulous and inspired reflections of Paul Forman are associated with the unsubstantiated claims of postmodern relativists.

17 Frank Lentricchia, 'Last Will and Testament of an Ex-Literary Critic', *Lingua Franca*, September/October 1996, pp. 59–67.

CHAPTER 1

1 Einstein, 'Der Angst-Traum [The Nightmare]', *Berliner Tageblatt*, 25 December 1917, Morgen-Ausgabe, Beiblatt, in *The Collected Papers of Albert Einstein, 6: The Berlin Years, Writings, 1914–1917*, ed. A. J. Kox, Martin J. Klein, and Robert Schulman (Princeton: Princeton University Press, 1996), p. 581.

2 John Theodore Merz, *Reminiscences* (Edinburgh: William Blackwood and Sons, 1922), p. 85.

CHAPTER 2

1 John Heilbron, *Electricity in the Seventeenth and Eighteenth Centuries: A Study of Early Modern Physics* (Berkeley: University of California Press, 1978), p. 139.

2 Henry Adams, *The Education of Henry Adams* (Boston: Houghton Mifflin, 1918), p. 78.

3 John Theodore Merz, *Reminiscences* (Edinburgh: William Blackwood and Sons, 1922), pp. 66–7.

CHAPTER 3

1 A. Rupert Hall, *The Revolution in Science, 1500–1750* (London: Longman, 1983), pp. 226–7.

2 Martha Ornstein, *The Role of Scientific Societies in the Seventeenth Century* (Chicago: University of Chicago Press, 1928), pp. 67–8; Roger Hahn, *Anatomy of a Scientific Institution: The Paris Academy of Sciences, 1666–1803* (Berkeley: University of California Press, 1971), pp. x–xi; James E. McClellan, *Science Reorganized: Scientific Societies in the Eighteenth Century* (New York: Columbia Unversity Press, 1985), p. 10.

3 Michael Hunter, *Establishing the*

New Science: The Experience of the Early Royal Society (Woodbridge, Suffolk: The Boydell Press, 1989), pp. 1–2.

4 Hall, *Revolution*, pp. 211–12; 14, Michael Hunter, 'First Steps in Institutionalization: The Role of the Royal Society of London', in Tore Frangsmyr, ed., *Solomon's House Revisited: The Organization and Institutionalization of Science* (Canton, MA: Science History, 1990), p. 14.

5 Hunter, *Establishing*, pp. 2–3.

6 Marie Boas Hall, *Promoting Experimental Learning: Experiment and the Royal Society, 1660–1727* (Cambridge: Cambridge University Press, 1991), pp. 10–11; Hall, *Revolution*, p. 224, and see review of this work in *British Journal for the History of Science*, 26 (1993), pp. 90–91.

7 Hunter in Frangsmyr, *Solomon's House*, p. 24; McClellan, *Science Reorganized*, p. 50.

8 On the Accademia del Cimento, see W. E. Middleton, *The Experimenters: A Study of the Accademia del Cimento* (Baltimore: Johns Hopkins University Press, 1971); Ornstein, *Scientific Societies*, pp. 76; 177–191; Hall, *Revolution*, pp. 212–13, 228–9.

9 Hall, *Revolution*, pp. 219–20, 223, 225–26; McClellan, *Science Reorganized*, p. 51; Hahn, *Anatomy*, 24; Harcourt Brown, *Scientific Organizations in Seventeenth Century France, 1620–1680* (New York: Russell & Russell, 1967), pp. 27, 147–8, 159–60.

10 Brown, *Scientific Organizations*, pp. 33, 162–66; 179, 181; Hall, *Revolution*, p. 214.

11 See the list in Maurice Crosland, *Science under Control: The French Academy of Sciences, 1795–1914* (Cambridge: Cambridge University Press, 1992), p. 71.

CHAPTER 5

1 The preceding section draws upon works by Krzysztof Pomian, *Collectioneurs, amateurs et curieux: Paris, Venice, XVIe–XVIIIe siècle* (Paris: Gallimard, 1987), pp. 61–65, 248–9, 252; Eilean Hooper-Greenhill, *Museums and the Shaping of Knowledge* (London: Routledge, 1992), pp. 115, 161–2; Simon Tait, *Palaces of Discovery: The Changing World of Britain's Museums* (London: Quiller Press, 1989), p. 2; and the essay by Michael Hunter in Oliver Impey and Arthur MacGregor, eds., *The Origins of Museums: The Cabinet of Curiosities in Sixteenth- and Seventeenth-Century Europe* (Oxford: Clarendon Press, 1985), pp. 163–5.

2 A. C. L. Günther, 'Objects and Use of Museums,' presidential address before the British Association for the Advancement of Science, Section D, Biology, Swansea, 1880, p. 593.

3 S. F. Markham and H. Hargreaves, *The Museums of India* (London: Museums Association, 1936), pp. 3, 21, 49, 57–8.

4 Mitchell Library (Sydney, Australia), McCoy

correspondence: William Denison to Frederick McCoy, 29 June [1859?]. S. F. Markham and H. C. Richards, *A Report on the Museums and Art Galleries of Australia* (London: Museums Association, 1933), p. 29. Gerard Krefft, 'The Improvements Effected in Modern Museums in Europe and Australia,' *Transactions of the Royal Society of New South Wales* (1868), p. 20. Ronald Strahan, *Rare and Curious Specimens: An Illustrated History of the Australian Museum, 1827–1979* (Sydney: Australian Museum, 1979), p. 31. Herbert M. Hale, 'The First Hundred Years of the Museum, 1856–1956,' *Records of the South Australian Museum* 12 (1956): p. 73.

5 Barrie Dyster, 'Argentine and Australian Development Compared,' *Past and Present, 84* (Aug. 1979): pp. 91–110, esp. p. 91. Warwick Armstrong and John Bradbury, 'Industrialisation and Class Structure in Australia, Canada, and Argentina: 1870 to 1980,' in *Essays in the Political Economy of Australia*, ed. E. L. Wheelwright and K. Buckley, 5 (Sydney: Australia and New Zealand Book Co., 1983), pp. 43–74.

6 William Swainson, *Taxidermy, Bibliography, and Biography* (London: Longman, 1840), pp. 77–8.

CHAPTER 6

1 Oliver Impey and Arthur MacGregor, eds., *The Origins of Museums: The Cabinet of Curiosities in Sixteenth- and Seventeenth-Century Europe* (Oxford: Clarendon Press, 1985), p. 1.

2 H. N. Wethered, *A Short History of Gardens* (London: Methuen, 1933), p. 168; Richardson Wright, *Story of Gardening* (New York: Dover, 1963), p. 195; William Howard Adams, *The French Garden, 1500–1800* (New York: George Braziller, 1979), pp. 75, 86.

3 Colloquium proceedings on *John Claudius Loudon and the Early Nineteenth Century in Great Britain* (Washington, DC: Dumbarton Oaks, 1980), p. 3; Wethered, p. 67; William Thomas Stearn, 'Botanical Gardens and Botanical Literature in the Eighteenth Century' in *Catalogue of Botanical Books in the Collection of Rachel McMasters Miller Hunt*, comp. Allan Stevenson, vol. 2 (Pittsburgh: The Hunt Botanical Library, 1961), p. liii; Wright, pp. 197–8; John Prest, *The Garden of Eden: The Botanic Garden and the Re-Creation of Paradise* (New Haven: Yale University Press, 1981), p. 49.

4 Lynn Barber, *The Heyday of Natural History* (London: Jonathan Cape, 1980), p. 48; Prest, pp. 54, 91–2, 102–4; Ann Leighton, *American Gardens in the Eighteenth Century* (Boston: Houghton Mifflin, 1976), pp. 18, 98–99; Arthur W. Hill, 'The History and Functions of Botanic Gardens', *Annals of the Missouri Botanical Garden, 2* (1915), p. 196.

5 Andrew Cunningham, 'The

Culture of Gardens', in N. Jardine *et al.*, eds., *Cultures of Natural History* (Cambridge: Cambridge University Press, 1996), pp. 38–56, on p. 53.

6 Stearn, pp. lxxii, lxxx–lxxxi; C. Stuart Gager, 'Botanic Gardens' in *The Standard Cyclopedia of Horticulture*, ed. L. H. Bailey (New York: Macmillan, 1914), pp. 528, 530; R. T. M. Pescott, *The Royal Botanic Gardens, Melbourne: A History from 1845 to 1970* (Melbourne: Oxford University Press, 1982), p. xi; David Stuart, *The Garden Triumphant: A Victorian Legacy* (New York: Harper & Row, 1988), p. 35; Wethered, p. 290.

7 Hill, 207, 209; Stearn, pp. c, cviii–cix; F. Nigel Hepper, *Kew: Gardens for Science & Pleasure* (London: Her Majesty's Stationery Office, 1982), pp. 10, 17; Edward Hyams, *A History of Gardens and Gardening* (London: Dent, 1971), p. 12; Gager, p. 527.

8 Gustave Loisel, *Histoire des ménageries de l'antiquité à nos jours*, 3 vols. (Paris: Octave Doin, 1912), pp. 1: 6–7; 2: 285–8, 290–2; Bill Jordan and Stefan Ormrod, *The Last Great Wild Beast Show: A Discussion on the Failure of British Animal Collections* (London: Constable, 1978), pp. 23–5, 27.

9 Loisel, 2: pp. 293–4, 296–99, 310, 317–21; 3: pp. 144–5; Jordan and Ormrod, pp. 30–31.

10 Bob Mullan and Garry Marvin, *Zoo Culture* (London: Weidenfeld & Nicolson, 1987), p. 51.

11. Solly Zuckerman, ed., *Great Zoos of the World: Their Origins and Significance* (Boulder, CO: Westview Press, 1980), p. 12; Loisel, 3: pp. 108–9; Jordan and Ormrod, pp. 33–4; Jon R. Luoma, *A Crowded Ark* (Boston: Houghton Mifflin, 1987), p. 15; Pescott, pp. x–xi.

12 *National Geographic, 184*, July 1993, p. 14.

CHAPTER 7

1 The number of degrees in a circle (360) also derives from Babylonian practice. Otto Neugebauer hypothesized that there were twelve standard length-units that could be travelled in one revolution of the sky; 360 celestial degrees stem directly from the fact that each length-unit divided into thirty parts.

2 Accuracy may be understood as precision related to a larger view of things. But one may be accurate without apparent precision ('The world will not end tomorrow') or precise without being accurate (as in supporting an assertion by a large number of irrelevant references).

3 In Japan, as in Europe, the length of the hours varied with the seasons. Hours in Japan, however, were counted backwards from 9 which served to denote both midnight and noon; half hours following odd hours were marked by one stroke, those after even hours by two strokes. The stroke sequence of early Japanese mechanical clocks, then, was 9–1–8–2–7–1–6–2–5–1–4–2–9.

CHAPTER 8

1 Joseph Needham has demonstrated that China knew movable type printing well before the European High Middle Ages. Nevertheless, we have no mention of it in Europe before Gutenberg. Printing may be an instance of the multiple-discovery phenomenon that attracted Robert Merton, Derek Price, and Thomas Kuhn.

2 James Burke, *The Day the Universe Changed* (London: BBC, 1985), pp. 116, 121.

3 Elizabeth Eisenstein, *Printing Press as an Agent of Change: Communications and Cultural Transformation in Early Modern Europe* (Cambridge: Cambridge University Press, 1980, pp. 66, 11.

4 Eisenstein, pp. 114, 379, 579.

5 Bernard Houghton, *Scientific Periodicals: Their Historical Development, Characteristics and Control* (Hamden, CT: Linnet Books, 1975), pp. 12–14.

6 Leona Rostenberg, *Library of Robert Hooke: The Scientific Book Trade of Restoration England* (Santa Monica, CA: Modoc Press, 1989, p. 110.

7 David A. Kronick, *A History of Scientific and Technical Periodicals: The Origins and Development of the Scientific and Technical Press, 1665–1790* (Metuchen, NJ: Scarecrow Press, 1976), pp. 121, 137; Houghton, pp. 18–19.

CHAPTER 9

1 Alfred W. Crosby, *The Columbian Exchange: Biological and Cultural Consequences of 1492*

(Westport, CT: Greenwood Press, 1972), and Crosby's magisterial *Biological Imperialism: The Biological Expansion of Europe, 900–1900* (Cambridge: Cambridge University Press, 1986).

2 A pioneering view of European history from the point of view of cultural mixing is Fernand Braudel's *The Mediterranean and the Mediterranean World in the Age of Philip II*, trans. Siân Reynolds (New York: HarperCollins, 1975). The mixing metaphor is a leitmotif in Joseph Needham's epic presentation of Chinese science, for example, in his *Moulds of Understanding: A Pattern of Natural Philosophy*, ed. Gary Werskey (London: George Allen and Unwin, 1976).

3 Joseph Needham, *Science and Civilization in China*, part IV, *3: Civil Engineering and Nautics* (Cambridge: Cambridge University Press, 1971), pp. 524–35.

4 *Ibid.*, p. 519.

5 This is the point emphasized by Lynn Thorndike in 1941. Thorndike, *A History of Magic and Experimental Science*, 5 and 6 (New York: Columbia University Press, 1941), pp. 9–10.

6 Bacon, in Carolyn Merchant, *The Death of Nature: Women, Ecology, and the Scientific Revolution* (San Francisco: Harper and Row, 1980), p. 170.

7 J. Heninger notes that the curators promised Boerhaave whichever of the five medical chairs (practical medicine, theoretical medicine, anatomy, chemistry, or botany) should first fall vacant. Heninger,

'Some Botanical Activities of Herman Boerhaave, Professor of Botany and Director of the Botanic Garden at Leiden', *Janus*, 47 (1971), pp. 1–78.

CHAPTER 10

1 Alfonso Leone, 'Maritime Insurance as a Source for the History of International Credit in the Middle Ages', *Journal of European Economic History*, 12 (1983), pp. 363–9; Charles Farley Trenerry, *The Origin and Early History of Insurance, including the Contract of Bottomry* (London: P. S. King & Son, 1926).
2 J. C. Riley, 'That Your Widows May Be Rich: Providing for Widowhood in Old Regime Europe', *Economisch en sociaal historisch jaarboek*, 45 (1982), pp. 58–76.
3 The emblematic writer about the limits of twentieth-century precision, Franz Kafka, resided in Prague, one of Europe's premier bilingual cities.
4 Cohen, *A Calculating People: The Spread of Numeracy in Early America* (Chicago: University of Chicago Press, 1982).
5 Robert E. Serfling, 'Historical Review of Epidemic Theory', *Human Biology*, 14 (1952), pp. 145–66.
6 John L. Heilbron, 'The Earliest Missionaries of the Copenhagen Spirit', *Revue d'histoire des sciences*, 38 (1985), pp. 195–230.
7 Fantappiè, *Principi di una teoria unitaria del mondo fisico e biologico* (Rome: Humanitas Nova, 1944).
8 Eddington cited in *ibid.*, p. 230.
9 Merz, *A History of European Scientific Thought in the Nineteenth Century* (Edinburgh: William Blackwood and Sons, 1904–14; New York: Dover, 1965), 2: pp. 552, 749.

CHAPTER 11

1 The word musket derives from the Italian word for sparrowhawk, and in the sixteenth century the Chinese referred to the European weapon as 'bird-beaked'. The common reference probably notes the pecking action of the cock.
2 H. D. Lasswell, 'The Garrison State', *American Journal of Sociology*, 46 (1941), pp. 455–68.
3 Clausewitz, *On War*, trans. J. J. Graham (London: K. Paul, French, Trubner & Co, 1940), pp. 117–21.
4 W. H. McNeill, *The Pursuit of Power: Technology, Armed Force, and Society since A. D. 1000* (Chicago: University of Chicago Press, 1982), pp. 133, 117.
5 Robert Gilpin, *France in the Age of the Scientific State* (Princeton: Princeton University Press, 1968), p. 258.
6 Jules Rouch, 'La prévision du temps', *Revue scientifique*, 56 (1918), 391–400, on p. 400.
7 Warriors outside the section of geography and navigation: Admiral Mouchez in the first division's astronomy section, Vice Admiral Abel Aubert du Petit-Thouars, Vice Admiral Jean-Philippe-Ernest de Fauque de Jonquières, Marshal Ferdinand Foch, and Marshal Jean-Baptiste Philibert, comte de Vaillant, all as *académiciens libres*. Membership in the Academy of Sciences is

given annually in the first number of the *Comptes rendus*. The discussion is based on the years 1860, 1880, 1900, 1920, and 1939; it excludes foreign associates and *correspondants*.

8 Robert Merton, *Science, Technology and Society in Seventeenth-Century England* (New York: Harper and Row, 1970), pp. xxi–xxii.

9 Walter McDougall, *The Heavens and the Earth: A Political History of the Space Age* (New York: Basic, 1985), p. 390.

10 Joseph de Maistre, cited in Pitirim A. Sorokin, *Contemporary Sociological Theories* (1928; New York: Harper and Row, 1964), p. 350.

CHAPTER 12

1 Derek J. de Solla Price, *Science since Babylon* (New Haven, CT: Yale University Press, 1961), p. 55.

2 H. B. Woodward, 'Geology in the Field and in the Study,' *Proceedings of the Geologists' Association, 13* (1894), p. 270.

3 J. W. Dawson, 'Things To Be Observed in Canada, and Especially in Montreal and Its Vicinity', *Canadian Naturalist, 3* (1858): pp. 1–12.

4 *Canadian Naturalist, new series 1* (1864), p. 52.

5 *Canadian Naturalist, new series 8* (1878), p. 450; *Canadian Naturalist, new series 10* (1883), p. 241.

6 Natalie Zemon Davis, *Women on the Margins: Three Seventeenth-Century Lives* (Cambridge, MA: Harvard University Press, 1995), p. 149.

7 *Ibid.*, pp. 163–5.

8 Evelyn Gordon Bodek, 'Salonières and Bluestockings: Educated Obsolescence and Germinating Feminism', *Feminist Studies, 3* (1975–76), p. 186.

9 Carolyn C. Lougee, *Le Paradis des Femmes: Women, Salons, and Social Stratification in Seventeenth-Century France* (Princeton: Princeton University Press, 1976), p. 170.

10 Dena Goodman, 'Enlightenment Salons: The Convergence of Female and Philosophic Ambitions', *Eighteenth-Century Studies, 22* (1988–9), pp. 329–50, on p. 338.

11 Joan B. Landes, *Women and the Public Sphere in the Age of the French Revolution* (Ithaca: Cornell University Press, 1988), p. 22.

12 Dena Goodman, *The Republic of Letters: A Cultural History of the French Enlightenment* (Ithaca: Cornell University Press, 1994), pp. 74, 76.

13 'Châtelet', *Dictionary of Scientific Biography, 3*, pp. 215–17.

14 Linda Gardiner Janik, 'Searching for the Metaphysics of Science: The Structure and Composition of Madame Du Châtelet's *Institutions de physique*, 1737–1740', *Studies on Voltaire and the Eighteenth Century, 201* (1982), pp. 60–2.

15 Londa Schiebinger, *The Mind Has No Sex? Women in the Origins of Modern Science* (Cambridge, MA: Harvard University Press, 1989), pp. 26 ff.

16 Landes, p. 51.

17 Linda Gardiner, 'Women in Science', in Samia I. Spencer, ed., *French Women and the Age of*

Enlightenment (Bloomington: Indiana University Press, 1984), p. 182.

CHAPTER 13

1 Edith Fountain Hussey to Orlando Fountain, 1911, in Lewis Pyenson, *Cultural Imperialism and Exact Sciences* (New York: Peter Lang, 1985), p. 190.
2 Tettje Clay-Jolles, the physicist wife of physicist Jacob Clay, complained in 1922 that it was not possible to live well on Java with a monthly income equal to a yearly income in the Netherlands. Lewis Pyenson, *Empire of Reason* (Leiden: E. J. Brill, 1989), p. 144.
3 Maclaurin, *The Theory of Light: A Treatise on Physical Optics, 1* (Cambridge: Cambridge University Press, 1908), pp. 1–15.
4 Jagdish N. Sinha, 'Science and the Indian National Congress', in Deepak Kumar, ed., *Science and Empire* (Delhi: Anamika Prakashan, 1991), pp. 161–81, on p. 167.

CHAPTER 14

1 P. M. Rattansi, 'Science and Religion in the Seventeenth Century' in Maurice Crosland, ed., *Emergence of Science in Western Europe* (New York: Science History, 1976), p. 80.
2 Charles Webster, 'Puritanism, Separatism, and Science' in D. C. Lindberg and R. L. Numbers, eds., *God and Nature: Historical Essays on the Encounter between Christianity and Science* (Berkeley: University of California Press, 1986), p. 213;

T. K. Rabb, 'Puritanism and the Rise of Experimental Science in England', in Leonard M. Marsak, ed., *The Rise of Science in Relation to Society* (New York: Macmillan, 1964), p. 55.
3 John Hedley Brooke, *Science and Religion: Some Historical Perspectives* (Cambridge: Cambridge University Press, 1991), pp. 64, 111; I. B. Cohen, ed., *Puritanism and the Rise of Modern Science* (New Brunswick: Rutgers University Press, 1990), p. 1.
4 Brooke, pp. 110, 112.
5 I. B. Cohen, 'Some Documentary Reflections on the Dissemination and Reception of the Merton Thesis', in J. Clark, et. al., eds., *Robert K. Merton: Consensus and Controversy* (London: Falmer Press, 1990), p. 316.
6 Richard L. Greaves, 'Puritanism and Science: The Anatomy of a Controversy', *Journal of the History of Ideas, 30* (1969), p. 353; Cohen, *Puritanism*, p. 2.
7 Cohen, *Puritanism*, p. 228; Greaves, pp. 349 ff.
8 Charles Webster, *The Great Instauration*, pp. 487–8, 491, 510, 505–6. Webster, 'Puritanism', p. 193.
9 Webster, pp. 485, 491, 493–8, 503–4, 510–11, 516; Webster, 'Puritanism', pp. 211–13.
10 In J. H. Brooke, 'Science and Religion', in R. C. Olby, et. al., eds., *Companion to the History of Science* (London: Routledge, 1990), pp. 773–4, and Brooke, pp. 167, 172–3, 175, 177.
11 Peter Burke, 'Religion and Secularisation', in *The New Cambridge Modern History*, Peter Burke, ed. (Cambridge:

Cambridge University Press, 1979), *13*, pp. 300–1.

12 Brooke, pp. 181–3, 189.

13 Olby, p. 775; Brooke, p. 193.

14 Brooke, pp. 198–200, 204, 209, 215–16; Frank M. Turner, *Contesting Cultural Authority: Essays in Victorian Intellectual Life* (Cambridge: Cambridge University Press, 1993), pp. 106–7, 111; Olby, pp. 775–6.

15 Olby, pp. 776–7.

16 McGill University Archives: J. W. Dawson to Charles Lyell, 26 April 1860.

17 Peter J. Bowler, *The Eclipse of Darwinism: Anti-Darwinian Evolution Theories in the Decades around 1900* (Baltimore: Johns Hopkins University Press, 1983), p. 44.

18 Olby, pp. 778–9.

19 Mead, *Twentieth Century Faith: Hope and Survival* (New York: Harper and Row, 1972), pp. 86–7.

CHAPTER 15

1 Peter M. Heimann, 'The Scientific Revolutions', in *The New Cambridge Modern History*, Peter Burke, ed. (Cambridge: Cambridge University Press, 1979), *13*, pp. 258–9; Brian P. Copenhaver, 'Natural Magic, Hermeticism, and Occultism in Early Modern Science', in David Lindberg and Robert Westman, eds., *Reappraisals of the Scientific Revolution* (Cambridge: Cambridge University Press, 1990), p. 265.

2 Christopher Hill, *The World Turned Upside Down* (London: Temple Smith, 1972), pp. 131–2.

3 Richard L. Greaves, 'Puritanism and Science: The Anatomy of a Controversy', *Journal of the History of Ideas*, *30* (1969), p. 366.

4 I. B. Cohen, 'Some Documentary Reflections on the Dissemination and Reception of the "Merton Thesis"', in J. Clark, *et al.*, eds., *Robert K. Merton: Consensus and Controversy* (London: Falmer Press, 1990), pp. 307–498.

5 R. F. Jones, *Ancients and Moderns: A Study of the Rise of the Scientific Movement in Seventeenth Century England* (Berkeley: University of California Press, 1965), p. 4.

6 Walter E. Houghton, 'The History of Trades: Its Relation to Seventeenth Century Thought', *Journal of the History of Ideas*, *2* (1941), pp. 34, 39, 50.

7 Jones, pp. viii–xi; Margaret C. Jacobs, *The Cultural Meaning of the Scientific Revolution* (Philadelphia: Temple University, 1988); Christopher Hill, *Intellectual Origins of the English Revolution* (Oxford: Oxford University Press, 1965); P. M. Rattansi, 'Science and Religion in the Seventeenth Century', in Maurice Crosland, ed., *Emergence of Science in Western Europe* (New York: Science History, 1976); Charles Webster, *The Great Instauration* (New York: Holmes & Meier, 1976), p. 515; John Bagnell Bury, *The Idea of Progress: An Inquiry into Its Origin and Growth* (London: Macmillan, 1920; New York: Dover, 1955), pp. 51–9.

8 Jacobs, p. 34; Bury, p. 59; Rattansi in Crosland, p. 83.

9 Frank A. Kafker, ed., *Notable*

Encyclopaedias of the Seventeenth and Eighteenth Centuries (Oxford: Voltaire Foundation, 1981), p. 8.

10 Kafker, pp. 223, 232; Thomas Hankins, *Jean d'Alembert, Science and the Enlightenment* (Oxford: Clarendon Press, 1970), pp. 66–7, 39–40, 83; P. N. Furbank, *Diderot: A Critical Biography* (London: Secker & Warburg, 1992), p. 84.

11 Stephen J. Gendzier, ed., *Denis Diderot's The Encyclopedia: Selections* (New York: Harper Torchbooks, 1967), p. 37.

12 Jacobs, p. 204.

13 David Knight, *Zoological Illustration*, (Folkestone: Dawson, 1977), p. 178.

14 Hankins, pp. 100–1

15 Maurice Mandelbaum, *History, Man, and Reason* (Baltimore: Johns Hopkins University Press, 1971), p. 11.

16 Bury, p. 284.

17 Robert Brown, 'Comte and Positivism', in C. L. Ten, ed., *The Nineteenth Century* [Routledge History of Philosophy, vol. 7] (London: Routledge, 1994), pp. 150–1.

CHAPTER 16

1 Robert Musil, trans. Eithne Wilkins and Ernst Kaiser, *The Man without Qualities*, vol. 1 (New York: Capricorn, 1953), p. 117.

2 Paul Forman, 'Truth and Objectivity', *Science*, 269 (1996), pp. 565–7, 707–10.

3 A. Toynbee, *A Study of History*, 9 (London: Oxford University Press, 1954), pp. 705–17.

4 Einstein to the Queen Mother of Belgium, 3 January 1952, in *Einstein on Peace*, Otto Nathan and Heinz Norden, eds (1960; New York: Avenel, 1981), p. 562.

FURTHER READING

INTRODUCTION

Robert Merton comments on his career, as well as Derek Price's and especially Thomas Kuhn's, in *The Sociology of Science: An Episodic Memoir* (Carbondale, IL: Southern Illinois University Press, 1979). Merton expands on his intellectual development in his collection, *On Social Structure and Science*, ed. Piotr Sztompka (Chicago: University of Chicago Press, 1996). Merton's persistent intellectual debt to George Sarton is found in his memoir, *On the Shoulders of Giants: A Shandean Postscript* (1965; New York: Harcourt Brace Jovanovich, 1985). Price's career is limned in Silvio A. Bedini, 'Memorial: Derek J. de Solla Price (1922–1983)', *Technology and Culture*, 15 (1984), pp. 701–5. Jed Z. Buchwald writes about Thomas Kuhn in 'Memories of Tom Kuhn', History of Science Society, *Newsletter, 25*, 4 (1996), pp. 3–4. A summing up may be found in Thomas Kuhn, 'Afterwords', in *World Changes: Thomas Kuhn and the Nature of Science*, ed. Paul Horwich (Cambridge, MA: Massachusetts Institute of Technology Press, 1993), pp. 311–41.

A case for the decline of science is made in Paul Horgan, *The End of Science: Facing the Limits of Knowledge in the Twilight of the Scientific Age* (Reading, MA: Addison-Wesley, 1996). Michael Sokal, 'Transgressing the Boundaries: Toward a Transformative Hermeneutics of Quantum Gravity', *Social Text*, nos. 46/47 (1996), pp. 217–52, offers a lampoon that was for a time accepted as a legitimate contribution to postmodernist thought (Sokal, 'A Physicist Experiments with Cultural Studies', *Lingua Franca*, May/June 1996, pp. 62–4). The journal *Social Text*, edited from Duke University, has been a forum for postmodernist ideologues. In the same issue of *Social Text*, another article concludes with an endorsement of one of our own studies: Steve Fuller, 'Does Science Put an End to History, or History to Science?' *Social Text*, nos. 46/47 (1996), pp. 27–42.

A recent review of approaches to scientific disciplines is found in Mary Jo Nye, *From Chemical Philosophy to Theoretical Chemistry: Dynamics of Matter and Dynamics of Disciplines, 1800–1950* (Berkeley: University of California Press, 1993), pp. 13–31. The discipline of the history of science is treated in Lewis Pyenson: 'Inventory as a Route to Understanding: Sarton, Neugebauer, and Sources', *History of Science, 33* (1995), pp. 253–82; 'Prerogatives of European Intellect: Historians of Science and the Promotion of Western Civilization', *History of Science, 31* (1993), pp. 289–315; 'What Is the Good of History of Science?' *History of Science, 27* (1989), pp. 353–89; ' "Who the Guys Were": Prosopography in the History of Science', *History of Science, 15* (1977), pp. 155–88.

CHAPTER 1

A classic treatment of the 'classical' world, placing science instruction in context, is Henri Irénée Marrou, *A History of Education in Antiquity*, trans. George Lamb (London: Sheed and Ward, 1956). Concise and still fresh is Thomas W. Africa, *Science and the State in Greece and Rome* (New York: Wiley, 1968). David Lindberg makes the case for a specifically European scientific culture in his *Beginnings of Western Science: The European Scientific Tradition in Philosophical, Religious, and Institutional Context, 600 B.C. to 1450* (Chicago: University of Chicago Press, 1992). Notable for its appeal to secondary sources in Eastern European languages is Jaroslav Krejčí's *Before the European Challenge: The Great Civilizations of Asia and the Middle East* (Albany: State University of New York Press, 1990).

A straightforward account may be found in Howard Spilman Galt, *A History of Chinese Educational Institutions* (London: A. Probsthain, 1951). A general comparative introduction, contrasting Western disputational education with Eastern scribal traditions and also elaborating on the evolution of science in Japan over the past thousand years, is Shigeru Nakayama, *Academic and Scientific Traditions, China, Japan, and the West*, trans. Jerry Dusenbury (Tokyo: University of Tokyo Press, 1984); this may be complemented with Nakayama's *A History of Japanese Astronomy: Chinese Background and Western Impact* (Cambridge, MA: Harvard University Press, 1969). Joseph Needham, the Erasmus of the twentieth century, has been identified in a British obituary as 'one of the greatest scholars in this or any country, of this or any century', a man who in his *Science and Civilisation in China* (Cambridge: Cambridge University Press, 1954–) undertook 'perhaps the greatest work of scholarship achieved by one individual since Aristotle'. Mansel Davies, 'Joseph Needham (1900–95)', *British Journal for the History of Science, 30* (1997), pp. 95–100. Needham's multivolume opus contains a wealth of information about Chinese educational institutions.

The chequered history of higher learning in Byzantine Constantinople is sketched in Robert Browning, 'The Patriarchal School at Constantinople in the Twelfth Century', *Byzantion, 32* (1962), pp. 167–202, in Paul Lemerle, *Cinq études sur le XIe siècle byzantin* (Paris: Centre National de la Recherche Scientifique, 1977), and especially in C. N. Constantinides, *Higher Education in Byzantium in the Thirteenth and Early Fourteenth Centuries (1204–ca.1310)* (Nicosia: Cyprus Research Centre, 1982). The case for a Byzantine 'university' is found in Friedrich Fuchs, *Die höheren Schulen von Konstantinopel im Mittelalter* (Leipzig: Teubner, 1926). For the situation under the Ottoman Turks see, in *Transfer of Modern Science and Technology to the Muslim World*, ed. Ekmeleddin Ihsanoglu (Istanbul: Research Center for Islamic History, Art and Culture, 1992): Ihsanoglu, 'Ottoman Science in the Classical Period and Early Contacts with European Science and Technology', pp. 1–48.

S. N. Sen has provided a perhaps overly optimistic survey of science education in South Asia before the European arrival in the first part of his *Scientific and Technical Education in India, 1781–1900*, published as nos. 1 and 2 of the *Indian Journal of History of Science* for 1988. Zaheer Baber summarizes much of the South Asian literature in *The Science of Empire: Scientific Knowledge, Civilization, and Colonial Rule in India* (Albany: State University of New York Press, 1996). A comprehensive treatment of the madrasa may be found in George Makdisi's *Rise of Colleges: Institutions of Learning in Islam and in the West* (Edinburgh: Edinburgh University Press, 1981). A. I. Sabra argues for the integrity of medieval Islamic scientific endeavour in 'The Appropriation and Subsequent Naturalization of Greek Science in Medieval Islam: A Preliminary Statement', *History of Science, 25* (1987), pp. 223–43. The unique features of European universities lie at the centre of Toby E. Huff, *Rise of Early Modern Science: Islam, China, and the West* (Cambridge: Cambridge University Press, 1993).

CHAPTER 2

A History of the University in Europe, vol. 1: Universities in the Middle Ages, ed. Hilde de Ridder-Symoens (Cambridge: Cambridge University Press, 1992) is the single best introduction to current scholarship; especially brilliant are general chapters by Walter Rüegg and

Jacques Verger and the sparkling discussion of students by Rainer Christoph Schwinges. In *Rebirth, Reform and Resilience: Universities in Transition, 1300–1700*, ed. James M. Kittelson and Pamela J. Transue (Columbus: Ohio State University Press, 1984), Lewis W. Spitz discusses the Renaissance and Reformation universities in 'The Importance of the Reformation for the Universities: Culture and Confessions in the Critical Years', pp. 42–67; Edward Grant makes a case for free scientific enquiry under medieval scholasticism in 'Science and the Medieval University', pp. 68–102; and John M. Fletcher evaluates university speciation in 'University Migrations in the Late Middle Ages, with Particular Reference to the Stamford Secession', pp. 163–189.

Nancy G. Siraisi discusses university medical faculties in *Medieval and Early Renaissance Medicine: An Introduction to Knowledge and Practice* (Chicago: University of Chicago Press, 1990). Joseph Needham makes the case for Chinese precedence in 'China and the Origin of Qualifying Examinations in Medicine', in Needham, *Clerks and Craftsmen in China and the West* (Cambridge: Cambridge University Press, 1970), pp. 379–95. Lu Gwei-Djen and Needham analyse the most striking of Chinese therapies – based on an enormous body of medical knowledge that circulated without state restriction – in a volume *hors série: Celestial Lancets: A History and Rationale of Acupuncture and Moxa* (Cambridge: Cambridge University Press, 1980). Paul Unschuld provides translations and commentary in *Medicine in China: A History of Pharmaceutics* (Berkeley: University of California Press, 1986).

Mordechai Feingold provides evidence that Oxford and Cambridge were far from scientific wastelands during the sixteenth and seventeenth centuries in *The Mathematicians' Apprenticeship: Science, Universities, and Society in England, 1560–1640* (Cambridge: Cambridge University Press, 1984), a contention disputed by H. Floris Cohen in *New Trends in the History of Science: Proceedings of a Conference Held at the University of Utrecht*, eds. R. P. W. Visser, H. J. M. Bos, L. C. Palm, and H. A. M. Snelders (Amsterdam: Rodopi, 1989), pp. 49–52. John Gascoigne summarizes the literature in 'A Reappraisal of the Role of the Universities in the Scientific Revolution', in *Reappraisals of the Scientific Revolution*, eds. David C. Lindberg and Robert S. Westman (Cambridge: Cambridge University Press, 1990), pp. 207–60.

John Heilbron, in his masterful *Electricity in the Seventeenth and Eighteenth Centuries: A Study of Early Modern Physics* (Berkeley: University of California Press, 1979), pp. 98–166, identifies the importance of university-based research in northern Europe and reminds us about the key role in science played by collegiate educational institutions, notably Jesuit colleges. Carl E. Schorske discusses *Wissenschaft* at Switzerland's oldest university in 'Science as a Vocation in Burckhardt's Basel', in *The University and the City: From Medieval Origins to the Present*, ed. Thomas Bender (Oxford: Oxford University Press, 1988), pp. 198–209. Joseph Ben-David focuses on the rise of research in nineteenth-century German universities in *The Scientist's Role in Society* (Englewood Cliffs, NJ: Prentice-Hall, 1971); a critical re-examination of Ben-David's German theses appears in R. Steven Turner, Edward Kerwin, and David Woolwine, 'Nineteenth-Century Physiology: Zloczower Redux', *Isis*, 75 (1984), pp. 523–9.

CHAPTER 3

The definitive work on scientific societies is Martha Ornstein, *The Role of Scientific Societies in the Seventeenth Century* (Chicago: University of Chicago Press, 1928). Harcourt Brown developed and extended her discussion of French institutions in *Scientific Organizations in Seventeenth Century France, 1620–1680* (1934; New York: Russell & Russell, 1967). For Italy, see W. E. Knowles Middleton, *The Experimenters: A Study of the Accademia del Cimento* (Baltimore: Johns Hopkins University Press, 1971). The subject has been reinvigorated and brought up to date in James E. McClellan III's *Science Reorganized: Scientific Societies in the Eighteenth Century* (New York: Columbia University Press, 1985). On specialized societies see Susan Sheets-Pyenson, 'Geological Communication in the Nineteenth Century: The Ellen S. Woodward Autograph Collection at McGill University', *Bulletin of the British Museum (Natural History)*, Historical Series, *10*, no. 6, 1982, pp. 179–226.

Histories of the Royal Society of London include Marie Boas Hall, *Promoting Experimental Learning: Experiment and the Royal Society, 1660–1727* (Cambridge: Cambridge University Press, 1991) and chapter 8 in A. Rupert Hall, *The Revolution in Science, 1500–1750* (London: Longman, 1983). Charles Webster is concerned with other issues in his *Great Instauration: Science, Medicine and Reform, 1626–1660* (1975; New York: Holmes & Meier, 1976), but his discussion of the early Royal Society in Part II, chapter 9 is fascinating. On the Royal Society as the institutionalization of science see

Michael Hunter, *Establishing the New Science: The Experience of the Early Royal Society* (Woodbridge, Suffolk: The Boydell Press, 1989)

On the French Academy of Science see Maurice Crosland, *Science under Control: The French Academy of Sciences, 1795–1914* (Cambridge: Cambridge University Press, 1992). This work expands on and carries forward Roger Hahn's pioneering *Anatomy of a Scientific Institution: The Paris Academy of Sciences, 1666–1803* (Berkeley: University of California Press, 1971).

CHAPTER 4

Good surveys of European instrumentation are provided in Anthony Turner, *Early Scientific Instruments; Europe, 1400–1800* (London: Sotheby's, 1987) and J. A. Bennett, *The Divided Circle: A History of Instruments for Astronomy, Navigation and Surveying* (Oxford: Phaidon/Christie's, 1987). A classic and still useful history is Henry C. King, *The History of the Telescope* (London: Charles Griffin & Co., and Cambridge, MA: Sky Publishing, 1955).

The institutional basis of Islamic astronomy is analysed intensively in Aydin Sayili's *Observatory in Islam and Its Place in the General History of the Observatory* (Ankara: Türk Tarih Kurumu Basimevi, 1960). More recent developments are treated in *Transfer of Modern Science and Technology to the Muslim World*, ed. Ekmeleddin Ihsanoglu (Istanbul: Research Center for Islamic History, Art and Culture, 1992): Ihsanoglu, 'Introduction of Western Science to the Ottoman World: A Case Study of Modern Astronomy', pp. 67–120; S. M. Razaullah Ansari, 'Modern Astronomy in Indo-Persian Sources', pp. 121–44; and George Saliba, 'Copernican Astronomy in the Arab East: Theories of the Earth's Motion in the Nineteenth Century', pp. 145–55. Addenda have appeared to Joseph Needham's volume on mathematics and astronomy in *Science and the Civilization in China*: Needham, Wang Ling, and Derek J. de Solla Price, *Heavenly Clockwork: The Great Astronomical Clocks of Medieval China* (1960; Cambridge: Cambridge University Press, 1986); Needham, Lu Gwei-Djen, John H. Combridge, and John S. Major, *The Hall of Heavenly Records: Korean Astronomical Instruments and Clocks, 1380–1780* (Cambridge: Cambridge University Press, 1986). The Tower of Winds appears with many illustrations in Price's 'Athens' Tower of Winds', *National Geographic, 131* (1967), pp. 586–96.

General accounts of timekeeping include David S. Landes, *Revolution in Time: Clocks and the Making of the Modern World* (Cambridge,

MA: Belknap Press of Harvard University Press, 1983); Kenneth F. Welch, *Time Measurement: An Introductory History* (Newton Abbot: David & Charles, 1972). Anthony Aveni's *Empires of Time: Calendars, Clocks, and Cultures* (New York: Basic, 1989) is useful for its account of indigenous New World developments. European calendrical science is surveyed in Arno Borst, *The Ordering of Time: From the Ancient Computus to the Modern Computer*, trans. Andrew Winnard (Chicago: University of Chicago Press, 1993). Rich in cultural history is Norbert Elias's *Time: An Essay*, trans. Edmund Jephcott (Cambridge: Blackwell, 1992).

Many modern observatories have been the object of a history. Typical examples are: William de Sitter, *A Short History of the Observatory of the University at Leiden, 1633–1933* (Haarlem: Joh. Enschede, 1933); Donald E. Osterbrock, John R. Gustafson, and W. J. Shiloh Unruh, *Eye on the Sky: Lick Observatory's First Century* (Berkeley: University of California Press, 1988). Astronomers have traditionally attracted biographers, as may be seen from bibliographies in the *Dictionary of Scientific Biography*. A systematic inventory appears in Derek Howse, *The Greenwich List of Observatories: A World List of Astronomical Observatories, Instruments and Clocks, 1670–1850* (Chalfont St Giles, 1986) [*Journal of the History of Astronomy*, 17, no. 4]. For the twentieth century there is *Astrophysics and Twentieth-Century Astronomy to 1950, Part A*, ed. Owen Gingerich (Cambridge: Cambridge University Press, 1984) [*General History of Astronomy* 4A, ed. Michael Hoskin]. Jesuit astronomy in the nineteenth and twentieth century is analysed in Lewis Pyenson, *Civilizing Mission: Exact Sciences and French Overseas Expansion, 1830–1940* (Baltimore: Johns Hopkins University Press, 1993). English-language observatories are surveyed in David S. Evans, *Under Capricorn: A History of Southern Hemisphere Astronomy* (Bristol: Adam Hilger, 1988). There are also national histories, for example, Richard Jarrell, *The Cold Light of Dawn* (Toronto: University of Toronto Press, 1988), on Canadian astronomy.

CHAPTER 5

On the early history of natural history museums and their evolution from 'curio cabinets' see Krzysztof Pomian, *Collectionneurs, amateurs et curieux: Paris, Venise, XVIe–XVIIIe siècle* (Paris: Gallimard, 1987). Eilean Hooper-Greenhill, unabashedly drawing upon the works of Michel Foucault, presents a somewhat different view of this evolution in her *Museums and the Shaping of Knowledge* (London: Routledge, 1992). Oliver Impey and Arthur MacGregor, eds., *The Origins of Museums: The Cabinet of Curiosities in Sixteenth- and Seventeenth-Century Europe* (Oxford: Clarendon Press, 1985) surveys these cabinets. Lorraine Daston's essay review of this work and others in *Isis*, 79 (1988), pp. 452–67, is especially interesting.

On early science museums in the United States, see Joel L. Orosz, *Curators and Culture: The Museum Movement in America, 1740–1870* (Tuscaloosa: University of Alabama Press, 1990). On museums of science and technology (including a useful bibliography on the topic) see Bernard S. Finn, 'The Museum of Science and Technology' in *The Museum: A Reference Guide*, ed. Michael S. Shapiro (Westport, CT: Greenwood Press, 1990), pp. 59–84, and Stella V. F. Butler, *Science and Technology Museums* (Leicester: Leicester University Press, 1992). Some historical discussion is interspersed in Simon Tait's *Palaces of Discovery: The Changing World of Britain's Museums* (London: Quiller Press, 1989), which emphasizes the contemporary situation.

Portions of this chapter are drawn from Susan Sheets–Pyenson's work; see especially her *Cathedrals of Science: The Development of Colonial Natural History Museums during the Late Nineteenth Century* (Montreal: McGill-Queen's University Press, 1988)

CHAPTER 6

On the history of gardens, chiefly ornamental, see Edward Hyams, *A History of Gardens and Gardening* (London: J. M. Dent & Sons, 1971); William Howard Adams, *The French Garden, 1500–1800* (New York: George Braziller, 1979); H. N. Wethered, *A Short History of Gardens* (London: Methuen & Co., 1933). Also see the introductory essay by Joseph Ewan in *Hortus Botanicus: The Botanic Garden and The Book*, comp. Ian MacPhail (Chicago: Morton Arboretum, 1972). Richardson Wright's *Story of Gardening* (1934; New York: Dover, 1963) is somewhat less useful.

For botanic gardens, see John Prest, *The Garden of Eden: The Botanic Garden and the Re-Creation of Paradise* (New Haven: Yale University Press, 1981); C. Stuart Gager, 'Botanic Garden', in *The Standard Cyclopedia of Horticulture*, ed. by L. H. Bailey (New York: Macmillan, 1914), pp. 526–32; Arthur W. Hill, 'The History and Functions of Botanic Gardens', *Annals of the Missouri Botanical Garden*, 2 (1915), pp. 185–240. Also see William Thomas Stearn,

'Botanical Gardens and Botanical Literature in the Eighteenth Century', in *Catalogue of Botanical Books in the Collection of Rachel McMasters Miller Hunt*, comp. Allan Stevenson, vol. 2 (Pittsburgh: The Hunt Botanical Library, 1961), pp. *xli–cxl*.

Edward Hyams provides a sketch of hundreds of botanical gardens, including their history, in *Great Botanical Gardens of the World* (London: Thomas Nelson & Sons, 1969). For a complete register of botanic gardens, see C. Stuart Gager, 'Botanic Gardens of the World', *Brooklyn Botanic Garden Record, 27* (July 1938), pp. 151–406.

For the case of Great Britain, see the colloquium proceedings on *John Claudius Loudon and the Early Nineteenth Century in Great Britain* (Washington, DC: Dumbarton Oaks, 1980). On the Royal Botanic Gardens at Kew, see Ronald King, *The World of Kew* (London: Macmillan, 1976) and F. Nigel Hepper, *Kew: Gardens for Science & Pleasure* (London: Her Majesty's Stationery Office, 1982). On Kew's role abroad, see Lucile H. Brockway, *Science and Colonial Expansion: The Role of the British Royal Botanic Gardens* (New York: Academic Press, 1979). For colonial botanic gardens, see Ray Desmond, *The European Discovery of the Indian Flora* (Oxford: Oxford University Press, 1992).

Gardens of Australia are discussed in R. T. M. Pescott, *The Royal Botanic Gardens, Melbourne: A History from 1845 to 1970* (Melbourne: Oxford University Press, 1982) and Lionel Gilbert, *The Royal Botanic Gardens, Sydney: A History, 1816–1985* (Melbourne: Oxford University Press, 1986). For the United States, see Ann Leighton, *American Gardens in the Eighteenth Century* (Boston: Houghton Mifflin, 1976) and her *American Gardens of the Nineteenth Century* (Amherst: University of Massachusetts Press, 1987).

The best general history of zoological gardens is still Gustave Loisel, *Histoire des ménageries de l'antiquité à nos jours*, 3 vols. (Paris: Octave Doin et Fils, 1912). The entry on 'Zoological Gardens', *Encyclopaedia Britannica*, 11th ed., surveys the world's zoos in 1910. More recent histories include Jon R. Luoma, *A Crowded Ark* (Boston: Houghton Mifflin, 1987); Solly Zuckerman, ed., *Great Zoos of the World: Their Origins and Significance* (Boulder, CO: Westview Press, 1980); and Bill Jordan and Stefan Ormrod, *The Last Great Wild Beast Show: A Discussion on the Failure of British Animal Collections* (London: Constable, 1978)

One of the few accounts of a particular zoo that devotes attention to its history is William Bridges, *Gathering of Animals: An Unconventional History of the New York Zoological Society* (New York:

Harper & Row, 1966). On the London Zoological Society see Lynn Barber, *The Heyday of Natural History* (London: Jonathan Cape, 1980).

CHAPTER 7

Systems of counting and measuring in diverse civilizations and languages are the subject of Karl Menninger, *Number Words and Number Symbols: A Cultural History of Numbers*, trans. Paul Broneer (Cambridge, MA: Massachusetts Institute of Technology Press, 1969). European mechanical computational devices are discussed and illustrated in Anthony Turner, *Early Scientific Instruments: Europe, 1400–1800* (London: Sotheby's, 1987). A general work by an English scholar is Samuel L. Macey's *Dynamics of Progress: Time, Method, and Measure* (Athens, GA: University of Georgia Press, 1989). Derek Price provides a summary treatment of the thesis that tactile crafts lie behind scientific discovery in 'Of Sealing Wax and String', *Natural History*, 1984, no. 1, pp. 49–56; Price surveys how measurements were used in 'A History of Calculating Machines', *Institute of Electrical and Electronics Engineers, Micro*, February 1984, pp. 23–52.

The standard source of measurement in classical antiquity is Friedrich Otto Hultsch, *Griechische und römische Metrologie* (Berlin, 1882; Graz: Akademische Druck, 1971). A survey of weights and measures appears in Mabel Lang and Margaret Crosby, *Weights, Measures, and Tokens* (Princeton, 1964) [The Athenian Agora: Results of Excavations Conducted by the American School of Classical Studies at Athens, 10]. An exhaustive survey appears in Erich Schilbach, *Byzantinische Metrologie* (Munich: C. H. Beck, 1970). A useful and well-illustrated compendium is Garo Kürkman, *Ottoman Weights and Measures* (Istanbul: Museum of Turkish and Islamic Art, 1991). See also: Feza Günergun, 'Introduction of the Metric System to the Ottoman State', in *Transfer of Modern Science and Technology to the Muslim World*, ed. Ekmeleddin Ihsanoglu (Istanbul: Research Center for Islamic History, Art, and Culture, 1992), pp. 297–316. A useful survey of North-African measures is found in Marcel Legendre, *Survivance des mesures traditionnelles en Tunisie* (Paris: Presses Universitaires de France, 1958).

The Antikythera machine is brilliantly analysed by Derek de Solla Price in *Gears from the Greeks: The Antikythera Mechanism, A Calendar Computer from ca 80 B.C.* (Philadelphia, 1974) [American

Philosophical Society, *Transactions, 64,* p. 7]. Su Sung's great astronomical clock-tower and much else receive masterly treatment in Joseph Needham, Wang Ling, and Derek J. de Solla Price, *Heavenly Clockwork: The Great Astronomical Clocks of Medieval China* (Cambridge: Cambridge University Press, 1960). Japanese horology is treated in J. Drummond Robertson, *The Evolution of Clockwork, with a Special Section on the Clocks of Japan* (London: Cassell, 1931), which discusses the clock of Ino Tadataka. A popular survey is Kenneth F. Welch, *Time Measurement: An Introductory History* (Newton Abbot: David and Charles, 1972), which discusses the Shortt free pendulum.

Kathryn Olesko analyses the origins of modern physics in Carl Neumann's school of measurement in *Physics as a Calling: Discipline and Practice in the Königsberg Seminar for Physics* (Ithaca: Cornell University Press, 1991). Nineteenth-century issues are masterfully treated in Christa Jungnickel and Russell McCormmach, *Intellectual Mastery of Nature: Theoretical Physics from Ohm to Einstein,* 2 vols. (Chicago: University of Chicago Press, 1986). The principal source for the history of measurement in modern times is the periodical *Historical Studies in the Physical and Biological Sciences,* edited by Russell McCormmach and published by Princeton University Press and Johns Hopkins University Press between 1969 and 1979, and since then edited by John L. Heilbron and published by the University of California Press.

CHAPTER 8

Chapter 8 owes a great debt to Elizabeth L. Eisenstein's fascinating account of the importance of the invention of the printing press for the emergence of modern science. See her *Printing Press as an Agent of Change: Communications and Cultural Transformation in Early Modern Europe* [vols. 1 and 2 in one vol.] (1979; Cambridge: Cambridge University Press, 1980). An especially witty and engaging expression of Eisenstein's thesis appears in James Burke's *The Day the Universe Changed* (London: British Broadcasting Corporation, 1985), chapter 4: 'Matter of Fact'. Interesting insights on printing, publishing, and science are put forward in Derek J. de Solla Price's *Science since Babylon* (1961; New Haven: Yale University Press, 1967), especially chapter 3, 'Renaissance Roots of Yankee Ingenuity', and his *Little Science, Big Science* (New York: Columbia University Press, 1963), chapter 1.

Bernard Houghton provides a useful overview of the development of scientific periodicals in his *Scientific Periodicals: Their Historical Development, Characteristics and Control* (Hamden, CT: Linnet Books, 1975), especially chapters 1 and 2. Also see David A. Kronick, *A History of Scientific and Technical Periodicals: The Origins and Development of the Scientific and Technical Press, 1665–1790* (Metuchen, NJ: Scarecrow Press, 1976).

Leona Rostenberg discusses the early years of the *Philosophical Transactions* in chapter 8 of her *Library of Robert Hooke: The Scientific Book Trade of Restoration England* (Santa Monica, CA: Modoc Press, 1989) and in her *Literary, Political, Scientific, Religious and Legal Publishing, Printing and Bookselling in England, 1551–1700: Twelve Studies* (New York: Burt Franklin, 1965). On the firm of Taylor & Francis see W. H. Brock and A. J. Meadows, *The Lamp of Learning: Taylor & Francis and the Development of Science Publishing* (London: Taylor & Francis, [1984]). (This book pays special attention to the history of the *Philosophical Magazine*.) More general but somewhat less useful is A. J. Meadows, ed., *Development of Science Publishing in Europe* (Amsterdam: Elsevier, 1980). For the *Annals of Natural History*, see Susan Sheets-Pyenson, 'From the North to Red Lion Court: The Creation and Early Years of the *Annals of Natural History*', *Archives of Natural History*, 10 (1981), pp. 221–49. On popular scientific periodicals, see Sheets-Pyenson, 'Popular Science Periodicals in Paris and London: The Emergence of a Low Scientific Culture', *Annals of Science*, 42 (1985), pp. 549–72.

On botanical illustration, see Wilfrid Blunt, *The Art of Botanical Illustration* (1950; London: Collins, 1967). For zoological illustration, David Knight, *Zoological Illustration: An Essay towards a History of Printed Zoological Pictures* (Folkestone, England: Dawson; Hamden, CT: Archon Books, 1977) and S. Peter Dance, *The Art of Natural History: Animal Illustrators and Their Work* (Woodstock, NY: Overlook Press, 1978). Magnificent illustrations appear in Handasyde Buchanan's *Nature into Art: A Treasury of Great Natural History Books* (New York: Mayflower Books, 1979).

CHAPTER 9

Augustine Brannigan discusses the various meanings of the discovery of America in *The Social Basis of Scientific Discovery* (Cambridge: Cambridge University Press, 1981). Genevieve Lloyd presents Bacon as a masculine dominator of female nature in *The*

Man of Reason: 'Male' and 'Female' in Western Philosophy (Minneapolis: University of Minnesota Press, 1984).

A recent reconsideration is a collection edited by Karen Ordahl Kupperman, *America in European Consciousness, 1493–1750* (Chapel Hill: University of North Carolina Press, 1995), notably the essay by Henry Lowood, 'The New World and the European Catalogue of Nature'.

Revised discussion of the early botanical accounts of the New World are found in José Pardo Tomás and María Luz López Terrada, *Las primeras noticias sobre plantas americanas en las relaciones de viajes y crónicas de Indias (1493–1553)* (Valencia, 1993) [Cuadernos Valencianos de Historia de la Medicina y de las Ciencia, 40, Series A]; José María López Piñero and José Pardo Tomás, *Nuevos materiales y noticias sobre la Historia de las plantas de Nueva España de Francisco Hernández* (Valencia, 1994) [Cuadernos Valenciano de Historia de la Medicina y de la Ciencia, 44, Series A]. From his chair at Valencia, José María López Piñero has done more than another other scholar to shed new light on science in Spanish America; his extensive bibliography defies abbreviation, although one may note his collaborative volume, *Medicinas, drogas y alimentos vegetables del Nuevo Mundo* (Madrid: Ministerio de Sanidad y Consumo, 1992). No acknowledgement of the great Valencia school appears in a recent general survey, Anthony Grafton, with April Shelford and Nancy Siraisi, *New Worlds, Ancient Texts: The Power of Tradition and the Shock of Discovery* (Cambridge, MA: Belknap Press, 1992), which focuses on the slow erosion of classical learning.

Significant reappraisals are found in the collections *De la ciencia ilustrada a la ciencia romántica: Actas de las II Jornadas sobre 'España y las expediciones científicas en América y Filipinas'*, ed. Alejandro R. Díez Torre, Tomás Mallo and Daniel Pacheco Fernández (Madrid: Doce Calles, 1995); *Las relaciones entre Portugal y Castilla en la época de los descubrimientos y la expansión colonial*, ed. Anna María Carabias Torres (Salamanca: Ediciones Universidad, 1994).

CHAPTER 10

The early social roots of mathematics are discussed in Jens Høyrup's *In Measure, Number, and Weight: Studies in Mathematics and Culture* (Albany: State University of New York Press, 1994). John Theodore Merz – chemist, philosopher, industrialist, and intellectual historian – traced the history of statistics from William Petty through Laplace to Maxwell, Darwin, Galton, and Pearson in the chapter, 'The Statistical View of Nature', in volume two of his magisterial *History of European Thought in the Nineteenth Century* (1904–14; New York: Dover, 1965). Merz is always rewarding, and he is especially relevant for the issues considered in the present chapter.

The past decades have seen increasing interest in the history of probability and statistics; not a little of the work is redundant. A useful and idiosyncratic discussion of the roots of a statistical understanding of nature appears in Ian Hacking, *The Emergence of Probability: A Philosophical Study of Early Ideas about Probability, Induction and Statistical Inference* (Cambridge: Cambridge University Press, 1975). This may be followed by Lorraine Daston, *Classical Probability in the Enlightenment* (Princeton: Princeton University Press, 1988), Stephen M. Stigler, *The History of Statistics: The Measurement of Uncertainty before 1900* (Cambridge, MA: Belknap Press of the Harvard University Press, 1986), and Theodore M. Porter, *The Rise of Statistical Thinking, 1820–1900* (Princeton: Princeton University Press, 1986). Late eighteenth-century numeracy is considered by John L. Heilbron in *Weighing Imponderables and Other Quantitative Science around 1800* (Berkeley: University of California Press, 1993) [Historical Studies in the Physical and Biological Sciences, 24, pt 1, supplement]. Ponderous and inelegant works may also be consulted: Anders Hald, *A History of Probability and Statistics and Their Applications before 1750* (New York: Wiley, 1990); L. E. Maistrov, *Probability Theory: A Historical Sketch*, trans. and ed. Samuel Kotz (New York: Academic Press, 1974). Useful and entertaining are Karl Pearson's lectures from the years 1921–1933: *The History of Statistics in the 17th and 18th Centuries against the Changing Background of Intellectual, Scientific and Religious Thought*, ed. E. S. Pearson (London: C. Griffin, 1978). A collection of articles is found in the two-volume *Probabilistic Revolution* (Cambridge, MA: Massachusetts Institute of Technology Press, 1987–89): vol. 1 (1989), ed. Lorenz Krüger, Lorraine J. Daston, and Michael Heidelberger; vol. 2 (1987), ed. Lorenz Krüger, Gerd Gegerenzer, and Mary S. Morgan; of particular interest in volume 1 is Ian Hacking's periodization of the development of probability and statistics during the nineteenth and early twentieth centuries. M. Norton Wise and Crosbie Smith consider early nineteenth-century ideology in "Work and Waste: Political Economy and Natural Philosophy in Nineteenth Century Britain,' *History of Science*, 27 (1989), pp. 263–

301, 391–449; *28* (1990), pp. 221–61. A good summary of the broader course of statistics in the nineteenth century is found in Theodore M. Porter, 'A Statistical Survey of Gases: Maxwell's Social Physics', *Historical Studies in the Physical and Biological Sciences,* *12* (1981), pp. 77–116. The finest discussion of Francis Galton, Karl Pearson, and R. A. Fisher is found in Daniel J. Kevles's *In the Name of Eugenics: Genetics and the Uses of Human Heredity* (New York: Knopf, 1985). Weimar uncertainty is the subject of Paul Forman's classic study, 'Weimar Culture, Causality, and Quantum Theory, 1918–1927: Adaptation by German Physicists and Mathematicians to a Hostile Intellectual Environment', *Historical Studies in the Physical Sciences, 3* (1971), pp. 1–115. A broad, impressionistic survey is available in Stephen G. Brush, 'Irreversibility and Indeterminism: Fourier to Heisenberg', *Journal of the History of Ideas, 37* (1976), pp. 603–30. Twentieth-century developments are covered in Sharon E. Kingsland's *Modeling Nature: Espisodes in the History of Population Ecology* (Chicago: University of Chicago Press, 1985). Gregory Bateson is quoted by James Lovelock in *The Ages of Gaia: A Biography of Our Living Earth* (New York: W. W. Norton, 1990), p. 218.

A general survey is found in Vardit Rispler-Chaim, 'Insurance and Semi-Insurance Transactions in Islamic History until the 19th Century', *Journal of the Economic and Social History of the Orient, 34* (1991), pp. 142–58. Chance in classical antiquity may be read in: Jacqueline Bordes, 'Le tirage au sort, principe de la démocratie athénienne', *Ethnologie française, 17* (1987), pp. 145–50; Didier Pralon, 'Les sorts des dieux et des héros,' *ibid.,* pp. 151–57.

CHAPTER 11

A portion of this chapter is adapted from Lewis Pyenson, *Civilizing Mission: Exact Sciences and French Overseas Expansion, 1830–1940* (Baltimore: Johns Hopkins University Press, 1993).

Francis Bacon is cited (in an expressive translation from Latin to English) in Joseph Needham, Ho Ping-Yü, Lu Gwei-Djen, and Wang Ling, *Science and Civilization in China, 5,* Pt 7: *Military Technology: The Gunpowder Epic* (Cambridge: Cambridge University Press, 1986), on p. *xxx*; p. 339 for the firepower of a medieval Chinese battalion.

Denis I. Duveen and Roger Hahn, 'Laplace's Succession to Bézout's Post of *Examinateur des élèves de l'Artillerie*', *Isis, 48* (1957), pp. 416–27, provide a bright and succinct discussion on the late eighteenth-century military scientists. On the early nineteenth-century Japanese reaction to European surgery: Ken Vos, *Assignment Japan: Von Siebold, Pioneer and Collector* (The Hague: SDU, 1989).

The recent literature on the American military-industrial complex, and its connection with the academic world, is largely silent about nineteenth-century antecedents and French precedent. No word is given to these issues in a recent symposium: Everett Mendelsohn, Merritt Roe Smith, and Peter Weingart, 'Science and the Military: Setting the Problem', in their edited volume, *Science, Technology and the Military* (Dordrecht: Reidel, 1988), pp. xi–xxix. For the army as a technological innovator: *Military Enterprise and Technological Change: Perspectives on the American Experience*, Merritt Roe Smith, ed. (Cambridge, MA: Massachusetts Institute of Technology Press, 1985), especially the historiographical remarks by the editor (pp. 1–37). For a corrective: Paul Forman and José Manuel Sánchez Ron, eds., *National Military Establishments and the Advancement of Science and Technology: Studies in Twentieth Century History* (Dordrecht: Kluwer, 1996).

A perceptive account of the rise of the American research university is Nathan Reingold, 'Graduate School and Doctoral Degree: European Models and American Realities', in *Scientific Colonialism: A Cross-Cultural Comparison*, Reingold and Marc Rothenberg, eds. (Washington: Smithsonian Institution Press, 1987), pp. 129–49. The research university rose dramatically between 1880 and 1900, just at the time that the military and naval slice of federal science funding fell from 27 per cent to 7 per cent. Nathan Reingold and Joel N. Bodansky, 'The Sciences, 1850–1900: A North Atlantic Perspective', *Biological Bulletin, 168* [supplement] (1985) [*The Naples Zoological Station and the Marine Biological Laboratory: One Hundred Years of Biology*], pp. 44–61, on pp. 53–4.

Japan, where the Samurai ethic pervaded Western science late in the nineteenth century, is a notable exception to twentieth-century generalizations. Shigeru Nakayama, trans. Jerry Dusenbury, *Academic and Scientific Traditions in China, Japan, and the West* (Tokyo: University of Tokyo Press, 1984), pp. 205–6.

Science and the military in revolutionary France have recently been illuminated in an exhaustive study about ordnance by Charles Coulston Gillispie, 'Science and Secret Weapons Development in Revolutionary France, 1794–1804: A Documentary History', *Historical Studies in the Physical and Biological Sciences, 23* (1992), pp. 35–152.

Elisabeth Crawford, *Nationalism and Internationalism in Science, 1880–1930: Four Studies of the Nobel Population* (Cambridge: Cambridge University Press, 1992), p. 31, for Jozef Pilsudski's contention that the state invents the nation, not the reverse. Walter A. McDougall, *The Heavens and the Earth: A Political History of the Space Age* (New York: Basic Books, 1985), p. 305 for the political significance of prestige.

CHAPTER 12

A preliminary census of scientific societies appears in John Cohen, C. E. M. Hansel and Edith F. May, 'Natural History of Learned and Scientific Societies', *Nature*, *173* (1954), pp. 328–33. A unique perspective on the important functions of learned societies for less noteworthy scientists appears in K. Theodore Hoppen, *The Common Scientist in the Seventeenth Century: A Study of the Dublin Philosophical Society, 1683–1708* (Charlottesville: University Press of Virginia, 1970). For a nuanced discussion of the rise of literary and philosophical societies see Arnold Thackray, 'Natural Knowledge in Cultural Context: The Manchester Model', *American Historical Review*, *79* (1974), pp. 672–709.

On the creation of peripatetic associations for the advancement of science, see Everett Mendelsohn, 'The Emergence of Science as a Profession in Nineteenth-Century Europe', in K. Hill, ed., *The Management of Scientists* (Boston: Beacon, 1964), pp. 3–48. Jack Morrell and Arnold Thackray provide exhaustive discussion in *Gentlemen of Science: Early Years of the British Association for the Advancement of Science* (New York: Clarendon Press, 1981). On the British Association meetings in Montreal and the Montreal Natural History Society see the relevant chapters of Susan Sheets-Pyenson, *John William Dawson: Faith, Hope, and Science* (Montreal: McGill-Queen's University Press, 1996).

On the rise of field clubs and other amateur scientific organizations see David Elliston Allen, *The Naturalist in Britain* (London: Penguin Books: 1976) and Lynn Barber, *The Heyday of Natural History* (London: Jonathan Cape, 1980). Susan Sheets-Pyenson treats organisations devoted to geology in 'Geological Communication in the Nineteenth Century: The Ellen S. Woodward Autograph Collection at McGill University', *Bulletin of the British Museum (Natural History)*, Historical Series, *10*, no. 6, 1982, pp. 179–226.

For fascinating discussions of the role of women in science see David F. Noble, *A World Without Women: The Christian Clerical Culture of Western Science* (New York: Knopf, 1992) and Evelyn Fox Keller, *Reflections on Gender and Science* (New Haven: Yale University Press, 1985).

Marina Benjamin, ed., *Science and Sensibility: Gender and Scientific Enquiry, 1780–1945* (Oxford: Blackwell, 1991) and Londa Schiebinger, *The Mind Has No Sex? Women in the Origins of Modern Science* (Cambridge, MA: Harvard University Press, 1989) emphasize the place of women within the natural history and craft traditions. Also see Sharon D. Valiant, 'Questioning the Caterpillar', *Natural History, 101* (Dec. 1992), pp. 46–59 for a discussion of the life of Maria Sibylla Merian. The fullest and most fascinating treatment of Merian appears in Natalie Zemon Davis's *Women on the Margins: Three Seventeenth-Century Lives* (Cambridge, MA: Harvard University Press 1995), p. 149.

On Mary Somerville, see Elizabeth C. Patterson, *Mary Somerville and the Cultivation of Science, 1815–40* (Boston: Martinus Nijhoff, 1983). On the role of British women in nineteenth century geology see Mary R. S. Creese and Thomas M. Creese, 'British Women Who Contributed to Research in the Geological Sciences in the Nineteenth Century', *British Journal for the History of Science, 27* (1994), pp. 23–54.

CHAPTER 13

Early syntheses include George Basalla, 'The Spread of Western Science', *Science, 156* (1967), pp. 611–22; Donald Fleming, 'Science in Australia, Canada, and the United States: Some Comparative Remarks', in *Proceedings of the 10th International Congress of the History of Science, 1962, 1* (Paris: Hermann, 1964), pp. 179–96. Later comparative treatments include Lewis Pyenson, 'Pure Learning and Political Economy: Science and European Expansion in the Age of Imperialism', in *New Trends in the History of Science*, R. P. W. Visser, *et al.*, eds. (Amsterdam: Rodopi, 1989), pp. 209–78.

European expansion is reconsidered in Alfred W. Crosby's *Ecological Imperialism: The Biological Expansion of Europe, 900–1900* (Cambridge: Cambridge University Press, 1986). A useful general history is Germán Arciniegas, *Latin America: A Cultural History*, trans. Joan McLean (New York: Alfred A. Knopf, 1967).

Colonial science in the eighteenth century is recounted in Arcienegas, *Latin America*, pp. 235–7 on Galileo's interest in the New World; John Tate Lanning, *Academic Culture in the Spanish Colonies* (Oxford: Oxford University Press, 1940); Robert Jonas Shafer, *The Economic Societies in the Spanish World, 1763–1821* (Syracuse: Syracuse University Press, 1958); Harry Woolf, *The Transits of Venus: A Study of Eighteenth-Century Science* (Princeton: Princeton University Press, 1959); Dirk J. Struik, *Yankee Science in the Making* (New York: Collier, 1962); Raymond Phineas Stearns, *Science in the British Colonies of North America* (Urbana: University of Illinois Press, 1970); Max Savelle, *Empires to Nations: Expansion in America, 1713–1824* (Minneapolis: University of Minnesota Press, 1974); James E. McClellan III, *Colonialism and Science: Saint Domingue in the Old Regime* (Baltimore: Johns Hopkins University Press, 1992) and in McClellan's *Science Reorganized: Scientific Societies in the Eighteenth Century* (New York: Columbia University Press, 1985); Kerjariwal, *The Asiatic Society of Bengal and the Discovery of India's Past, 1784– 1838* (Delhi: Oxford University Press, 1988). A fitting conclusion to the eighteenth-century colonial effort is Charles C. Gillispie, 'Scientific Aspects of the French Egyptian Expedition, 1798–1801', *Proceedings of the American Philosophical Society, 133* (1989), pp. 447– 74.

Aspects of colonial science in the nineteenth century are analysed in Lucile H. Brockway, *Science and Colonial Expansion: The Role of the British Royal Botanic Gardens* (New York: Academic Press, 1979); Susan Sheets-Pyenson, *Cathedrals of Science: Colonial Natural History Museums in the Late Nineteenth Century* (Montreal: McGill-Queen's University Press, 1988); Deepak Kumar, *Science and the Raj, 1857–1905* (Delhi: Oxford University Press, 1995); S. N. Sen, *Scientific and Technical Education in India, 1781–1900*, appearing as nos. 1 and 2 of the *Indian Journal of History of Science, 23* (1988); Nathan Reingold, 'Graduate School and Doctoral Degree: European Models and American Realities', in Reingold, *Science, American Style* (New Brunswick: Rutgers University Press, 1991), pp. 171–89.

For twentieth-century developments in the United States, John L. Heilbron and Robert W. Seidel, *Lawrence and His Laboratory: A History of the Lawrence Berkeley Laboratory, 1* (Berkeley: University of California Press, 1989), p. 310 for the geographical diffusion of cyclotrons; Stanley Coben, 'The Scientific Establishment and the Transmission of Quantum Mechanics to the United States, 1919– 32', *American Historical Review, 76* (1971), pp. 442–66, on American doctorates in physics.

On Australia: R. W. Home, 'Origins of the Australian Physics Community', *Historical Studies*, 20 (1983), pp. 383–400; R. W. Home and Masao Watanabe, 'Physics in Australia and Japan to 1914: A Comparison', *Annals of Science*, 44 (1987), pp. 215–35; R. W. Home and Masao Watanabe, 'Forming New Physics Communites: Australia and Japan, 1914–1950', *Annals of Science*, 47 (1990), pp. 317–45. Further on Japan, John Z. Bowers, *When the Twain Meet: The Rise of Western Medicine in Japan* (Baltimore: Johns Hopkins University Press, 1980); Kenkichiro Koizumi, The Emergence of Japan's First Physicists, 1868–1900', *Historical Studies in the Physical Sciences*, 6 (1975), pp. 3–108; Kiyonobu Itakura and Eri Yagi, 'The Japanese Research System and the Establishment of the Institute of Physical and Chemical Research', in *Science and Society in Modern Japan*, eds. Shigeru Nakayama, David L. Swain, and Eri Yagi (Tokyo: University of Tokyo Press, 1974), pp. 158–201; Chikara Sasaki, 'Science and the Japanese Empire, 1868–1945: An Overview', in *Science and Empires*, Patrick Petitjean, Catherine Jami, and Anne Marie Moulin, eds. (Dordrecht: Kluwer, 1992), pp. 243–6.

On India: Prahlad Singh, *Jantar-Mantars of India* (Jaipur: Holiday Publications, 1987); R. K. Kochhar, 'The Growth of Modern Astronomy in India, 1651–1960', *Vistas in Astronomy*, 34 (1991), pp. 69–105; R. K. Kochhar, 'Science in British India, II: Indian Response', *Current Science*, 64 (1993), pp. 55–62; Satpal Sangwan, 'Science Education in India under Colonial Constraints, 1792–1857', *Oxford Review of Education*, 16 (1990), pp. 81–95; In *Science and Empire*, Deepak Kumar, ed. (Delhi: Anamika Prakashan, 1991): Aparnaj Basu, 'The Indian Response to Scientific and Technical Education in the Colonial Era, 1820–1920', pp. 126–38; Chittabrata Palit, 'Mahendra Lal Sircar, 1833–1904: The Quest for National Science', pp. 152–60. Recent summaries may be found in Deepak Kumar, *Science and the Raj, 1857–1905* (Delhi: Oxford University Press, 1995) and Zaheer Baber, *The Science of Empire: Scientific Knowledge, Civilization, and Colonial Rule in India* (Albany: State University of New York Press, 1996).

Collections include Nathan Reingold and Marc Rothenberg, eds., *Science and Colonialism: A Cross-Cultural Comparison* (Washington: Smithsonian Institution Press, 1987); Patrick Petitjean, Catherine Jami, and Anne Marie Moulin, eds., *Science and Empires*, noted above; Antonio Lafuente, Alberto Elena, and María Luisa Ortega, eds., *Mundialización de la ciencia y cultura nacional* (Madrid: Doce

Calles, 1993); Alejandro R. Díez Torre, Tomás Mallo, and Daniel Pacheco Fernández, eds., *De la ciencia ilustrada a la ciencia romántica* (Madrid: Doce Calles, 1995); the *Journal of the Netherlands Institute in Japan* for 1991 contains a fine summary of recent opinion about science across cultures over the past millennium.

A recent trilogy is Lewis Pyenson's *Cultural Imperialism and Exact Sciences: German Expansion Overseas, 1900–1930* (New York: Peter Lang, 1985), notably for Argentina (Hussey quotation on p. 190); *Empire of Reason: Exact Sciences in Indonesia, 1840–1940* (Leiden: E. J. Brill, 1989); *Civilizing Mission: Exact Sciences and French Overseas Expansion, 1830–1940* (Baltimore: Johns Hopkins University Press, 1993).

CHAPTER 14

An excellent overview of the role of religious faith in the history of science appears in John Hedley Brooke, *Science and Religion: Some Historical Perspectives* (Cambridge: Cambridge University Press, 1991). Also see his 'Science and Religion', in R. C. Olby *et. al.*, eds., *Companion to the History of Science* (London: Routledge, 1990), pp. 763–82. Another fascinating interpretation is Amos Funkenstein, *Theology and the Scientific Imagination from the Middle Ages to the Seventeenth Century* (Princeton: Princeton University Press, 1986). Also see Peter Burke's discussion of the eighteenth century in 'Religion and Secularisation', in *The New Cambridge Modern History*. I. Bernard Cohen, ed., *Puritanism and the Rise of Modern Science* (New Brunswick: Rutgers University Press, 1990) summarizes the literature on the relationship between Puritanism and science. He supplies a critical bibliography of works on the Merton thesis on pp. 89–111. Also see his 'Some Documentary Reflections on the Dissemination and Reception of the "Merton Thesis"' in J. Clark, *et al.*, eds., *Robert K. Merton: Consensus and Controversy* (London: Falmer Press, 1990), pp. 307–498. Other recent discussions include Charles Webster's *Great Instauration* and his 'Puritanism, Separatism, and Science', in D. C. Lindberg and R. L. Numbers, eds., *God and Nature: Historical Essays on the Encounter between Christianity and Science* (Berkeley: University of California Press, 1986), pp. 192–217. Also see P. M. Rattansi, 'Science and Religion in the Seventeenth Century', pp. 79–87 in Maurice Crosland, ed., *Emergence of Science in Western Europe* (New York: Science History, 1976).

Earlier works that helped to crystallize this debate are Christopher Hill, *The World Turned Upside Down* (London: Temple Smith, 1972) and T. K. Rabb, 'Puritanism and the Rise of Experimental Science in England', in Leonard M. Marsak, ed., *The Rise of Science in Relation to Society* (New York: Macmillan, 1964), pp. 54–67. Also see Richard L. Greaves, 'Puritanism and Science: The Anatomy of a Controversy', *Journal of the History of Ideas, 30* (1969), pp. 345–68.

The voluminous literature on Puritanism and science is summarized in Lotte Mulligan, 'Puritans and English Science: A Critique of Webster', *Isis, 71* (1980), pp. 456–69. The articles from *Past and Present* have been reprinted in Charles Webster, ed., *Intellectual Revolution of the Seventeenth Century* (London: Routledge & Kegan Paul, 1974).

For the nineteenth century, see Frank M. Turner, 'The Victorian Conflict between Science and Religion: A Professional Dimension', *Isis, 69* (1978), pp. 356–76 and his *Contesting Cultural Authority: Essays in Victorian Intellectual Life* (Cambridge: Cambridge University Press, 1993). On aspects of the conservative reaction to evolution see Peter J. Bowler, *The Eclipse of Darwinism: Anti-Darwinian Evolution Theories in the Decades around 1900* (Baltimore: Johns Hopkins University Press, 1983).

CHAPTER 15

The literature on the Scientific Revolution is enormous. Frances A. Yates was the first to identify the cultivation of natural magic with an interest in experimental science; see 'The Hermetic Tradition in Renaissance Science', in *Art, Science, and History in the Renaissance*, ed. Charles S. Singleton, (Baltimore: Johns Hopkins University Press, 1967), pp. 255–274. Brian P. Copenhaver, 'Natural Magic, Hermeticism, and Occultism in Early Modern Science', updates the literature in David C. Lindberg and Robert Westman, eds., *Reappraisals of the Scientific Revolution* (Cambridge: Cambridge University Press, 1990). Also see Keith Thomas, *Religion and the Decline of Magic* (New York: Scribner's, 1971).

Other recent treatments include Peter M. Heimann, 'The Scientific Revolutions', *The New Cambridge Modern History*, vol. 13, ed. Peter Burke (Cambridge: Cambridge University Press, 1979), pp. 248–70 and Charles Webster, ed., *The Intellectual Revolution of the Seventeenth Century* (London: Routledge & Kegan Paul, 1974).

More dated but nonetheless useful are Harcourt Brown, 'The Utilitarian Motive in the Age of Descartes' *Annals of Science, 1* (1936), pp. 182–92; R. F. Jones, *Ancients and Moderns: A Study of the Rise of the Scientific Movement in Seventeenth Century England* (1936; Berkeley: University of California Press, 1965); and Walter E. Houghton, 'The History of Trades: Its Relation to Seventeenth Century Thought', *Journal of the History of Ideas, 2* (1941), pp. 33–60.

For eighteenth-century developments see Rom Harré, 'Knowledge', in G. S. Rousseau and Roy Porter, eds., *The Ferment of Knowledge: Studies in the Historiography of Eighteenth-Century Science* (Cambridge: Cambridge University Press 1980), pp. 11–54, and Margaret C. Jacob, *The Cultural Meaning of the Scientific Revolution* (Philadelphia: Temple University Press, 1988). For the notion that things continually improve, see John Bagnell Bury, *The Idea of Progress: An Inquiry into Its Origin and Growth* (London: Macmillan, 1920; New York: Dover, 1955).

On the role of science in the Enlightenment see Thomas Hankins, *Jean d'Alembert, Science and the Enlightenment* (Oxford: Clarendon Press 1970) and P. N. Furbank, *Diderot: A Critical Biography* (London: Secker & Warburg 1992). For the *Encyclopédie*, see Frank A. Kafker, ed., *Notable Encyclopedias of the Seventeenth and Eighteenth Centuries: Nine Predecessors of the Encyclopédie* (Oxford: Voltaire Foundation 1981) and Stephen J. Gendzier, ed. and trans., *Denis Diderot's The Encyclopedia: Selections* (New York: Harper Torchbooks, 1967).

The classic overview of nineteenth-century philosophy is John Theodore Merz, *A History of European Thought in the Nineteenth Century* (1904–14; New York: Dover, 1965), which supplements his two volume history of scientific thought, first published between 1898 and 1904. Also see Maurice Mandelbaum, *History, Man and Reason: A Study in Nineteenth-Century Thought* (Baltimore: Johns Hopkins University Press, 1971) and Basil Willey, *Nineteenth Century Studies* (London: Chatto & Windus, 1949). All of these works emphasize the important role of positivism.

On positivism, see W. M. Simon, *European Positivism in the Nineteenth Century* (Ithaca: Cornell University Press, 1963). Frank Manuel emphasizes the social activism of the French positivists in *Prophets of Paris* (Cambridge: Harvard University Press, 1962). Also see the article on Auguste Comte by Larry Laudans in the *Dictionary of Scientific Biography*, ed. Charles C. Gillispie (New York: Charles Scribner's Sons, 1970–80).

CHAPTER 16

A general introduction to the place of twentieth-century physics in a number of societies is found in José Manuel Sánchez Ron, *El poder de la ciencia: Historia socio-económica de la física (siglo XX)* (Madrid: Alianza Editorial, 1992). An interesting discussion of physics and international politics, where Einstein appears in the background, is found in Elisabeth Crawford, *Nationalism and Internationalism in Science, 1880–1939: Four Studies of the Nobel Population* (Cambridge: Cambridge University Press, 1992). A major, general assessment of mathematical physics, where relativity is prominent, is Chikara Sasaki, *The Historical Structure of Scientific Revolutions*, 2 vols. (Tokyo: Iwanami Shoten, 1985) [in Japanese]. A useful survey is found in Vladimir P. Vizgin, *Unified Field Theories: In the First Third of the Twentieth Century*, trans. Julian B. Barbour (Boston: Birkhäuser, 1994). An ambitious project by Princeton University Press to issue and comment on Einstein's published and unpublished work has been under way for nearly two decades; two volumes have appeared under the general editorship of John Stachel and four under the general editorship of Martin J. Klein (all volumes so far deal with the period up to the end of the First World War).

Relativity continues to generate an extraordinary amount of general commentary, most of it unworthy of serious reflection. A typical *mélange*, better than most, is Themistocles M. Rassias and George M. Rassias, eds., *Selected Studies: Physics-Astrophysics, Mathematics, History of Science. A Volume Dedicated to the Memory of Albert Einstein* (Amsterdam: North-Holland, 1982). A useful collection is B. Bertotti, R. Balbinot, S. Bergia, and A. Messina, eds., *Modern Cosmology in Retrospect* (Cambridge: Cambridge University Press, 1990). A recent volume in a continuing and uneven series of conference proceedings is Jean Eisenstaedt and Anne J. Kox, eds., *Studies in the History of General Relativity* (Boston: Birkhäuser, 1992). A global cross-cultural attempt at understanding the era of relativity appears in the contributions to Thomas F. Glick, ed., *The Comparative Reception of Relativity* (Dordrecht: Reidel, 1987). An idiosyncratic, philosophical encyclopaedia of sorts appears in Klaus Hentschel, *Interpretationen und Fehlinterpretationen der speziellen und der allgemeinen Relativitätstheorie durch Zeitgenossen Albert Einsteins* (Basel: Birkhäuser, 1990). It would serve no useful purpose to list treatises associating special and general relativity with musical, artistic, or literary concerns.

Recent articles of note relating to Einstein include: Lewis Pyenson, 'Einstein's Natural Daughter', *History of Science*, 28 (1990), pp. 365–79; Senta Troemel-Ploetz 'Mileva Einstein-Marić: The Woman Who Did Einstein's Mathematics', *Women's Studies International Forum*, 13 (1990), pp. 415–32; Seiya Abiko, 'On the Chemico-Thermal Origins of Special Relativity', *Historical Studies in the Physical and Biological Sciences*, 22 (1991), pp. 1–24; Klaus Hentschel, 'Grebe/Bachems photometrische Analyse der Linienprofile und die Gravitations-Rotverschiebung: 1919 bis 1922', *Annals of Science*, 49 (1992), pp. 21–46; Makoto Katsumori, 'The Theories of Relativity and Einstein's Philosophical Turn', *Studies in the History and Philosophy of Science*, 23 (1992), pp. 557–592; Setsuko Tanaka, 'Mach, Einstein, and Kuwaki', in John Blackmore, ed., *Ernst Mach: A Deeper Look* (Dordrecht: Kluwer, 1992), pp. 297–332; Seiko Yoshida and Seiji Takata, 'Uzumi Doi, An Anti-Relativist and the Physics Circle in the Taisho Era', *Kagakusi Kenkyu*, 31 (1992), pp. 19–26 [in Japanese]; Robert Schulmann, 'Einstein at the Patent Office: Exile, Salvation, or Tactical Retreat?' *Science in Context*, 6 (1993), pp. 17–24; Andreas Kleinert, 'Paul Weyland, der Berliner Einstein-Töter', in Helmuth Albrecht, ed., *Naturwissenschaften und Technik in der Geschichte* (Stuttgart: Verlag für Geschichte der Naturwissenschaften und der Technik, 1993), pp. 199–232; Adel Ziadat, 'Early Reception of Einstein's Relativity in the Arab Periodical Press', *Annals of Science*, 51 (1994), pp. 17–35.

PICTURE CREDITS

The author and publishers are grateful to the following for permission to reproduce illustrative material: Plate 1, courtesy of the Archives, California Institute of Technology; Plate 2, courtesy of the Westinghouse Electric Corporation; Plate 3, courtesy of the National Air and Space Museum, Washington, DC; Plate 4, courtesy of the Bakken Library and Museum, Minneapolis. Plate 5, taken from Jack Morrell and Arnold Thackray, *Gentlemen of Science: Early Years of the British Association for the Advancement of Science* (Oxford: Oxford University Press, 1981), plate no. 25, courtesy of Oxford University Press; Plate 7, courtesy of Chuou-Koron-sha Co., Ltd, Tokyo; Plate 8, courtesy of the Consejo Superior de Investigaciones Científicas, Madrid; Plates 9 and 11, Jan van der Straet, *New Discoveries: The Sciences, Inventions, and Discoveries of the Middle Ages and Renaissance as Represented in 24 Engravings Issued in the Early 1580s by Stradanus* (Norwalk, 1953) [Burndy Library Publication no. 8], courtesy of the Burndy Library, Dibner Institute for the History of Science and Technology, Massachusetts Institute of Technology; Plate 10, courtesy of the Library of the Academy of Natural Sciences, Philadelphia; Plate 12 from Geoffrey Wakeman, *Victorian Book Illustration* (Newton Abbot: David & Charles, 1973), p. 13, by kind permission of the publishers.

INDEX